Causes and Effects of Stratospheric Ozone Reduction: An Update

A report prepared by the
Committee on Chemistry and Physics of Ozone Depletion
and the
Committee on Biological Effects of Increased Solar Ultraviolet Radiation

Environmental Studies Board
Commission on Natural Resources
National Research Council

NATIONAL ACADEMY PRESS
Washington, D.C. 1982

NOTICE: The project that is the subject of this report was approved by the Governing Board of the National Research Council, whose members are drawn from the councils of the National Academy of Sciences, the National Academy of Engineering, and the Institute of Medicine. The members of the committees responsible for the report were chosen for their special competences and with regard for appropriate balance.

This report has been reviewed by a group other than the authors according to procedures approved by a Report Review Committee consisting of members of the National Academy of Sciences, the National Academy of Engineering, and the Institute of Medicine.

The National Research Council was established by the National Academy of Sciences in 1916 to associate the broad community of science and technology with the Academy's purposes of furthering knowledge and of advising the federal government. The Council operates in accordance with general policies determined by the Academy under the authority of its congressional charter of 1863, which establishes the Academy as a private, nonprofit, self-governing membership corporation. The Council has become the principal operating agency of both the National Academy of Sciences and the National Academy of Engineering in the conduct of their services to the government, the public, and the scientific and engineering communities. It is administered jointly by both Academies and the Institute of Medicine. The National Academy of Engineering and the Institute of Medicine were established in 1964 and 1970, respectively, under the charter of the National Academy of Sciences.

This study was supported by the U.S. Environmental Protection Agency under Contract No. 68-02-3701.

Library of Congress Catalog Card Number 82-81229

International Standard Book Number 0-309-03248-2

Available from

NATIONAL ACADEMY PRESS
2101 Constitution Avenue, N.W.
Washington, D.C. 20418

Printed in the United States of America

DSZM
N

83 006333

COMMITTEE ON CHEMISTRY AND PHYSICS OF OZONE DEPLETION

CHARLES H. KRUGER, JR. (Chairman), Stanford University
ROBERT E. DICKINSON, National Center for Atmospheric Research
JAMES P. FRIEND, Drexel University
DONALD M. HUNTEN, University of Arizona
MICHAEL B. McELROY, Harvard University

COMMITTEE ON BIOLOGICAL EFFECTS OF INCREASED SOLAR ULTRAVIOLET RADIATION

RICHARD B. SETLOW (Chairman), Brookhaven National Laboratory
JAMES P. FRIEND, Drexel University
MAUREEN M. HENDERSON, University of Washington
JOHN JAGGER, University of Texas at Dallas
RICHARD M. KLEIN, University of Vermont
JOHN A. PARRISH, Massachusetts General Hospital/Harvard University
HOWARD H. SELIGER, Johns Hopkins University
WILLIAM B. SISSON, U.S. Department of Agriculture/New Mexico State University

Staff

ADELE KING MALONE, Staff Officer
ELIZABETH G. PANOS, Administrative Assistant
MYRON F. UMAN, Senior Staff Officer

ENVIRONMENTAL STUDIES BOARD

DAVID PIMENTEL (Chairman), Cornell University
DANIEL A. OKUN (Vice-Chairman), University of North Carolina
ALVIN L. ALM, Harvard University
RALPH C. D'ARGE, University of Wyoming
ALFRED M. BEETON, University of Michigan
JOHN CAIRNS, JR., Virginia Polytechnic Institute and State University
JAMES A. FAY, Massachusetts Institute of Technology
MARGARET HITCHCOCK, Yale University School of Medicine
JULIUS E. JOHNSON, Dow Chemical Company
CHARLES H. KRUGER, JR., Stanford University
KAI N. LEE, University of Washington
CARL M. SHY, University of North Carolina
EDITH BROWN WEISS, Georgetown University Law Center

RAPHAEL G. KASPER, Executive Secretary
MYRON F. UMAN, Associate Executive Secretary

COMMISSION ON NATURAL RESOURCES

ROBERT M. WHITE (Chairman), University Corporation for Atmospheric Research
TIMOTHY ATKESON, Steptoe & Johnson
STANLEY I. AUERBACH, Oak Ridge National Laboratory
NEVILLE P. CLARK, Texas A&M University
NORMAN A. COPELAND, E.I. du Pont de Nemours and Company, Inc., retired
GEORGE K. DAVIS, University of Florida, retired
JOSEPH L. FISHER, Virginia Office of Human Resources
EDWARD D. GOLDBERG, Scripps Institution of Oceanography
KONRAD B. KRAUSKOPF, Stanford University
CHARLES J. MANKIN, Oklahoma Geological Survey
NORTON NELSON, New York University Medical Center
DANIEL A. OKUN, University of North Carolina
DAVID PIMENTEL, Cornell University
JOHN E. TILTON, Pennsylvania State University

WALLACE D. BOWMAN, Executive Director
THEODORE M. SCHAD, Deputy Executive Director

CONTENTS

PREFACE

The Clean Air Act, as amended in August 1977 (Part B, Title I), is intended, in part, to foster understanding of how human activities may affect the stratosphere, in particular the ozone layer, and how changes in the stratosphere, especially changes in ozone concentrations, may affect public health and welfare. The act requires the U.S. Environmental Protection Agency (EPA) and other agencies to conduct studies to increase our knowledge on these topics. The agencies must report to Congress biennially on the results of relevant research.

In the spring of 1981, EPA asked the National Research Council (NRC) for assistance in carrying out its responsibilities under the act. The NRC was asked to provide an assessment of the state of knowledge on ozone depletion and its effects, to be used by EPA in preparing its biennial report to Congress, due in January 1982.

The NRC had prepared earlier reports on these topics, _Environmental Impact of Stratospheric Flight: Biological and Climatic Effects of Aircraft Emissions in the Stratosphere_ (1975), _Halocarbons: Environmental Effects of Chlorofluoromethane Release_ (1976a), _Halocarbons: Effects on Stratospheric Ozone_ (1976b), _Nitrates: An Environmental Assessment_ (1978), _Protection Against Depletion of Stratospheric Ozone by Chlorofluorocarbons_ (1979a), and _Stratospheric Ozone Depletion by Halocarbons: Chemistry and Transport_ (1979b). The purpose of the current study was to update these previous reports by assessing the most recent scientific information. The study was assigned to the Environmental Studies Board within the Commission on Natural Resources of the NRC.

The study was divided into two parts: first, an assessment of changes in understanding of the atmospheric chemistry and physics of ozone depletion, and, second, an examination of current knowledge about environmental and human health effects of the increased intensities of solar ultraviolet radiation that would result from reductions in stratospheric ozone. EPA asked that the study emphasize the assessment of biological effects.

In May 1981, the Committee on Chemistry and Physics of Ozone Depletion and the Committee on Biological Effects of Increased Solar Ultraviolet Radiation were established under the auspices of the Environmental Studies Board. (Biographical data on the members of the committees appear in Appendix I.) In October 1981, EPA requested that the committees take whatever additional time may be necessary beyond the original contract deadline of December 21, 1981, to ensure that sufficient time was available for consideration of the report of an international workshop held in May 1981 under the auspices of the World Meteorological Organization (WMO) and the U.S. National Aeronautics and Space Administration (NASA), Federal Aviation Administration, and National Oceanic and Atmospheric Administration (Hudson et al. 1982). Accordingly, the period of study was extended through March 1982.

The two committees approached their tasks in different ways. The Committee on Chemistry and Physics of Ozone Depletion commissioned six consultants to review current developments in three areas: (1) laboratory measurements and modeling, (2) measurements in the stratosphere, and (3) understanding of stratospheric perturbations and trends. The consultants' papers were reviewed by independent peer reviewers. The commissioned papers and peer reviews along with successive drafts of the report of the NASA/WMO workshop provided the base of information from which the committee's report was developed. The commissioned papers are included as Appendixes A through F.

The Committee on Biological Effects of Increased Solar Ultraviolet Radiation organized a workshop that was held on July 30-31, 1981, at the National Academy of Sciences in Washington, D.C. Approximately 30 scientists who were active in research or familiar with the current literature participated, including the committee. Participants presented and assessed the information that had become available since the NRC (1979a) report, covering three topics: (1) molecular and cellular studies, (2) ecosystem effects, and (3) human health effects. The committee members drew on the presentations and discussions at the workshop, the work of the Panel to Review Statistics on

Skin Cancer (funded by the Department of Energy and the National Institute of Environmental Health Sciences) of the NRC Committee on National Statistics, and their own knowledge to develop their report. Workshop participants and several additional scientists active in the field were asked to review the report. The workshop participants are listed in Appendix H.

The report consists of a joint summary followed by Part I and Part II, which are from the Committee on Chemistry and Physics of Ozone Depletion and the Committee on Biological Effects of Increased Solar Ultraviolet Radiation, respectively. Of the material included in this volume, only the summary and Parts I and II have been critically reviewed by the NRC. Views expressed in the commissioned papers in the appendixes are not necessarily those of the committees.

The two committees wish to express their appreciation to Adele King Malone, Elizabeth G. Panos, and Myron F. Uman of the National Research Council for their contributions in managing our study and preparing this report. Other staff members providing assistance include Raphael Kasper, Estelle Miller, Roseanne Price, Robert Rooney, and Christina Shipman. We also want to thank the members of our two committees and consultants for the diligence and enthusiasm with which they approached our task. We are grateful, too, for the cheerful cooperation of personnel from EPA and NASA and for the helpful critiques provided by those who reviewed drafts of our consultants' papers and our report.

Our report is, as were the ones that preceded it, an attempt to describe the current state of knowledge in fields that are rapidly developing. The goal is to give policy makers an independent and objective assessment of what we know now, what we do not know, and the prospects for resolving current uncertainties. We hope our efforts will prove useful.

Charles H. Kruger, Jr., Chairman
Committee on Chemistry and Physics of Ozone Depletion

Richard B. Setlow, Chairman
Committee on Biological Effects of
Increased Solar Ultraviolet Radiation

SUMMARY

INTRODUCTION

This report reviews current knowledge about man-made causes of changes in concentrations of stratospheric ozone and the effects of those changes. Recent reports of the National Research Council (NRC 1975, 1976a,b, 1978, 1979a,b) have treated the chemical and physical aspects of potential reductions of stratospheric ozone in detail. Part I of this report reviews recent developments on that subject. Part II deals with the effects of reduction of stratospheric ozone on humans, other animals, and plants, independently of what might cause the reduction.

CHEMISTRY AND PHYSICS OF OZONE REDUCTION

The abundance of ozone in the stratosphere is determined by a dynamic balance among processes that produce and destroy it and transport it to the troposphere. According to current understanding, the most important photochemical reactions regulating ozone involve molecular and atomic oxygen and various radicals containing nitrogen, hydrogen, and chlorine. All of these compounds have natural sources, but their concentrations in the stratosphere can be significantly altered by human activities. The human activities that have thus far been identified as potentially influencing stratospheric ozone are as follows:

- The release of gaseous chlorinated carbon compounds, mainly chlorofluorocarbons (CFCs) and methyl chloroform (CH_3CCl_3). CFCs are used as foam-blowing agents, as working fluids in refrigeration systems, and as propellants in aerosol sprays. Methyl chloroform is an industrial solvent. These gases decompose in the stratosphere providing a significant source of radicals that contain chlorine.

- The release of nitrous oxide (N_2O) from combustion and its enhanced release from soils and waters as a result of various agricultural and waste management practices. Nitrous oxide decomposes in the stratosphere, introducing radicals that contain nitrogen.
- The direct input of nitrogen radicals to the stratosphere due to nitrogen oxides (NO_x) in aircraft engine exhausts.
- The increased abundance of carbon dioxide (CO_2) in the atmosphere due to combustion of fossil fuels and deforestation. Increased carbon dioxide has a subtle influence, causing the temperature of the stratosphere to decrease, which leads to increased stratospheric ozone, and changing stratospheric concentrations of water vapor.

Key Findings and Conclusions

Over the past several years, research, driven by discrepancies between theory and observation, has led to considerable improvement in our understanding of the effects on stratospheric ozone of releases of CFCs and oxides of nitrogen. As a result, previous discrepancies between the estimates of models of stratospheric processes and observed concentrations of certain important species have been reduced. Important discrepancies still remain, however, which means that there are still uncertainties inherent in the results of modeling exercises.

Current scientific understanding, expressed in both 1- and 2-dimensional models, indicates that if production of two CFCs, CF_2Cl_2 and $CFCl_3$, were to continue into the future at the rate prevalent in 1977, the steady state reduction in total global ozone, in the absence of other perturbations, could be between 5 percent and 9 percent. Comparable results from models prevalent in 1979 ranged from 15 percent to 18 percent. The differences between current findings and those reported in 1979 are attributed to refinements in values of important reaction rates. Also, as an example, if the atmospheric concentration of N_2O were doubled in the absence of other perturbations, total ozone would be reduced by between 10 percent and 16 percent. Although atmospheric concentrations of N_2O appear to be increasing, we cannot reliably project the future course of N_2O emissions. Steady state reductions in both these cases would be reached asymptotically in times on the order of a century, although the assumption of doubling N_2O concentrations is unrealistic on such a time scale. The effects of perturbations by CFCs and N_2O are not additive, so the estimates of effects of combined perturbations require investigation of specific cases.

These results should be interpreted in light of the uncertainties and insufficiencies of the models and observations. For example, other chemicals released from human activities are understood to have the potential for affecting stratospheric ozone. Examples are methyl chloride (CH_3Cl), carbon tetrachloride (CCl_4), and particularly methyl chloroform. Observations of critical species need to be extended and confirmed by a number of measurements using independent techniques. Important assumptions in the models about rate constants, distributions of certain species, and the reactions taking place need to be tested. Furthermore, three important discrepancies between models and observations remain to be resolved: More chlorine monoxide (ClO) is observed at altitudes above 35 km than is predicted, the behavior of NO_x in winter at high latitudes is unexplained, and concentrations of CFCs in the lower stratosphere are lower than the models suggest.

We anticipate that research on these problems in the field, in the laboratory, and in theory currently under way, planned, and proposed will lead to continued improvement in understanding, resulting in further reduction of the remaining discrepancies between theory and observation. In particular, simultaneous measurement of the important chemical species as a function of altitude and latitude by various methods should prove critical to improving understanding during the next several years.

Examination of the historical record of measurements of ozone does not reveal a significant trend in total ozone that can be ascribed to human activities. This observational result is consistent with those of current models, since no detectable trend would be expected on the basis of current theory.

Because data on total global ozone cannot be analyzed to distinguish among causes of ozone changes, total ozone data alone cannot be relied upon for early detection of an anthropogenic change. Measurement of the spatial and temporal distribution of critical trace species and ozone, together with theoretical modeling taking into account all the major influences on stratospheric ozone, offers promise of understanding the causes of ozone changes and the consequences of alternative actions in response.

Recommendations

1. The national research program, including atmospheric observation, laboratory measurements, and theoretical modeling, should maintain a broad perspective with emphasis on areas of disagreement between theory and observation. Highest priority in research should be given to a coordinated program to understand the spatial and temporal distributions of important species, such as ClO and the hydroxyl radical (OH).

2. The global monitoring effort should include both ground-based and satellite observations of total ozone and concentrations of ozone above 35 km, where theory indicates the largest reductions might occur. Sound, satellite-based systems for stratospheric observations are essential.

3. Potential emissions of N_2O, CO_2, CH_3CCl_3, and other relevant gases should be assessed and their consequences for stratospheric ozone evaluated. Models should be developed to describe the consequences for stratospheric ozone of future emissions of these gases.

BIOLOGICAL EFFECTS OF INCREASED SOLAR ULTRAVIOLET RADIATION

Stratospheric ozone acts as a shield to screen out much of the short-wavelength ultraviolet (UV) in sunlight. Slight changes in this ozone layer may result in large changes in the amount of damaging UV striking the surface of the earth. Living creatures have adapted to the present level of UV and to its fluctuations from season to season and during the day. Part II of this report gives the current state of knowledge about the effects on biological systems of an increase in UV resulting from a decrease in stratospheric ozone concentration.

Each of the findings and conclusions summarized below has important implications for future research--either in efforts to decrease the uncertainty in concepts or in efforts to increase quantitative knowledge. These research implications are spelled out in our list of major recommendations. Recent advances in knowledge since the last NRC report on the subject (NRC 1979a) have clarified our view of the problem but have also pointed out scientific areas not emphasized in earlier reports that confound the simple prediction of the effects of ozone depletion on biological systems. The unraveling of these difficulties will be accomplished only by a research effort directed by knowledgeable scientists, especially photobiologists. In many instances, we are still not sure of the scientific questions to be asked. Similar comments were made in earlier NRC reports (NRC 1975). The fact that they have not been acted on with any reasonable financial commitment accounts for a large part of our inability to make better predictions.

It seems certain that more than 90 percent of skin cancer other than melanoma in the United States is associated with sunlight exposure and that the damaging wavelengths are in the UV-B region (290 nm to 320 nm) of the spectrum. A decrease in ozone will be accompanied by a well-predicted increase in UV-B. We estimate that there will be a 2 percent to 5 percent increase in basal cell skin cancer incidence per 1 percent decrease in stratospheric ozone. The increase in squamous cell skin cancer incidence will be about double that. Where in this range the value falls depends on which theory is used to make the estimate and on the appropriate dosimetric data used. The predicted increases are appreciably greater at lower latitudes than at higher.

Although the incidence of malignant melanoma increases with a decrease in latitude, the degree to which sunlight is responsible is not apparent, and there are few data implicating UV-B as the only responsible wavelength region. Therefore it is not appropriate to make quantitative predictions about the increase in the incidence of this disease associated with a decrease in ozone.

Some of the difficulty in making quantitative predictions about humans comes from uncertainties (even in simple cellular systems) about the effects of interactions among single wavelengths in a broad band, such as in the ultraviolet of sunlight, in producing antagonistic or synergistic effects. Moreover, it has been learned only recently that rapid repair of sunlight damage to human skin takes place during irradiation. An appreciable fraction is photorepair mediated by visible light, and a similar phenomenon seems to take place in anchovy populations. The quantitative magnitudes of such effects are not known.

The effects of ozone depletion on other animals and plants in the biosphere are as important as the direct effects on human health. However, scientists are still not able to predict quantitative effects on crop plants or ecosystems.

The details of our findings and recommendations are spelled out in Chapters 3, 4, and 5. Key findings and conclusions and major research recommendations have been extracted from the chapters and are listed below. Estimates are given, where possible, of how long the recommended research might take under ideal circumstances.

Key Findings and Conclusions

Molecular and Cellular Studies (Chapter 3)

1. Deoxyribonucleic acid (DNA) is probably the primary target in animal cells for most deleterious effects of UV-B, especially effects involving mutagenesis and neoplastic transformation. Other targets of possible biological significance for UV-B effects include membranes, ribonucleic acid (RNA), and proteins.

2. The spectrum for absorption of energy by DNA for wavelengths in the UV-B region and the spectra for biological damage to DNA as a function of wavelength (action spectra) are known. The absorption spectrum and the action spectra are similar but not identical, probably because long-wavelength light is absorbed in some components of this genetic material that are not effective in changing the structure of DNA. The action spectra in the UV-B region for affecting mammalian cells (killing, mutation, and neoplastic transformation) are similar to those for damaging DNA.

3. The formation of pyrimidine dimers (bonds between pyrimidine residues in one of the two strands of DNA that distort the normal DNA helical structure) appears to be the major injury to DNA from UV-B irradiation.

4. There are major interactions between the effects of UV-A (320 nm to 400 nm) and those of UV-B on DNA in cells. Some of these are antagonisms, whereby UV-A effects significantly reduce or repair the UV-B damage. Except for photoreactivation, which involves enzymic splitting of pyrimidine dimers back to normal single

residues mediated by UV-A and visible light, these interactions are still poorly understood.

5. In excision repair, dimers are removed from one strand of a DNA double helix by enzymes that work in the dark, leaving the unaltered strand as a template for reconstitution of a new normal strand. Photoreactivation and excision repair of pyrimidine dimers occurs rapidly in human skin.

Ecosystems and Their Components (Chapter 4)

6. Both UV-A and UV-B have been reported to be detrimental to plant growth and development and to a number of physiological processes of plants, when examined under non-field conditions. The adaptability of plant species appears to be sufficient, under current ambient levels of UV-B, to maintain food crop yields. The potential for further adaptation to predicted increases in ambient UV-B is not known.

7. Ambient UV-B at present levels or similar levels in the laboratory can damage sensitive aquatic organisms or stages in their lifecycles that occur at the water's surface. Natural populations of aquatic organisms have adapted to current UV-B levels so as to maximize reproduction potential. In the case of anchovy larvae, it has been demonstrated that photorepair of UV-B damage is effective even at UV-B levels significantly higher than those that would result from predicted ozone depletions. Photorepair may be a general adaptive mechanism of organisms evolving in the presence of UV-B. Currently, there is no information from which to predict the magnitude of adverse effects of enhanced UV-B on aquatic organisms.

8. From limited field experiments on terrestrial plants and laboratory experiments with captured or cultured aquatic organisms, it appears that different species of both plants and animals have different sensitivities to increases in UV-B above current levels. Changes in species compositions and abundances of organisms have been observed in simulated aquatic ecosystems subjected to enhanced UV-B. Mathematical models show that in systems subject to large natural oscillations in the size of the population, there are severe limitations on the minimum population density needed to maintain a species. However, the data currently available on food chains in the natural

ecosystem are not precise enough or complete enough to be used to predict population dynamics or the displacement of an individual species under current environmental conditions. It is doubtful therefore that a statistically significant causal relationship between increased UV-B levels and food chain success can be predicted in the near future.

9. Only minor effects of increased UV-B levels are predicted for animals used for human food.

Direct Human Health Hazards (Chapter 5)

10. A reduction in the concentration of stratospheric ozone will not create new health hazards, but will increase existing ones.

Effects Other Than Cancer

11. There is evidence that direct acute effects of UV on humans, such as sunburn (acute erythema) and corneal inflammation (photokeratitis), are linked more strongly to UV-B than to UV-A.

12. Acute erythema and photokeratitis can be predicted accurately for a given dose and spectrum of UV-B, since the action spectra, dose-response curves, and intensity-time reciprocity relationships are known.

13. Ultraviolet radiation affects many aspects of the immune system of animals and humans. Allergic contact dermatitis, skin graft rejection, tumor susceptibility, and function and viability of individual circulating and noncirculating cells of the immune system can be altered, primarily by UV-B.

Skin Cancer Other Than Melanoma

14. Data on the relative incidence rates of basal and squamous cell cancers in highly pigmented (black) versus lightly pigmented (white) persons indicate that more than 90 percent of skin cancers other than melanoma in U.S. whites are attributable to sunlight.

15. Molecular, cellular, and whole animal data all implicate UV-B as the major carcinogenic component of sunlight for skin cancers other than melanoma. The evidence is stronger for squamous than basal cell cancers because animals rarely get basal cell cancers. In

humans, basal cell cancers are virtually all related to sunlight.

16. Based on animal studies, UV-B is implicated not only as an initiator of carcinogenesis but also as a promoter (in the general sense and via indirect effects) of chemical carcinogenesis. With the current state of knowledge, it is not possible to assess the extent to which increasing exposures to chemicals would result in increases in skin cancers due to synergism, over and above any increase because of increased UV-B exposure alone.

17. A 1 percent reduction in the amount of stratospheric ozone is predicted to give an approximate 2 percent increase in biologically effective UV-B. Epidemiological data suggest that a 2 percent increase in UV-B would give a 2 percent to 5 percent increase in basal cell skin cancers. For squamous cell skin cancers the increase would be about twice these values (4 percent to 10 percent).

18. The risk of developing skin cancers other than melanoma and the increased risk due to increased exposure to UV-B could be mitigated by individuals through changes in lifestyle that would reduce exposure.

Melanoma

19. The incidence of skin melanoma appears to depend on latitude, an indication that sunlight is a contributing factor. Circumstantial evidence such as occupational differences and location of the cancers on the body suggests, however, that exposure to sunlight is only one of several factors. The association between sunlight and melanoma is not strong enough to make a prediction of increased incidence due to increased exposure to UV based on epidemiological data.

20. The only evidence that suggests UV-B causes melanoma in humans comes from studies of people with the inherited disease xeroderma pigmentosum. These people have a known defect in the mechanism that would repair UV-B damage to DNA, and they also have a very high incidence of skin cancers, including melanoma.

21. There are no reliable animal models for light-induced melanoma. The only models currently available are animals with chemically induced, preexisting pigmented lesions that can be made to look like melanoma after UV irradiation.

Major Research Recommendations

The estimates following each recommendation of how long the research might take are educated guesses based on the experience of individual committee members. The estimates provide only a rough idea of how long the research might take under ideal circumstances.

Molecular and Cellular Studies (Chapter 3)

1. An understanding is needed of why broad bands of UV (heterochromatic radiation) often do not act on DNA in vivo and on in vitro cell systems as a simple sum of monochromatic wavelengths.

(a) Studies of interactive effects between UV-A and UV-B are fundamental to understanding the mechanisms of cancer induction by sunlight. Such studies, employing bacteria or cultured mammalian cells, would take about two to five years.
(b) An understanding is needed of UV-A-induced repair systems in bacteria, as a first step in understanding possible similar systems in higher organisms. This would take about two to five years.
(c) Experiments should be conducted to determine the rate and extent of photoreactivation in humans in sunlight. Data are needed on how the level of dimers depends on the relative amounts of UV-A and visible light compared with the amount of dimer-producing UV-B. These experiments would take about two to five years.

2. Data are needed on the rates of repair, in the dark and in laboratory light, of UV-irradiated human skin cells as a function of UV dose. The differences, if any, between acute and chronic irradiations should be determined. One might be able (with informed consent) to study individuals who are exposed to high levels of UV-B as part of phototherapy for psoriasis. The aim of such experiments would be to determine whether the kinetics of dark repair of damage from pyrimidine dimers in human skin show two components, a slow one and a fast one, as is true for human cells irradiated in vitro. The two components represent repair of DNA in different regions of the DNA strands. Equally important questions are, what other types of biologically important damages occur

in skin, what are their lifetimes, and are any of them persistent? These data could be obtained in about four or five years.

Ecosystems and Their Components (Chapter 4)

3. Techniques must be developed for simulating changes in UV-B under natural ambient conditions. Only in this way can dose-response relationships be obtained. If these techniques cannot be developed for studies at temperate latitudes, they might best be achieved in a low-latitude (subtropical), minimal-cloud-cover, multiuser facility, which would provide UV-B radiation corresponding to reduced ozone concentrations at more northern latitudes. Priority should be given to screening representative species of important food plant systems for identifying possible adverse effects on crop productivity. Dosimetry and environmental regulation techniques must be developed to ensure optimum experimental conditions--conditions equivalent to the higher latitude ambient field conditions of the plants being tested. Without strict attention to these control conditions, studies will have limited potential for extrapolation or prediction. It would take about three years to develop the facility and another three years to conduct the species screening experiments.

4. The effects of UV dose on elements of aquatic food chains cannot be determined unless (a) the underwater spectral irradiances are integrated over the varying positions of organisms in water columns to obtain the exposures that simulate spectral intensities in the natural systems, and (b) damage to individuals can be related to population dynamics in the natural ecosystem. This would require an integrated research approach involving physical hydrography, physical optics, and organism physiology. It would take about five years to develop this approach and obtain results. Unless UV-B studies are made as a part of an ecosystem study, effects on populations and interactions among populations cannot be predicted. (Testing for whole ecosystem effects is addressed in another NRC report, Testing for Effects of Chemicals on Ecosystems (NRC 1981).)

An attempt to incorporate such an integrated approach was made for anchovy larvae. The interdisciplinary approach used in the anchovy study to assess UV-B damage to food chains, together with the specific laboratory measurements, should serve as a model for future research proposals.

Direct Human Health Hazards (Chapter 5)

5. Studies (animal and human) should be conducted in the developing field of photoimmunology to determine the magnitude of UV effects on the human immune system, the effective wavelengths, and the dose-response relationship. Results may increase understanding of skin cancer mechanisms, other effects of UV on skin, and certain other diseases. These studies would take about two to five years.

6. Animal studies of UV-induced skin cancers other than melanoma are needed to understand interactions among parameters such as intermittent exposures, different wavelengths, dose rates, chemical carcinogens and promoters, and agents that modify cellular responses to irradiation. These studies would take about two to five years.

7. The Surveillance, Epidemiology, and End Results program of the National Cancer Institute routinely collects data on incidence of melanoma. The incidence of skin cancers other than melanoma should be surveyed every decade at a time coinciding with the population census, so as to determine trends in time. Only a few locations are necessary, but these should be the same as past survey locations. Data should be collected in a way that permits cohort as well as cross-sectional analysis.

8. Animal models for UV- or light-induced melanomas are needed. They would allow studies of action spectra, dose-response curves, waveband interactions, and other parameters. It is not possible to predict how long it would take to develop such models.

9. To determine the association between UV and melanoma, it would be useful to determine the incidence of the various subtypes of melanoma and their dependence on latitude. Although this will be difficult because the majority of melanomas are of the superficial spreading type, the methodology is available. Careful epidemiological studies that are based on reliable clinical and histological studies of subtypes of melanoma are needed.

PART I: CHEMISTRY AND PHYSICS OF OZONE REDUCTION

Chapter 1

CURRENT STATUS

INTRODUCTION

This chapter reviews recent changes in the state of understanding of the chemical and physical processes that determine the effect of human activities on concentrations of stratospheric ozone. The report is motivated by a continuing need to assess the potential effects on stratospheric ozone of chlorofluorocarbons (CFCs) and other chemicals, as prescribed in the Clean Air Act, as amended (42 USC 7450). The topic has been the subject of intense study during the past decade; our report builds on that work, most notably on previous studies by the National Research Council (NRC 1975, 1976b, 1977, 1978, 1979b) and the National Aeronautics and Space Administration (NASA) (Hudson and Reed 1979). To prepare our assessment, we relied on our professional knowledge, on a concurrent technical review prepared under the auspices of NASA, the Federal Aviation Administration, the National Oceanic and Atmospheric Administration, and the World Meteorological Organization (WMO) (Hudson et al. 1982), and on a series of topical reviews prepared at our request by technical consultants. The consultants' reports are contained in Appendixes A to F.

PROCESSES DETERMINING OZONE CONCENTRATIONS

Ozone (O_3) is formed in the stratosphere by reaction of atomic oxygen (O) with diatomic molecular oxygen (O_2). The process is initiated by photolysis of O_2, that is, the dissociation of O_2 into atomic oxygen by absorption of solar ultraviolet radiation at wavelengths below 240 nanometers (nm). Photolysis of O_2 occurs mainly at altitudes above 25 km.

According to current understanding, approximately 1 percent of the ozone created in the stratosphere is removed by transport to the troposphere; the remaining 99 percent is destroyed by chemical reactions in the stratosphere that re-form ozone into O_2. The net effect of these chemical reactions is either the combination of ozone with atomic oxygen to form O_2, represented by the equation

$$O + O_3 \rightarrow 2O_2, \qquad \text{I}$$

or the combination of two ozone molecules represented by

$$O_3 + O_3 \rightarrow 3O_2. \qquad \text{II}$$

These equations represent the net results of a number of complex sets of reactions catalyzed by a variety of gases and chemical radicals present in the stratosphere in trace amounts.

Important examples of sets of reactions summarized by process I are

$$Cl + O_3 \rightarrow ClO + O_2 \qquad (1a)$$
$$ClO + O \rightarrow Cl + O_2, \qquad (1b)$$

$$NO + O_3 \rightarrow NO_2 + O_2 \qquad (2a)$$
$$NO_2 + O \rightarrow NO + O_2, \qquad (2b)$$

$$OH + O_3 \rightarrow HO_2 + O_2 \qquad (3a)$$
$$O + HO_2 \rightarrow OH + O_2. \qquad (3b)$$

Process I may also proceed by the direct path

$$O + O_3 \rightarrow 2O_2. \qquad (4)$$

These reactions are limited by the availability of oxygen atoms and therefore occur mainly at altitudes above 25 km. The reactions that limit the rates at which chains 1, 2, and 3 proceed are (1b), (2b), and (3b), respectively.

Process II summarizes reaction schemes in which atomic oxygen is not limiting, for example,

$$OH + O_3 \rightarrow HO_2 + O_2 \qquad (5a)$$
$$HO_2 + O_3 \rightarrow OH + 2O_2. \qquad (5b)$$

Reactions (5) account for most of the ozone lost below 25 km in current models. The chemistry of the lower stratosphere is complex, however (Appendix A), and one cannot exclude additional reaction schemes involving oxides of nitrogen and chlorine (NO_x, ClO_x) and oxidation products of hydrocarbons such as methane (CH_4).

Ozone removed from the stratosphere by transport to the troposphere is ultimately lost by chemical reactions in the gas phase or at the earth's surface.

The spatial and temporal distribution of the concentration of ozone reflects a dynamic balance among the processes that form and remove ozone (Figure 1.1). According to current understanding, photolysis of O_2 provides a global source of ozone of 50,000 million metric tons per year, with more than 90 percent of this amount formed above 25 km. Most of this ozone is removed by reactions represented by process I. At altitudes between 25 km and 45 km, reaction (2b) accounts for roughly 45 percent of the ozone removed while reactions (1b) and (4) each account for about 20 percent and reaction (3b) for 10 percent (S.C. Wofsy, Harvard University, private communication, 1982). About 1 percent of stratospheric ozone, 600 million metric tons per year, is removed below 25 km by process II, with a similar amount being lost by physical transport to the troposphere.

Only 30 percent of global ozone is stored at altitudes above 25 km, reflecting the relatively short chemical lifetime of ozone at high altitudes. The rest is contained in the region below 25 km, and more than 70 percent of the amount below 25 km is found at latitudes above 30°. The abundance of ozone below 25 km is determined by the balance between transport from the chemically more active region at higher altitudes and losses to the troposphere; its distribution is regulated by atmospheric motions.

Adding to the stratosphere substances that destroy ozone has the effect of creating a new balance between production and removal processes in which the total abundance of ozone is reduced. For example, stratospheric concentrations of chlorine monoxide (ClO) and nitrogen dioxide (NO_2) may be increased as a result of emissions of CFCs and nitrous oxide (N_2O) from human activities. The effects are persistent. A typical CFC molecule, CF_2Cl_2 for example, survives for approximately 75 years in the atmosphere before it is decomposed by sunlight releasing its constituent chlorine atoms in the

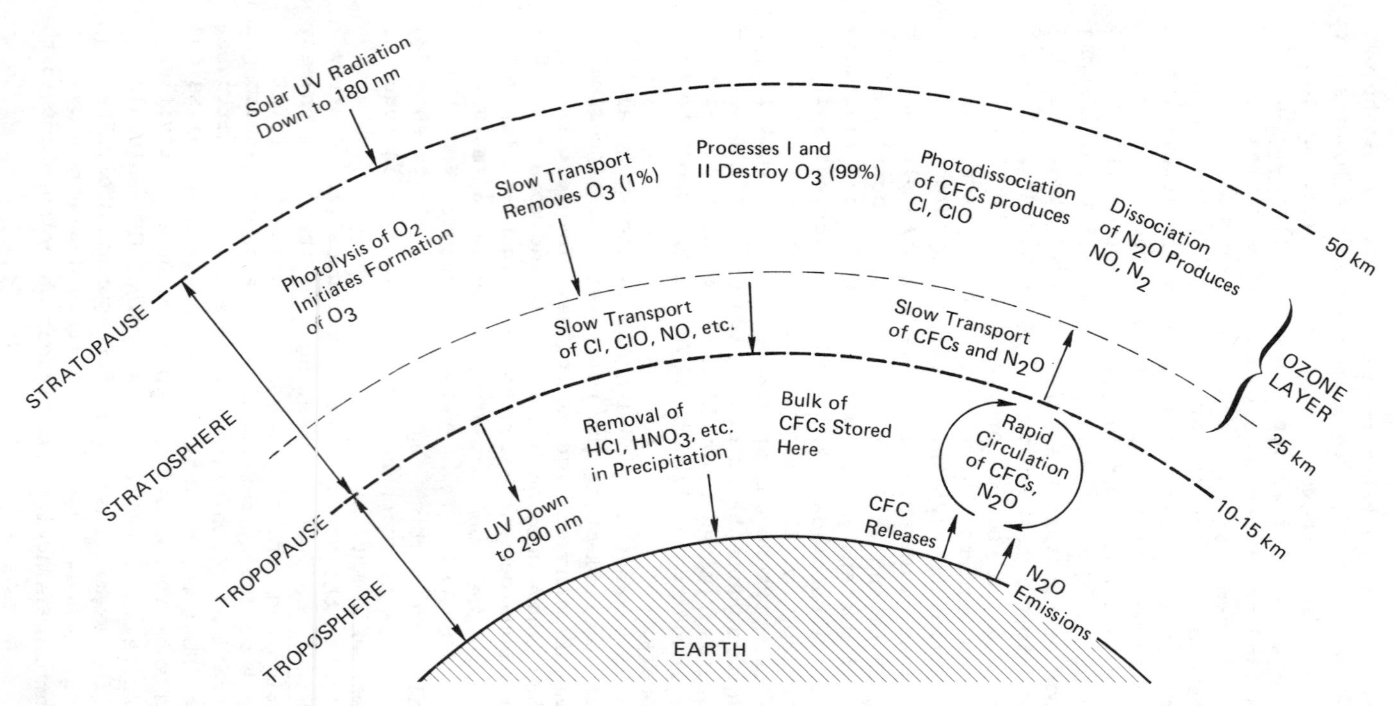

FIGURE 1.1 Representation of the processes that determine the concentration of ozone in the stratosphere.

stratosphere. A chlorine atom can affect recombination of between 10^4 and 10^5 ozone molecules during its lifetime in the stratosphere (on the order of two years) before it returns to the troposphere, mainly as hydrochloric acid (HCl). A similar situation holds for N_2O. Approximately 10 percent of N_2O molecules released to the atmosphere decompose by paths leading to production of stratospheric nitric oxide (NO), and subsequently NO_2, by reaction (2a). The average NO_x molecule also removes between 10^4 and 10^5 ozone molecules before it returns to the troposphere, after its typical two-year residence in the stratosphere. Current theoretical models lead us to conclude that the dependence of ozone concentration on altitude will also change, the net effect being a redistribution of ozone from higher to lower altitudes. Quantitative estimates of these effects have varied somewhat over the past decade (Appendix A).

Perturbations by Chlorine

Currently, approximately 3 parts per billion (ppb) of the lower stratosphere consists of chlorine bound in organic molecules such as methyl chloride (CH_3Cl), carbon tetrachloride (CCl_4), and CFCs (Hudson and Reed 1979, Hudson et al. 1982). Table 1.1 indicates the abundances of the more prevalent species; only methyl chloride is known to have natural origins. The table also shows estimates of current rates of release of man-made compounds found in the lower stratosphere.

Halocarbons decompose under the influence of sunlight at altitudes above 20 km; the fractional abundances (mixing ratios) of halocarbons (in ppb) are observed to decrease with increasing altitude (Appendix C). The chlorine produced by decomposition of halocarbons is converted to inorganic species, including HCl, chlorine nitrate ($ClNO_3$), ClO, and atomic chlorine (Cl). Hydrochloric acid is the major reservoir for chlorine at altitudes above 25 km (Appendix C). Concentrations of Cl, ClO, and HCl have been observed in the stratosphere; observations and predictions of theoretical models are in general agreement, although some difficulties remain (Appendix D), as we shall see.

Computer calculations using current understanding and incorporating new data on rates of several important reactions (Appendixes C and D) suggest that continued release of the CFCs, CF_2Cl_2 and $CFCl_3$, at rates

TABLE 1.1 Concentration in the Lower Stratosphere and Release Rates of Major Sources of Chlorine in the Stratosphere

Compound	Concentration (ppb)[a]		Rate of Release (million metric tons of Cl per year)
	Molecular	Chlorine	
Methyl chloride (CH_3Cl)	0.62	0.62	2[b]
F-12 (CF_2Cl_2)	0.30	0.60	0.19[c]
F-11 ($CFCl_3$)	0.18	0.54	0.20[c]
Carbon tetrachloride (CCl_4)	0.13	0.52	0.053[d]
Methyl chloroform (CH_3CCl_3)	0.11	0.33	0.35[e]

[a]Hudson et al. (1982).
[b]About 85 to 90 percent of CH_3Cl is naturally produced, the remainder being attributed to industrial sources (Cicerone 1981). The total release rate varies slowly in time because of the large contributions of natural sources.
[c]1980 release rate from "World Production and Release of Chlorofluorocarbons 11 and 12 through 1980," Chemical Manufacturers Association Fluorocarbon Program Panel, July 29, 1981. Release rate has decreased by about 20 percent from the peak rate of 1974.
[d]1976 release rate (NRC 1979b). The release rate is apparently relatively constant, although somewhat uncertain.
[e]Neely and Plonka (1978).

prevalent in 1977 would ultimately cause a net decrease of total global ozone roughly between 5 percent and 9 percent assuming no other perturbations (Hudson et al. 1982). We regard a representative result to be 7 percent (Appendix C). This would result in a smaller steady state reduction in ozone than reported in NRC (1979b), which was 16.5 percent with a 95 percent probability that the true value lies between 5 percent and 28 percent. (Other models current in 1979 gave reductions ranging from 15 percent to 18 percent (Hudson and Reed 1979). Estimates have fluctuated between roughly 5 percent and 20 percent over the past eight years as models have been refined (Appendix A).) The steady state reduction would be reached asymptotically in times on the order of a century. Calculations now indicate that the reduction would occur almost entirely at altitudes above 35 km, in the region of the stratosphere where the ozone concentration is determined primarily by chemical processes, with a smaller, partially compensating increase in ozone concentrations at lower altitudes. The current result obtains for both 1- and 2-dimensional models and further differs from that prevalent in 1979 in that earlier calculations showed regions of reduction both above and below 35 km.

The differences between current findings and those reported in 1979 are the result of refinements in the values for the rates of several reactions affecting the concentration of the hydroxyl radical (OH) (Appendixes C and D). The refinements are the result of improved laboratory measurements (Hudson et al. 1982). OH is important because the concentration of ClO in the lower stratosphere is particularly sensitive to it. Results of model calculations using current values for these reaction rates are in good agreement with observations of ClO for altitudes below 35 km (Appendixes C and D), whereas models using the reaction rates favored in 1979 give concentrations of ClO a factor of 3 higher than observed values in this range. The new reaction rates have not changed greatly the results of calculations for altitudes above 35 km, however, so that the amounts of reduction in ozone above 35 km obtained in the 1979 and current models are about the same. The models continue to indicate lower concentrations of ClO in the stratosphere above 40 km than are observed. We shall return to this discrepancy.

Increased attention to effects of releases of methyl chloroform (CH_3CCl_3) on stratospheric ozone is warranted because of the growing use of this compound, an industrial solvent. The release rate increased by a factor of about 50 between 1958 and 1978 (Neely and Plonka 1978).

Perturbations by Oxides of Nitrogen

The chemically active oxides of nitrogen in the stratosphere (such as NO_2) are thought to arise mainly from photooxidation of N_2O. N_2O is formed naturally by bacteria in soil and water. As indicated earlier, reactions involving NO_2 account for about 45 percent of the ozone removed in the stratosphere between 25 km and 45 km.

The human influence on the global cycle of fixed nitrogen is thought to be significant and increasing (NRC 1978). The global atmospheric concentration of N_2O appears to have increased by 2.7 percent (from 292 ppb in 1964 to 300 ppb in 1980) over the past 16 years (Weiss 1981, Weiss and Craig 1976). The concentration of N_2O in the atmosphere is likely to continue to increase with increases in emissions associated with agricultural practices, disposal of human and animal wastes, and possibly combustion; but we cannot say how or on what time scale.

An increase in N_2O concentrations of about 30 percent in the absence of other perturbations could cause a reduction in global ozone of an amount comparable with the 7 percent reduction currently estimated due to continued emissions of CF_2Cl_2 and $CFCl_3$ at 1977 rates, also taken as the sole perturbation. This estimate is based on current model calculations that indicate that, should the concentration of N_2O double in the absence of other perturbations, total global ozone would decline between 10 percent and 16 percent (Hudson et al. 1982).

Early attention to human influences on the stratosphere focused on effects of NO_x released by high-flying aircraft (NRC 1975). Models then and now suggest that an input of NO_x at altitudes above about 20 km should lead to reduction in stratospheric ozone. A source of NO_x at lower altitude, associated for example with subsonic commercial aviation, can modify local chemistry such as to cause an increase in tropospheric ozone. It has been suggested that reductions in the column of ozone above the earth's surface due to reductions in stratospheric ozone may be masked to some extent by increases in tropospheric ozone attributable to subsonic jets and urban smog.

Assessment of the impact on stratospheric ozone due to a combination of perturbations requires investigation of specific cases since the effects are not simply additive. Hudson et al. (1982) report the results of several studies of the effects of doubling atmospheric N_2O concentrations and continuing releases of CFCs at 1977 rates, both separately and in combination. The Lawrence Livermore National Laboratory (LLNL) model, for example, indicated a reduction of 12.5 percent due to doubling N_2O with a reduction of 12.9 percent due to the combination of perturbations. The LLNL model gives a reduction of 5.0 percent for CFC releases alone. Another model, from Atmospheric and Environmental Research, Inc., gives reductions of 9.5 percent for doubling N_2O, 6.1 percent for continuing CFC releases, and 13.0 percent for the combination. The results may be misleading, however, since current trends suggest a considerably longer time scale for doubling atmospheric concentrations of N_2O than for reaching the steady state reduction due to continued emissions of CFCs.

Perturbations by Other Species

Stratospheric ozone may be affected by human activity in a number of other ways. Of greatest potential concern are changes in concentrations of carbon dioxide (CO_2), water vapor (H_2O), and perhaps methane (CH_4).

The well-documented increase in atmospheric concentrations of CO_2 is directly attributable to combustion of fossil fuels and wood. This increase is expected to lead to a global warming of the atmosphere near the surface of the earth but is expected to cause a reduction in the temperature of the stratosphere (Fels et al. 1980).

Lower stratospheric temperatures would have at least two effects. First, the chemical removal processes affecting ozone that were described earlier are sensitive functions of temperature, being less efficient at lower temperature. Consequently, with lower temperature the equilibrium concentration of ozone would be higher. Current models incorporating this effect suggest that the steady state reduction in total ozone due to continuing emissions of CFCs at 1977 rates would change from 5 percent to 9 percent to between 4 percent and 6 percent if global CO_2 were doubled concurrently (Hudson et al. 1982). (Global CO_2 has increased by about 6 percent in the past 22 years.)

The possible second effect of lower statospheric temperatures resulting from increased CO_2 is a thermally driven change in stratospheric water vapor (H_2O) caused by a change in the temperature of the tropical tropopause. Dissociation of H_2O provides the source of hydrogen radicals, and these radicals play a key role in stratospheric chemistry regulating abundances of both active NO_x and Cl_x species in addition to their contributions to reactions (3) and (5). A complete model for stratospheric chemistry should include a description of H_2O interactions, a requirement beyond current capability.

Stratospheric ozone may also vary in response to changes in concentrations of CH_4, which plays an important role in reaction (1b) by regulating the partitioning of chlorine between HCl and ClO (Hudson et al. 1982). Recent reports (Rasmussen and Khalil 1981) suggest increases in global concentrations of CH_4, but likely future changes and their consequences are unknown.

CURRENT STATUS OF MODELS OF THE STRATOSPHERE

Theoretical models of stratospheric chemistry cannot be validated by measurements of total ozone only, owing to the diversity of factors, natural and man-made, that may affect ozone concentrations. Comparison of calculated and observed values for the concentrations of important trace species and radicals--such as OH, ClO, NO_2, and atomic oxygen--must play a central role in any orderly strategy for validating models.

In general terms, agreement in detail between the predictions of theoretical models and observations is excellent. For example, changes in reaction rates since 1979 have resulted in substantial agreement between theory and observation for ClO below 35 km. There are, however, three areas in which discrepancies remain. The discrepancies may or may not point to significant difficulties in modeling. Similarly, agreement between modeling results and observation of ClO, while encouraging, need not imply validity of the model at lower altitudes.

The improved agreement between observed and calculated concentrations of ClO in the lower stratosphere may be attributed mainly to changes in rate constants for reactions affecting OH. Concentrations of OH in current models are lower than values obtained in 1979, with the result that a larger fraction of Cl_x is now found as HCl. The chemistry of the lower stratosphere is complex, however. Agreement between model and observed values of ClO in the lower stratosphere should be considered necessary but not sufficient for validation. A more extensive and demanding test would require comparison of theoretical and observed profiles of other radicals, particularly OH.

To improve understanding of stratospheric chemistry also requires that attention be directed to the assumptions of the models and to the measurements against which models are tested. High-quality measurements are obviously prerequisite to validation of models. Confidence in observations of critical species is enhanced by using a number of independent, inter-calibrated techniques, each relying on different physical properties. Validation of measurement technique is difficult since concentrations of the important atmospheric species may vary in time and space on scales that are not well understood. Validation procedures involve coordinated studies in the field requiring considerable logistical support.

The assumptions of models are of two types: (1) input data on environmental conditions, reaction rates, and other parameters, and (2) the reaction schemes incorporated into the model. There are still uncertainties about the appropriateness of some assumptions common in current models. For example, the rate for reaction of OH with HO_2, an important path for removal of hydrogen radicals, remains uncertain despite extensive and continuing efforts in the laboratory. There are other reactions in need of similar clarification. Models are sensitive to assumptions about the abundance and distribution of stratospheric H_2O; the underlying physical and chemical processes that regulate this key parameter are not well understood. It is difficult to rule out the possibility of an important role for species not now included in models, and, if history is a guide, there may well be future surprises in this area. Models for the stratosphere have been adjusted over the past decade in just this manner to include gases such as ClO (1974), $ClNO_3$ (1976), and HOCl (1978) (Appendix A, Figure A.1), and there is current discussion of a possible participation of sodium (Kolb and Elgin 1976, Murad et al. 1981). Progress in recognition of missing species or reactions occurs through a combination of laboratory, field, and theoretical studies, the normal practice of validating models and resolving discrepancies.

As was noted earlier, there is reasonable agreement between model calculations and observations for ClO in the lower stratosphere. Currently, however, there is a discrepancy between theory and observation for ClO in the region above 35 km, where chlorine-mediated catalysis is most important. The average value for the concentration of ClO measured by Anderson and co-workers (see Appendix D) near 40 km is almost a factor of 2 larger than the value calculated from models. Furthermore, theory and experiment give different dependences of the concentration of ClO on altitude in the upper stratosphere. The ClO discrepancy is particularly important because it occurs at altitudes where ozone is most sensitive to perturbations caused by CFCs (Appendix D, Figure D.52).

Extensive ground-based observations of NO_2 have been made over a range of latitudes by J. Noxon (see Appendix C), revealing a sharp spatial discontinuity in concentration in the winter with very low concentrations poleward of the discontinuity. Thus far, no theroretical model has been able to explain this phenomenon.

A third area of discrepancy between current models and observations is in concentrations of $CFCl_3$ and CF_2Cl_2 at altitudes above about 20 km (see Appendix C, Figure C.11). Observed values are substantially lower than predicted values. The difference could be due to errors in model simulation of ultraviolet radiance in the lower stratosphere, which, if true, would imply that the CFCs have shorter residence time in the lower stratosphere. The issue is not resolved and requires continuing attention.

Nevertheless, the extent of agreement between measurement and theory is encouragingly good.

MONITORING AND ASSESSMENT OF TRENDS

Measurements of the total amount of ozone above a unit area of the earth's surface (called total column ozone) are essential for assessing the human influence on ozone (as well as the potential effects of changes in ozone on humans and other organisms). As detailed in Appendix A, total column ozone fluctuates on a variety of spatial and temporal scales owing to natural causes; these fluctuations tend to mask possible systematic changes due to man-made perturbations. For example, current models for single and combined perturbations predict a reduction of total column ozone over the past decade of less than 1 percent, but a change of this magnitude cannot be distinguished from fluctuations due to other causes (Appendixes D and E).

Models of the stratosphere predict that the largest reductions in ozone due to releases of CFCs should occur near 40 km. Reductions should therefore be most readily detectable at this altitude. Current models suggest that ozone concentrations at 40 km should have decreased by several percent over the past decade. There have been reports in the press that an effect of this order has been detected in data from satellite experiments (see, for example, Science, Sept. 4, 1981, pp. 1088-1089). The community of atmospheric scientists has not yet had the opportunity to scrutinize this evidence, which must therefore be regarded as preliminary (Appendix F).

Our ability to detect trends in ozone in the future will depend on the availability of consistent, high-quality data taken over long time intervals. Improvements in the current monitoring systems are feasible and clearly needed. For example, it is vitally important to improve and enhance systems for monitoring ozone profiles in the upper stratosphere that could provide a valuable early indication of systematic changes in ozone due to emissions of CFCs or N_2O, but existing data in the upper stratosphere are inadequate for this purpose. It is also imperative to continue, and desirable to expand, the high-quality monitoring of total ozone by Dobson spectrophotometers.

THE QUESTION OF EARLY DETECTION

A notable feature of the ozone issue is that a reduction due to increases in the tropospheric concentrations of CFCs or N_2O, once it has taken place, is expected to persist for more than 100 years even if the practices that caused it are stopped immediately. It is therefore important to detect an anthropogenic effect at the earliest possible time. Three methods currently exist for this purpose.

1. Measurement of total ozone. Relying on measurement of total ozone has the following advantages (Appendix E): There exists a relatively long historical base (30 to 50 years) of data. Ground-based instrumentation is available and may be readily complemented by observations from satellites. Finally, total ozone is most directly related to one of the consequences of depletion that is of concern, the possibility of enhanced exposure to ultraviolet radiation at the ground (Part II). Since, however, the reduction due to CFCs is expected to be concentrated at high altitudes, measurements of total column ozone are less sensitive indicators of an anthropogenic effect than are measurements of ozone profiles.

2. Measurement of ozone at high altitudes. The advantages of this method derive from the theoretical result that changes in ozone due to CFCs are predicted to be largest at high altitudes. Changes in the spatial distribution of ozone may be important for understanding the second major consequence of depletion that is of concern, the possibility of climate change (Appendixes B and C). The disadvantages stem from the difficulty of making the measurements, whose quality and stability are inferior to those of total ozone (Hudson and Reed 1979, Hudson et al. 1982). Satellite data are particularly subject to changes in calibration of instruments, which

cannot be refurbished; ground-based measurements by the Umkehr method give poor height resolution and are subject to perturbations by hazes and stratospheric particulate matter. Partly because of these difficulties, the data base is relatively small and somewhat fragmented (Appendix F). Ozone measurements using satellites would have the desirable attribute of obtaining temporal and spatial distributions that would be useful in validating 2- and 3-dimensional models.

3. Measurement of key radicals involved in chemical removal processes. Measurements of spatial and temporal profiles of important species such as ClO and OH may be combined with chemical models for assessment of trends and their causes, such that the dependence on specific models can be relatively slight. This method is in principle the most sensitive, but it is also the least direct.

The last approach is regarded by many experts as having already shown the effect of chlorine of human origin, mainly connected with emissions of CFCs. But this conclusion would be more firmly established with more direct confirmation, as discussed in the previous section. Ideally, all three types of measurement should be integrated (with due regard to their sensitivity) in a strategy for early detection of anthropogenic effects.

UNCERTAINTY

Quantitative estimates of the uncertainties inherent in current estimates of reductions in ozone due to emissions of CFCs and N_2O are difficult to obtain. The ability to make quantitative estimates of uncertainty depends both on what we know and on what we do not know. Such estimates employ professional judgments about the importance of various factors and the sensitivity of the results to potential changes in understanding.

Our major concern in estimating uncertainties in our understanding of stratospheric ozone is with the possibility that some key process or processes may be missing from current models. In an orderly scientific strategy, continuing development of models on the basis of an ongoing comparison with observational data is expected. Progress is stimulated by the existence of discrepancies or uncertainties and tends to occur in more or less discrete steps rather than uniformly. Our understanding of the lower stratosphere has improved over the past two years as a result of developments that may be attributed at least in part to efforts to resolve earlier (and larger) discrepancies between observed and computed values for the concentration of ClO. Agreement between observed and computed values of ClO is now satisfactory below 35 km, but, as noted earlier, there continues to be a serious discrepancy at higher altitudes. This disagreement illustrates the difficulty of estimating limits of uncertainty for current estimates for reduction in ozone due to CFCs.

For example, observed values of ClO at higher altitudes are larger than calculated values, suggesting that the long-term reduction in ozone could be correspondingly larger. One can, however, conceive of speculative chemical schemes that could suggest a stratosphere less vulnerable to perturbations.

In circumstances such as this, the usual ways of estimating uncertainty (using mathematically rigorous procedures) are not applicable. Instead we rely on professional judgment. The predictions of the current chemical scheme have been cross-checked against observed atmospheric data in many ways, and the agreement in general is quite good. As stated earlier, a representative estimate of potential steady state reduction of global ozone due to continued releases of CFCs at the 1977 rate in the absence of other perturbations is 7 percent. There continue to be, however, important discrepancies between theory and observation.

Our opinions are divided on whether there are sufficient scientific grounds to estimate the effect of resolving one of the discrepancies, that of ClO in the upper stratosphere, on calculations of ozone reduction. We agree that we do not know enough at this time to make a quantitative judgment of the uncertainty associated with the other major discrepancies, NO_2 at high latitudes and lifetime of CFCs in the stratosphere above 20 km.

Those of us who believe there are grounds to judge the effect of resolving the ClO issue conclude that our estimate of ozone reduction from CFC emissions should not change by more than a factor of 2.

Those of us unwilling to offer quantitative estimates of uncertainty hold the conviction that no rigorous scientific basis exists for such statements. We are concerned by implications of the discrepancies noted earlier. These discrepancies should be resolved in the next few years by orderly application of the scientific method with appropriate interaction between theory and observation. We see no reason to prejudge the result of this process.

Research during the past several years has enhanced our understanding of the factors affecting stratospheric ozone. Development of the field is progressing rapidly. We anticipate further developments in both observation and modeling in the next few years that will result in considerable improvement in our understanding, both clarifying and reducing uncertainties.

FINDINGS

1. Our understanding of the stratosphere has advanced considerably in the past two years. Progress is significant in all areas with improvements in our ability to model the system in more than 1 dimension, with impressive achievements in techniques for measurement of chemical reactions in the laboratory, and with major advances in our ability to measure concentrations of important trace species in the atmosphere. We note here that the success of the research is due in no small part to the breadth of the scientific effort involving scientists from many countries with support from both private and governmental sources. We expect continued improvement in understanding of the chemistry and dynamics of ozone reduction to result from research currently under way, planned, and proposed.

2. The concern regarding the possibility of reduction in stratospheric ozone due to CFCs remains, although current estimates for the effect are lower than results given in NRC (1979b). The change in estimates of ozone reduction reflects improvements in our understanding of chemical processes in the stratosphere below 35 km. There has been no significant change in results obtained by models for the stratosphere above 35 km. The major impact of CFCs is predicted for the height range of 35 km to 45 km.

3. The chlorine species Cl and ClO participate in a series of chemical reactions that destroy ozone. The radical ClO has been measured in the stratosphere in significant amounts and is believed to be primarily of human origin. Our current understanding indicates that if production of CFCs continues into the future at the rate existing in 1977, the steady state reduction in

total ozone, in the absence of other perturbations, would be between 5 percent and 9 percent. Previous estimates fluctuate between roughly 5 percent and 20 percent, with those current in 1979 ranging from 15 percent to 18 percent. Latest results also suggest that CFC releases to date should have reduced the total ozone column by less than 1 percent.

4. According to current understanding, increases of N_2O in the stratosphere would result in reductions in total ozone, with the largest effects occurring in the lower stratosphere. Although concentrations of N_2O in the stratosphere appear to be increasing, we cannot reliably project the future course of N_2O sources. If, however, the concentrations of N_2O in the atmosphere were to double, in the absence of other perturbations, current models suggest that the steady state reduction in the total ozone would be between 10 percent and 16 percent.

5. On the whole, there have been substantial improvements in the agreement between model predictions and observed profiles of trace species in the past several years. Three exceptions are still a cause for concern: Above 40 km, more ClO is observed than is predicted by current theory; the behavior of NO_x in winter at near-polar latitudes is unexplained; and concentrations of CFCs in the stratosphere above 20 km are lower than predicted by the models.

6. Examination of historical data (extending back 30 to 50 years) has not yet shown a significant trend in total ozone that can be ascribed to human activities. Current models of combinations of pollutants suggest that a reduction of total ozone to date from human activities would be less than 1 percent. No detectable trend would be expected on the basis of these results.

7. Data on total ozone should not be used alone to guide decisions on whether to take action to prevent future changes in stratospheric ozone. Although an important guide, analysis of trends in total ozone cannot by itself reveal causes of ozone reductions or increases. Such analysis, together with measurement of altitude profiles of trace species and ozone and theoretical modeling, offers promise of understanding causes of ozone changes and the consequences of alternative actions in response.

8. The impact of CFCs should be assessed in the context of a broad understanding of the variety of ways in which human activity can alter stratospheric

composition. Ozone may be reduced by increasing levels of CFCs and N_2O, but reductions might be offset in part by higher concentrations of CO_2 and perhaps CH_4. Human activities have already increased the amounts of CO_2 and CFCs in the atmosphere, and from the known release rates, further increases can be confidently expected. In addition, there is evidence that N_2O and CH_4 concentrations are also increasing. A special reason for concern about perturbations potentially caused by CFCs and N_2O is the long lifetime of these gases in the atmosphere, of the order of 50 to 150 years. Even if the releases of these gases were reduced, the atmosphere would not recover until far in the future.

RECOMMENDATIONS

In light of our findings, we believe it is important to maintain a competent, broadly based research program that includes a long-term commitment to monitoring programs. The research effort should extend over at least two solar cycles (of 11 years each) to distinguish between changes induced by variations in the sun from those associated with man. Accordingly, we make the following recommendations:

1. The national research program, including atmospheric observations, laboratory measurements, and theoretical modeling, should maintain a broad perspective with some focus on areas of discrepancy between theory and observation. A coordinated research program to understand the spatial and temporal distributions of key species and radicals merits highest priority. Observations should be extended to include studies of the equatorial and polar regions.
2. The global monitoring effort should include both ground-based and satellite observations of total ozone and of concentrations of ozone above 35 km, where theory indicates the largest reductions might occur. We also need data to define the variability of stratospheric temperature and water vapor. We regard sound, satellite-based systems for stratospheric observations as essential.
3. Potential emissions of a number of relevant gases, in addition to CFCs and N_2O, and their consequences for stratospheric ozone should be thoroughly evaluated and assessed. It is important that we understand current and

potential rates of emissions of these compounds and the effects these emissions might have on ozone in addition to understanding emissions and effects of CFCs. There is observational evidence that atmospheric concentrations of N_2O and CO_2 are increasing. Models should be developed to describe the combined effects on stratospheric ozone of future changes in releases of all relevant gases, such as CFCs, N_2O, CO_2, CH_4, CH_3Cl, and CH_3CCl_3.

PART II: BIOLOGICAL EFFECTS OF INCREASED SOLAR ULTRAVIOLET RADIATION

Chapter 2

INTRODUCTION

It is well known that ultraviolet radiation (UV) can be harmful to plants and animals including humans. The effects of UV on living cells and organisms depend on the wavelength of the radiation. The ultraviolet portion of the electromagnetic spectrum is conventionally divided into three parts--UV-A, UV-B, and UV-C--in order of decreasing wavelength (Figure 2.1). The divisions are somewhat arbitrary, based largely on our understanding of how UV affects humans. For the purposes of this report, UV-A is the wavelength region from 320 nanometers (nm) to 400 nm (near-UV); UV-B, wavelengths from 290 nm to 320 nm (mid-UV); and UV-C, wavelengths from 190 nm to 290 nm (far-UV).

The known harmful effects per unit dose of the shorter wavelengths, UV-C and UV-B, are greater than those of the longer wavelengths, UV-A (Blum 1959; Harm 1980b; NRC 1975, 1976a, 1979a; Parrish et al. 1978). A familiar effect of UV on humans is sunburn (Figure 2.1). UV also affects the metabolism of, kills, and mutates cells in culture, and is carcinogenic for animals, including humans.

The ozone layer provides protection from UV by absorbing the most harmful wavelengths. The spectrum of solar radiation reaching the surface of the earth for the current atmospheric distribution of ozone is shown schematically in Figure 2.1. Radiation in the UV-C band is essentially completely absorbed by stratospheric ozone and does not reach the surface of the earth; even with large reductions (tens of percents) in the concentration of stratospheric ozone almost no UV-C would be transmitted to the earth. Most of the solar UV-B also does not reach the surface of the earth. Absorption in the UV-B band is a sensitive function of the amount of ozone,

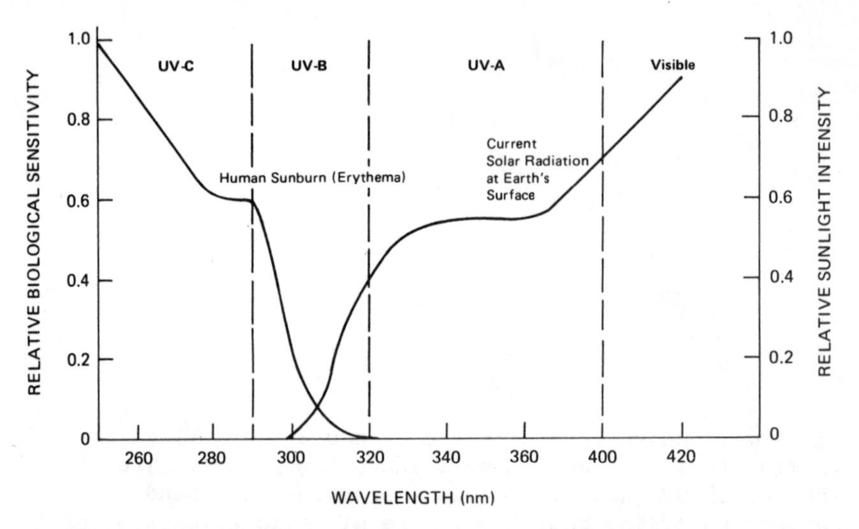

FIGURE 2.1 Schematic plots of the UV portion of the solar electromagnetic radiation currently reaching the surface of the earth and the biological sensitivity curve for human sunburn (erythema) are shown as functions of wavelength.

however, and so if ozone concentrations decrease, either as a result of natural causes or as a result of human activity, the amount of UV-B reaching the surface of the earth will increase and the harmful effects of UV will also increase. The amount of UV-A reaching the surface of the earth is not sensitive to changes in ozone concentration.

Changes in ozone abundance resulting from the release of chlorofluorocarbons, and other human activities, would take place only over a long period of time, probably decades. It is conceivable that many living creatures with relatively fast reproductive cycles could adapt biologically to a slow increase in the average intensity of UV, because they would go through many generations in the time it takes for the intensity to reach some new steady state value. Humans, on the other hand, could not adapt biologically nearly as rapidly. Furthermore, if an increase in UV gives rise to an increased incidence of skin cancer, the increased cancer incidence is not likely to be detected for many years after the increase in UV. Thus, the continued release of chlorofluorocarbons may lead to reductions in stratospheric ozone some time in the future, and that may lead to increases in the incidence of cancer in humans even farther in the future. The effects of human activities on stratospheric ozone are of concern for the long term, but the effects of current events on the future will not readily be reversible. The second part of this report addresses the biological effects to be expected from changes in solar UV. The uncertainties in understanding are large in spite of substantial advances in basic knowledge. These advances have not answered all of the important questions. A long-term commitment to research designed to answer the remaining critical questions is needed to facilitate predictions about the effects of enhanced UV on biological systems.

With new knowledge comes the possibility of reduced or increased concerns about ozone reduction, either from changes in understanding of the effects currently recognized or from previously unknown effects. Continuous surveillance of the problem by knowledgeable photobiologists is highly desirable, not only directly but also indirectly via basic research. For example, a number of years ago the fact that visible light can ameliorate the damaging effects of UV on human cells was not suspected. Now, as a result of experiments of a basic nature on cells in culture (Harm 1980a, Sutherland et al. 1974), this amelioration is recognized as an important factor (D'Ambrosio et al. 1981b, Sutherland et al. 1980b).

THE PROBLEM

At the surface of the earth the intensity of sunlight is a strong function of wavelength, decreasing rapidly for wavelengths below 320 nm (Figure 2.1). Intensities at wavelengths below 320 nm are affected most by changes in stratospheric ozone. Figure 2.2 shows the effect of large reductions in ozone on the spectrum of light reaching the earth. The net effect is a shift in the entire spectrum of UV at the surface of the earth toward shorter wavelengths; that is, the intensity of the short-wavelength UV increases. While the reductions of ozone illustrated in Figure 2.2 are much larger than is generally anticipated, the figure illustrates the point. For example, an approximate 50 percent decrease in stratospheric ozone gives rise to a change in intensity that increases from a factor of about 2 at 305 nm to a factor of about 50 at 295 nm. In general, for any change

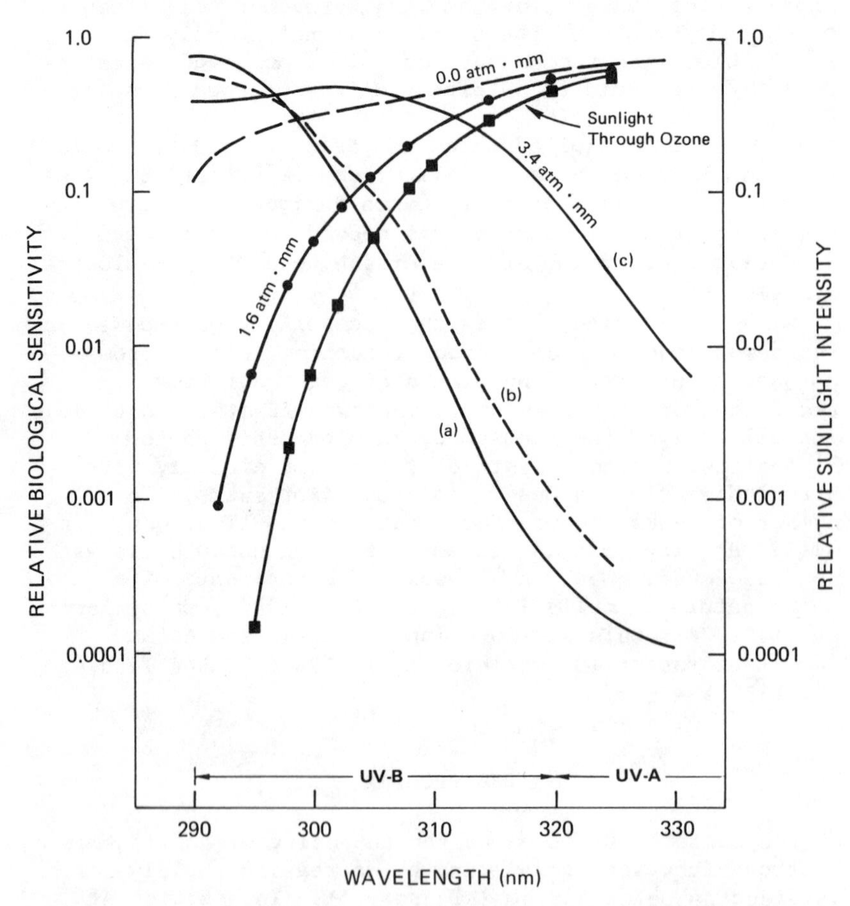

FIGURE 2.2 The relative intensity of sunlight (solar elevation of 60°) reaching the surface of the earth for different amounts of stratospheric ozone (the normal amount is close to 3.4 atmosphere · mm). The shapes of two biological sensitivity curves are also shown: (a) damage to DNA multiplied by the transmission of human epidermis, and (b) human erythema or sunburn. Curve (c) is the response of the Robertson-Berger meter (discussed in Chapter 5). (Source: The three curves of sunlight intensity are from U.S. Congress, Senate (1975); the two biological sensitivity curves are from Setlow (1974) and Scott and Straf (1977); the Robertson-Berger meter curve is from Berger et al. (1975).)

in ozone concentration, one can compute with reasonable confidence the change in the UV spectrum striking the surface of the earth. Hence a predicted decrease in stratospheric ozone will give rise to predicted increases in intensity as a function of wavelength of solar UV (Johnson et al. 1976).

The extent of the known deleterious effects of UV also depends strongly on wavelength and, as a rule, increases rapidly for wavelengths below 320 nm. Figure 2.2 shows two curves of biological sensitivity (Scott and Straf 1977, Setlow 1974). The figure illustrates the findings that UV-A wavelengths are much less biologically effective for damaging DNA or causing sunburn than UV-B, and that in the UV-B region the biological sensitivity per unit dose is an extremely sensitive function of wavelength. Thus, even if the increase in the absolute amount of UV penetrating the ozone layer is small, the changes will occur in a region of the spectrum that is very effective biologically.

Plots of biological sensitivity as a function of wavelength--so-called action spectra--are obtained experimentally. These experiments are difficult to do on simple biological systems and even more difficult to do on animals, plants, and ecosystems. Thus there are uncertainties in our understanding of the dependence of effects on wavelength. Furthermore, ethical considerations prevent the controlled investigation of some action spectra, specifically those for various types of skin cancer in humans.

Most of the available data do not derive from direct experiments on the biological systems of interest. For example, the basic data on human skin cancer are epidemiological: incidence, prevalence, and mortality at a relatively small number of locations in the United States. The locations differ in many ways, for example, in the average UV-B exposure during the year, the maximum UV-B exposure at any time during the year, the amount of visible light, and the ethnic and occupational backgrounds and life styles of the populations. Without data from many more locations that differ widely in the variables that might affect skin cancer incidence, it is not possible to use epidemiological data alone to determine the important variables or the action spectrum responsible for skin cancer. Thus we must draw inferences about the action spectrum for human skin cancer from animal experiments and molecular theories. Without knowing the action spectrum for a particular effect, that is, without knowing the biological sensitivity curve, it is not possible to make even rough predictions. For example, if curve (a) in Figure 2.2 were not the proper one to use because the major effect arose from wavelengths in the UV-A region, there would be no real consequences of ozone depletion on the biological system of interest. But if the sensitivity curve were as given in curve (a) of Figure 2.2, the depletion of ozone would have a large effect.

There are two general approaches to measuring and predicting the effects of increased UV on biological systems.

1. A straightforward approach is to irradiate a system with solar simulators, which mimic the spectrum of the sun, as a function of time and for various concentrations of ozone. This approach is useful for studying effects on crop plants and small animals, but, even if there were large numbers of such simulators available, it is impractical for studying effects on ecosystems because they are too large. In addition, the experimental irradiation of people is not ethical, even though large segments of the U.S. population willingly participate in a natural experiment of a sort through their propensity for sunbathing.

2. A second approach is to expand and apply photobiological theories of effects on molecules, microorganisms, cells in culture, plants, and animals in order to improve the data base and our understanding. Predictions about the effects of ozone depletion on complex biological systems, such as humans and ecosystems, can then be made from fundamental principles.

THE UNDERLYING BIOLOGICAL QUESTIONS

Part II of this report builds on the large amount of photobiological data accumulated in the U.S. Department of Energy's Climatic Impact Assessment Program and in two extensive National Research Council reports: _Environmental Impact of Stratospheric Flight_ (NRC 1975) and _Protection Against Depletion of Stratospheric Ozone by Chlorofluorocarbons_ (NRC 1979a). Since those reports were written, there have been important additions to the basic knowledge of photobiological processes and some modest increases in basic epidemiological data. The changes and refinements in knowledge are summarized in the chapters that follow.

Because straightforward solar-simulation experiments cannot be used to estimate most of the biological effects likely to result from a change in stratospheric ozone, the problem must be approached by determining directly, or indirectly by extrapolation from simpler biological systems, the answers to four key questions. With the answers to these questions, models and theories can be constructed from which reasonable predictions of photobiological responses can be made.

1. What is the shape of the dose-response curve? An increasing dose of UV produces an increasing biological effect, but the effect is usually not linearly proportional to the dose. The quantitative relationship between dose and response may be described by a dose-response curve. Figure 2.3 shows the general shapes of three possible dose-response curves. If the dose-response curve were a straight line, a 10 percent change in dose would give a 10 percent increase in effect. If, on the other hand, it were curved sharply upward, as in curve (c) in Figure 2.3, a 10 percent increase in dose would give rise to different increases in the biological effect, depending on the initial dose. For an initial dose of 2.5 arbitrary units per year, a 10 percent increase in dose would give a 70 percent increase in the biological effect. If the dose-response curve were curved downward, as in curve (a) in Figure 2.3, a 10 percent increase in the same initial dose would give rise to only about a 4 percent increase in the biological effect. Hence it is necessary to know the actual form of this relationship for the shortest wavelengths of UV that penetrate the ozone layer, say 290 nm, to the longest that have an important biological effect on the system being investigated. For the induction of cancer in mice, for example, this longest wavelength is near 320 nm. If the dose-response curves have similar shapes for all wavelengths investigated, one can have confidence that the fundamental photobiological processes are the same at all wavelengths. On the other hand, if the curves do not have the same shape at all wavelengths, different types of photochemical or photobiological mechanisms must operate at different wavelengths.

2. Is there a reciprocal relationship between intensity and duration of exposure in responses? In a number of biological systems, low intensities delivered for a long time give the same result as high intensities delivered for a short time, as long as the same total

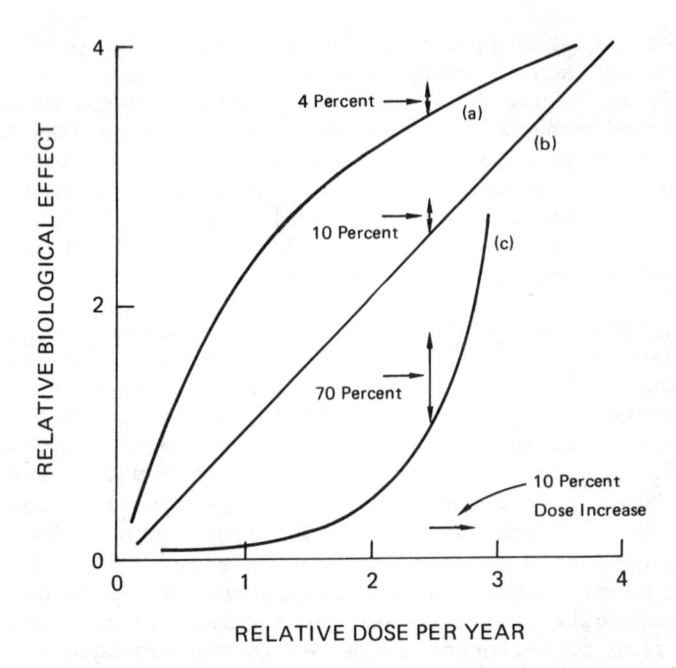

FIGURE 2.3 Hypothetical dose-response curves (a), (b), and (c), illustrating the effect on the changes in anticipated biological effects resulting from a 10 percent dose increase. (In the UV-B spectral region, a 5 percent change in ozone concentration will probably produce an approximate 10 percent change in dose.)

dose is given--the response is simply the product of intensity and time, or the total, time-integrated dose. Exposure time and dose rate are then said to be related reciprocally, and the reciprocity law holds. If the reciprocity law does not hold, one must know not only the dose-response relationship, but also the dependence of the response on the exposure time and dose rate. For example, in simple cellular systems, a given effect usually requires a higher dose at low intensities than at high intensities, presumably because during low-intensity irradiation repair processes take place and little damage accumulates (Harm 1980b). In rats a single dose is more tumorigenic than an equal dose fractionated over 12 weeks (Strickland et al. 1979). On the other hand, to produce tumors in 50 percent of mice by UV irradiation, a higher dose is required at high intensities than at low intensities (see Chapter 5). It is not known how intermittent exposures, as might actually be experienced by humans at work or during recreation, affect the dose-response relationship.

A difficulty in extrapolating the effects of laboratory-type experiments to the outside world is the fact that most laboratory experiments involve acute exposures usually taking only a fraction of a cell cycle time. In sunlight, however, many biological systems are exposed to low intensities for long times (chronic exposure). Since exposures to sunlight, and in particular to UV-B, may often be weak, early or late in the day, or during the winter, many chronic exposures may be at dose rates well below those used in the laboratory to determine whether reciprocity holds.

3. <u>How does biological sensitivity depend on wavelength</u>? It is clear from descriptions given above that the specific biological effects resulting from a change in amount of ozone depends critically on the action spectra. If these curves are not known from direct experiment or cogent theory, there is no theoretical basis for making a prediction of the effects of ozone depletion. The answers to questions (1) and (2) above must be known before the shape and the wavelength dependence of each action spectrum, that is, the relative effectiveness of different monochromatic wavelengths in producing the observed effect, can be determined. The product, wavelength by wavelength, of the action spectrum and the spectrum of sunlight at the surface of the earth gives the relative effectiveness of sunlight in producing the specific biological effect (Caldwell 1971, NRC 1979a, Setlow 1974), provided that interactive effects (see question (4) below) are small. Any ozone depletion will change the spectrum of sunlight at the surface of the earth. This change, when multiplied by the action spectrum, will give the radiation amplification factor, i.e., the percentage increase in biologically damaging UV per percentage decrease in ozone. The radiation amplification factor depends on the action spectrum.

4. <u>Are there effects at different wavelengths that interact</u>? The studies that have been conducted in the three areas discussed above have used single wavelengths of UV. An extrapolation to the effects of sunlight on crops, ecosystems, and humans from experiments in which the effects of single wavelengths are studied can be made only if the effects of the isolated wavelengths are purely additive, and not synergistic or antagonistic. Hence it is crucial to determine whether biological systems irradiated with a range or band of wavelengths

act as one would predict from the sum of the effects at discrete wavelengths. In addition, there may be other synergistic or antagonistic agents in the environment to consider, such as visible light, temperature, and chemicals. As will be discussed in Chapter 3, there are large synergistic effects between UV-B and longer wavelengths in many simple photobiological systems.

Despite the present uncertainties in understanding, there has been impressive progress in the extent of knowledge and in the delineation of the types of questions that can be answered easily. Certain questions may take several years and much data accumulation to answer, and some appear at present to be unanswerable but perhaps could be answered in the future with the help of a strong program of basic research.

Chapter 3

MOLECULAR AND CELLULAR STUDIES

SUMMARY

A decrease in stratospheric ozone has biological consequences that reflect two processes: (1) the increase in the intensity of UV-B reaching the earth as a function of decreasing wavelength, and (2) the increase in biological sensitivity with decreasing wavelength. The latter function is called an action spectrum. Knowledge of action spectra is important in evaluating the hazards of ozone depletion but is not sufficient for making quantitative predictions.

In the past few years, extensive advances have been made in understanding the effects of single-wavelength (monochromatic) light on simple biological systems such as bacteria and cultured mammalian cells. It is now known that the biological effects of UV-B are quite similar to those of UV-C (with which almost all experiments in the literature have been done), and that DNA is the target for many deleterious effects of UV. The action spectrum for damage to DNA is well characterized. However, despite the similarity between the DNA action spectrum and the action spectra for killing, mutating, or transforming mammalian cells in culture, irradiation of such cellular systems with broad bands of radiation does not give results that would be predicted from the sum of the effects at discrete wavelengths. There are good indications that the longer wavelengths in UV-B, or UV-A, may modify the effects of UV-B. For example, in a process called photoreactivation, cellular enzymes use longer wavelength light energy to reverse the effects of shorter wavelength light. Longer wavelengths may delay the growth of cells, leaving more time for repair processes to act on the short-wavelength damage. Thus since sunlight comprises a broad band of wavelengths, simple action spectra do not permit reliable predictions of the responses of biological systems to sunlight.

One of the more important recent findings concerns the fate of photoproducts in skin cells of animals and humans. Photoproducts result from the absorption of photons (quanta of light energy) by molecules. Sensitive ways have been designed to measure the most important UV-C photoproducts, pyrimidine dimers in DNA, in intact skin irradiated by sunlamps. The presence of these dimers has severe biological consequences, including mutation and cell death. Studies have shown that there are two processes that remove dimers from cellular DNA in vivo; one of these repair mechanisms acts in the dark (excision repair), and the other acts in the presence of longer wavelength UV and visible radiation (photoreactivation). Photoreactivation is highly specific for pyrimidine dimers. Preliminary estimates indicate that photoreactivation is very rapid in humans and takes place to an appreciable extent even while human skin is being irradiated by the UV-B in sunlight. Consequently, the level of pyrimidine dimers in cellular DNA in vivo depends upon the relative intensities of the dimer-forming wavelengths in UV-B and the dimer-splitting wavelengths in UV-A and visible radiation. This argument suggests that the effects of sunlight exposure on people might depend upon their exposure habits. For example, exposure in the afternoon might be much less deleterious than exposure in the morning. This is because the intensity of UV-B in relation to that of UV-A and visible light increases in the morning and is greatest at noon. If exposure ended at noon, dimer formation would be at a peak relative to dimer splitting. If exposure occurred in the afternoon, more photoreactivating (dimer splitting) activity would occur as the relative intensity of UV-A and visible light to UV-B increased. These findings underscore the importance of obtaining quantitative measures of life style and effective exposure.

Both excision repair and photoreactivation have now been found to occur more slowly in mice than in humans; this fact must be taken into account in extrapolating data from rodents to humans.

An added complication in attempting to estimate the effects of UV on, say, skin cancer induction is the effect of UV-B on the immune systems of animals and humans. A preliminary action spectrum has been determined for UV-induced immunosuppression in mice. If this spectrum were the correct one to use in estimating the carcinogenic effects of UV on humans, the predicted effects of ozone depletion would be significantly smaller than those obtained by using an action spectrum determined for DNA alteration or for production of erythema in human skin.

INTRODUCTION

The studies described in this chapter aim at supplying the basic knowledge required to estimate the shapes of dose-response curves, to extrapolate from high to low dose rates, to determine the appropriate action spectra, and to assess synergistic and antagonistic phenomena in analyzing biological effects of ozone depletion. Without such data it is not possible to extrapolate from cellular studies to predict the effects on ecosystems and humans.

Damage to DNA

Many of the deleterious effects of UV arise from the damage it does to DNA. Thus a great deal of effort has gone into understanding the biological effects of specific damages (photoproducts) in DNA. Early work in this field established definitively the shape of the action spectrum for damage to viral or bacterial DNA. It was supposed that this action spectrum was the appropriate one to use to calculate the biologically effective solar UV dose for humans and the change in this dose as a result of ozone depletion (Setlow 1974). This presumption was a big extrapolation. There were no reliable action spectra for UV effects on mammalian cells, and there was a good possibility that the correct action spectrum might be different from that for bacterial DNA, because in higher organisms DNA is not a naked polymer but is closely associated with proteins. This association might change the action spectrum and might give rise to other deleterious photoproducts, for example, cross-links between protein and DNA.

The effects of solar UV on DNA are emphasized here because in much of the current work on carcinogenesis it is assumed with reasonable assurance that DNA is an important target for initiating carcinogenic events. We enumerate in Appendix G a number of reasons for this presumption. It is important to realize, however, that other external or internal cellular factors, such as tumor promoters, hormones, or immunosuppressors, may be important in the development of cancer. In addition, UV could be both an initiator and a promoter, which would complicate the interpretation of dose-response curves.

DNA Repair

Although dosimetry for UV damage to cells is accurate and one can enumerate easily the DNA photoproducts existing immediately after irradiation (Setlow and Setlow 1972), the biological effects depend not only on the presence of such photoproducts but also on their lifetimes in cells. Most cells have repair mechanisms that remove the photoproducts or permit cells to ignore them (Friedberg et al. 1979, Hanawalt et al. 1979, Setlow 1978). A great deal is known about two of these repair mechanisms, photoreactivation and excision repair.

Photoreactivation is a process by which visible light or UV-A may reverse the effects of UV-B or UV-C radiation. In this process, an enzyme binds to a DNA molecule containing pyrimidine dimers. The complex of enzyme and damaged DNA can absorb UV-A or visible light, which causes the dimer to split, thereby repairing the damage. When the photoproducts are removed, the enzyme dissociates from the DNA. Photoreactivation is highly specific for pyrimidine dimers (Setlow and Setlow 1972), and when it is observed for a particular biological effect (such as survival, chromosome breaks, or mutation), it is taken as evidence that dimers caused that effect.

Many cells, including normal human cells, contain repair systems such as excision repair that operate in the dark. In excision repair, products of UV irradiation are removed from one strand of a DNA double helix by specific enzymes. The opposite, unaltered strand is then used as a template on which a new, unaltered strand is reconstituted. Excision repair is a very active process in normal human cells. Cells from individuals with a genetically inherited, sunlight-sensitive, cancer-prone disease called xeroderma pigmentosum are in almost all cases defective in excision repair. The high prevalence of skin cancer in such individuals is ascribed to the defect (Kraemer 1980). Mouse cells in culture also are defective in excision repair, and this defect must be taken into account in attempting to extrapolate from mice to humans.

ADVANCES IN KNOWLEDGE

Transformation of Cells in Culture

Transformation, an inheritable alteration of cells, can lead to cancer formation. UV-C and UV-B are able to transform mouse, hamster, and human cells in vitro, so that their growth on surfaces is no longer inhibited by contact with neighboring cells. The cells grow into piled-up clumps of cells instead of monolayers and can grow without being attached to a surface (Chan and Little 1976, DiPaolo and Donovan 1976, Sutherland et al. 1980a). The fraction of transformed cells per surviving cell increases with dose. In numerous experiments, colonies of UV-transformed rodent cells are usually tumorigenic when injected into certain mouse strains, but no tumorigenicity has been shown for the UV-transformed human cells described in the experimental results shown in Figure 3.1. Transformation of mouse and hamster cells is accomplished by single acute doses of UV, but transformation of human cells thus far has only been effected by several small UV doses (Sutherland et al. 1980a) or after a single dose under rather special conditions (Milo et al. 1981).

Photoproducts in DNA

Evidence that pyrimidine dimers are one of the major UV-C photoproducts in cells of higher organisms comes from studies on photoreactivation. About 65 percent of the lethal damage to frog cells, which have high levels of photoreactivating activity, is photoreversible (Rosenstein and Setlow 1980). Since under experimental conditions only 80 percent of the pyrimidine dimers in the cells are returned to monomers, the results indicate that approximately 0.65/0.8 (≈80 percent) of the lethal damage can be ascribed to pyrimidine dimers. In other systems with high levels of photoreactivating activity (frog cells and chicken embryonic fibroblasts), there is extensive photoreactivation of UV-C-induced chromosome aberrations and sister-chromatid exchanges (Griggs and Bender 1973, Natarajan et al. 1980). Between 75 percent and 95 percent of the dimers are reduced to monomers, and the effective reduction in sister-chromatid exchanges was calculated to be between 65 percent and 80 percent, indicating again that a major fraction of this particular

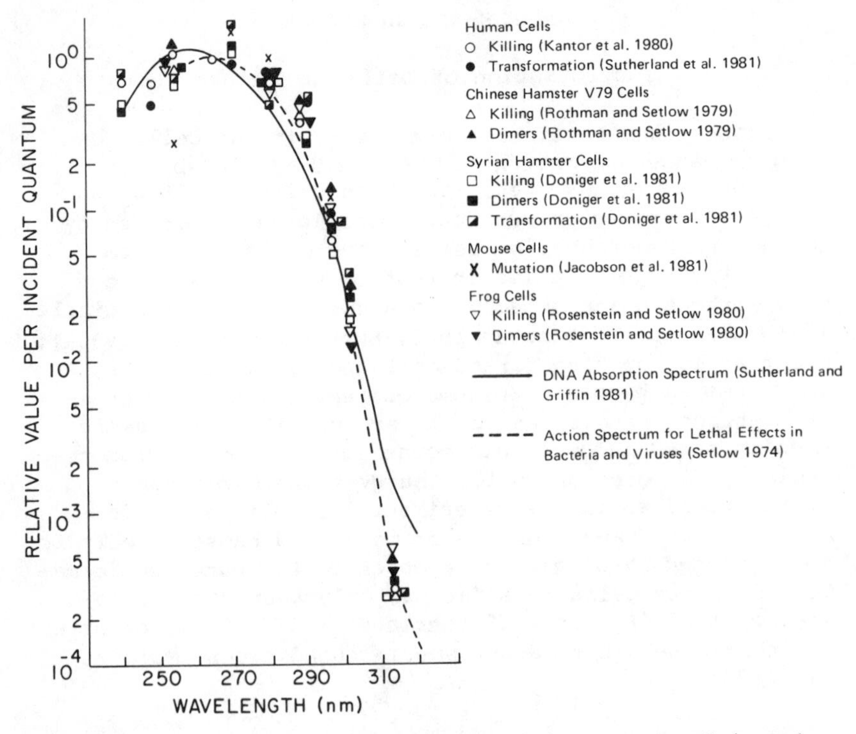

FIGURE 3.1 Points on action spectra of a variety of higher organism cells (normalized to 1.00 at 265 nm). Also shown are the absorption spectrum of purified mammalian DNA (solid line) and the action spectrum for lethal effects in bacteria and viruses (dashed line).

type of chromosomal damage arises from pyrimidine dimers in DNA (Natarajan et al. 1980). The initiation by UV-C irradiation of the transformation of human cells has now also been found to be photoreversible (Sutherland et al. 1980a).

That these conclusions may be extended to the UV-B region is strongly indicated for certain lesions. The killing of frog cells is photoreversed with equal effectiveness at each damage-producing wavelength tested between 252 nm and 313 nm (Rosenstein and Setlow 1980), and transformation of human cells by UV-B is photoreactivable (Sutherland et al. 1980a).

UV-C makes other photoproducts in DNA in addition to pyrimidine dimers, but these products have generally not been analyzed for their biological consequences. Examples include other photoproducts of thymine, DNA-protein cross-links, and single-strand breaks (but of a different character than those made by ionizing radiation) (Erickson et al. 1980, Setlow and Setlow 1972). The ratio of other photoproducts to pyrimidine dimers appears to be a function of the wavelength of irradiation (Hariharan and Cerutti 1977). If other products were important, they might distort an action spectrum for affecting DNA from the shape of the action spectrum for dimer formation. The action spectrum for dimer formation itself is a complicated function of wavelength since all types of pyrimidine dimers may be formed, such as thymine-thymine, thymine-cytosine, and cytosine-cytosine. The ratio of cytosine-thymine to thymine-thymine dimers appears to increase with wavelength from 290 nm to 313 nm (Ellison and Childs 1981).

It should be recognized that cell killing by UV-A is produced mostly by mechanisms that are quite different from those produced by UV-C, although the lethal lesion is still primarily damage to DNA. This is evidenced by the fact that bacterial mutants lacking DNA repair systems are very sensitive to UV-A. Roughly 90 percent of the UV-A killing requires oxygen (Webb 1977), whereas UV-C killing does not. Further evidence that UV-A damage is different is that the action spectrum for aerobic killing has a specific structure that suggests absorption of UV by dyelike molecules (Webb 1977). The fraction of UV-A killing that does not require oxygen (10 percent) may be due to direct production of damage in DNA, as evidenced by the fact that the action spectrum below 350 nm is without structure (Webb 1977) and similar to the absorption spectrum of DNA (Cabrera-Juarez et al. 1976, Peak and Tuveson 1979, Sutherland and Griffin 1981). Thus a small fraction of the lethal damage produced by UV-A may be similar in mechanism to that produced by UV-C.

There are many indications that the mechanisms causing effects on DNA from UV-B irradiation represent a mixture of the UV-C and the UV-A mechanisms, although the UV-C mechanisms clearly predominate, as shown, for example, by the data of Figure 3.1. Thus, even though damage to DNA from UV-B (the waveband critical in ozone depletion effects) is quite similar to that produced by UV-C, it is not identical to it.

Action Spectra for Affecting Simple Cellular Systems

Action spectra for killing a wide variety of higher organism cells in culture have been obtained for wavelengths up to 313 nm. Action spectra have been obtained for killing human cells, mutating mouse cells, and transforming hamster cells up to 313 nm and for transforming human cells up to 297 nm. The data points for a number of action spectra, normalized to a value of 1.00 at 265 nm, are shown in Figure 3.1. All the spectra that could be drawn for the various sets of points are very similar and coincide with the spectrum for forming pyrimidine dimers in the DNA of these cells. Figure 3.1 illustrates that (1) the action spectra for effects on mammalian and frog cells are very similar to those for effects on bacteria and viruses, (2) from 297 nm to 313 nm, the shapes of the spectra that could be drawn for the effects on cells of higher organisms are all approximately the same, and (3) there is a substantial deviation at long wavelengths between the values for all the action spectra and the absorption spectrum of purified mammalian DNA. At long wavelengths, the biological effects indicated in Figure 3.1 are smaller than would be predicted by the DNA absorption spectrum, probably because much of the UV is absorbed in the purine (guanine) residues of DNA whereas the lethal photoproducts are primarily in the pyrimidine (cytosine and thymine) residues that absorb little at long wavelengths.

Possible effects at wavelengths longer than 313 nm have not yet been determined in higher biological systems because the energies needed are higher than those achievable with the monochromatic UV sources used in past studies. The lethal responses of a large number of normal human cell strains have been examined at 254 nm and 313 nm (Smith and Paterson 1981), and the ratios of their sensitivities are close to those shown in Figure 3.1. A similar ratio is obtained for xeroderma pigmentosum cells (Smith and Paterson 1981), indicating that these repair-deficient human cells show equally enhanced sensitivity to 254 nm and to 313 nm. Thus for monochromatic radiation sources and the effects shown in Figure 3.1, almost all cells follow the DNA action--not absorption--spectrum. The primary conclusion drawn from the current understanding of action spectra is that all have similar shapes and hence the DNA action spectrum for mammalian cells may be taken to represent an average spectrum (not drawn in the figure).

Effects of UV on Photoproducts in Animal and Human Skin

Sensitive enzymatic techniques have recently been developed to investigate the amounts of pyrimidine dimers in irradiated animal and human skin. These experiments indicate that excision repair (Ley et al. 1977) and photoreactivation (Ley et al. 1978) are negligible in mice, except for neonatal mice, in which there is a low level of photoreactivation (Ananthaswamy and Fisher 1980). For human skin irradiated with one minimal erythemal dose by a sunlamp, it is possible to measure the number of pyrimidine dimers immediately after irradiation. If incubation is continued in vivo in the dark, there is an appreciable loss of dimers within 20 minutes, presumably by excision repair (D'Ambrosio et al. 1981a, Sutherland et al. 1980b). If instead the skin is illuminated with light from an incandescent lamp (predominantly visible wavelengths) for 20 minutes immediately after the sunlamp, there is more loss of dimers than from incubation in the dark. With even higher photoreactivating illumination, 80 percent of the dimers are lost in 4 minutes (D'Ambrosio et al. 1981b). These data indicate that normal human skin has both an active excision repair process and an active photoreactivation process, as was inferred from experiments on cells in culture (see the section below, "Mitigation and Enhancement of UV-B Effects by Light at Other Wavelengths"). The experiments imply that the illumination of human skin by sunlight results in a rather complex set of reactions for exposures that cover an appreciable period of time. UV-B exposure makes pyrimidine dimers, but during the exposure dimers are being excised, and the UV-A and visible components of sunlight are reversing the dimers by enzymatic photoreactivation. Thus in some situations low, chronic UV exposures might have little effect, especially if the exposure continues into the later parts of the day when the UV-B component of sunlight is relatively low and the photoreactivating (UV-A) component is relatively high.

Action Spectra for Immune Responses

Irradiation of certain strains of mice with sunlamps (UV-A and UV-B) suppresses two immune responses, rejection of UV-induced tumors (DeFabo and Kripke 1979,

Fisher and Kripke 1977) and contact hypersensitivity to the chemical trinitrochlorobenzene (TNCB) (see Chapter 5). Such systemic effects in mice irradiated at low doses (less than 1 minimal erythemal dose) could well be a contributing factor to the efficacy of UV as a carcinogenic agent.

The abolition of contact hypersensitivity to TNCB requires much lower doses than those needed for the lowered rejection of UV-induced tumors. This makes it experimentally possible to obtain an action spectrum for inhibition of contact hypersensitivity. Preliminary descriptions of this spectrum (De Fabo and Noonan 1980) indicate that it is in the UV-B region and falls off rapidly as wavelength increases above 290 nm. The action spectrum matches closely the absorption spectrum of several compounds known to be in mammalian skin, such as urocanic acid, and its values are appreciably greater than the absorption spectrum of DNA (Figure 3.1) at wavelengths greater than 290 nm. However, the target(s) (as yet unknown) for this effect may lie below the surface of the skin, and light absorption of skin is greater at shorter wavelengths. Hence, if this action spectrum were expressed in terms of quanta incident on the target as they are in Figure 3.1, rather than on the surface of animals, the values for the shorter wavelengths would be increased in relation to the values for the longer ones. As a result, the shape of the action spectrum for the depression of contact hypersensitivity would be closer to, but probably not identical to, the DNA absorption spectrum.

If the suppression of these immune responses is important in UV carcinogenesis, and if their biological sensitivities at longer wavelengths are greater than that for damage to DNA, the effects of ozone depletion would be less than those computed for a DNA action spectrum. This is because it is the steepness of the action spectrum in the UV-B that makes ozone depletion important (see Figure 2.2).

Experiments using broad-band sunlamps have shown suppression of contact hypersensitivity in mice. The shapes of the dose-response curves are similar to those of the narrow-band UV used to determine the action spectrum (Noonan et al. 1981a). It is important to determine how the effects of heterochromatic radiation on these immune responses compare quantitatively with the sum of the effects of monochromatic radiation.

Mitigation and Enhancement of UV-B Effects by Light at Other Wavelengths

Sunlight and UV-B in Bacterial Systems

Much of what is known concerning photobiological effects in animal cells originates from studies of bacteria. Bacteria are very small, have large populations in a small volume, and have a very short generation time, permitting efficient study of rare events, such as mutation. Although bacteria are different in many ways from human cells, the fundamental biochemistry and genetics are analogous. It is generally true therefore that advances in bacterial photobiology have preceded and have pointed to advances in the photobiology of cells of higher organisms.

It is now evident that UV-B acts on bacteria in much the same way as do shorter wavelengths (UV-C), namely, through absorption of energy in DNA. Sunlight, however, contains much more of the longer wavelength UV-A than UV-B. Although UV-A by itself can kill and mutate bacteria, it does this with only very low efficiency; the primary effect of the UV-A in full spectrum sunlight is a modification of the action of UV-B. This modification may be either antagonistic or synergistic. Because of these modifications, it is clear that most actions of sunlight on biological systems cannot be understood from experimental work using monochromatic light alone.

Among the known antagonistic processes are photoreactivation and photoprotection. Photoreactivation was defined earlier as a process in which UV-A or visible light eliminates the pyrimidine dimers produced by UV-B. The great excess of UV-A in sunlight suggests that much of the UV-B damage will be repaired in the same exposure to sunlight that produced the damage (Webb 1977). Such effects have been observed in bacterial systems, which are normally killed (to a 10 percent survival level) by about 30 minutes of exposure to bright sunlight, but which have been shown to be more sensitive to killing by sunlight under conditions where photoreactivation is prevented (by low temperature or by using a system possessing defective photoreactivating enzyme). It is not easy to demonstrate such effects in animal cells irradiated with sunlight, but one may confidently expect that any animal tissue (such as human skin) that contains photoreactivating enzyme will in fact have some of its lethal damage repaired in this way (see the section above, "Effects of UV on Photoproducts in Animal and Human Skin").

Another way in which UV-A may decrease the damage caused by UV-B is through the phenomenon of photoprotection, that is, protection resulting from a preceding illumination with UV-A (see the review by Webb (1977)). Photoprotection in bacteria may be induced by as little as 10 minutes of exposure to bright sunlight. It involves the induction by UV-A of a delay in growth, allowing more time after UV-B irradiation is completed for error-free dark-repair systems to repair the damaged DNA (Tsai and Jagger 1981). In addition, the UV-A effective in photoprotection may actively inhibit error-prone repair (Turner and Webb 1981). In wild-type *Escherichia coli* bacteria, for example, it has been shown that sunlight does not induce mutations. The UV-A wavelengths effective in reducing mutations are the same as those that delay growth, so this process may be a photoprotection effect (Tyrrell 1980). Photoprotection and growth delay in *E. coli* are produced by the absorption of light by an unusual base in transfer RNA, 4-thiouracil (Tsai and Jagger 1981), but this base has not yet been shown to exist in the cells of higher animals. However, growth delay in *Bacillus subtillis* has been shown to be due to the absorption of light by quinones in the bacterial membrane (Taber et al. 1978). The process could also occur in the cells of higher organisms, although it is not yet tested experimentally.

Synergistic effects have been observed in *E. coli* between UV-A and UV-B wavelengths, at high doses of UV-A (Turner and Webb 1981, Tyrrell 1978, Webb et al. 1978). Some of these synergisms appear to be due to the destruction of error-free DNA repair systems by the UV-A radiation. Error-prone recombination repair is responsible for mutation induced by all UV wavelengths in bacteria. At high UV-A doses, the destruction of error-free repair systems results in an enhancement of mutation; at low doses, enhanced mutation is seen only in cells that are defective in error-free repair systems and thus are not capable of photoprotection. Such synergisms may operate in the sunlight induction of skin cancer in those humans whose skin is defective in error-free repair systems, such as those with excision-defective xeroderma pigmentosum (Maher and McCormick 1976).

Finally, consideration must be given to UV-induced repair systems. In the UV-C region, for example, UV-C itself induces the error-prone repair system that is responsible for most of the mutation produced by UV-C in bacteria (Witkin 1976). It has recently been discovered that UV-A will induce a repair system in bacteria that is capable of repairing damage caused by UV-A (Peters and Jagger 1981). It is not yet known if the UV-A system will repair damage caused by UV-B, or if it is error-prone and would therefore produce mutation.

Sunlight or Broad-Band Radiation and Mammalian Cells

In mammalian cells, comparisons have been made between 254-nm radiation, sunlamp radiation in the range from 290 nm to 365 nm, sunlamps filtered to remove wavelengths below 300 nm, and sunlamps filtered to remove radiation below 310 nm. The results depend on the response being studied. For example, in Chinese hamster ovary cells, sister-chromatid exchanges are proportional to the amounts of pyrimidine dimers made by 254-nm radiation or by sunlamp radiation at wavelengths greater than 290 nm (Reynolds et al. 1979). However, the ratio of killing efficiency to dimer production, or mutation efficiency to dimer production, increases as the shorter wavelengths are removed from the radiation bands with mutation per dimer increasing more rapidly (Zelle et al. 1980). Similar results are obtained with Chinese hamster V79 cells, where the longest wavelength band used (greater than 310 nm) produces, in the time of a typical irradiation, no cell killing but appreciable mutation, and with mouse cells, where it produces very little killing but considerable transformation (Elkind et al. 1978, Suzuki et al. 1981).

Thus it seems as if heterochromatic light in the longer wavelength regions of UV-B does not act as the sum of a series of monochromatic wavelengths. On the other hand, it is not clear which wavelengths are interacting to give the apparent synergistic effects for mutation and transformation. The interaction may arise between wavelengths in the 310-nm to 315-nm region and longer wavelengths such as UV-A. At present, the quantitative response to an enhancement in UV intensity in the region of 305 nm to 310 nm as a result of ozone depletion, with the other, longer UV wavelengths remaining constant, is not known.

There is some evidence for the existence of UV-induced repair systems in mammalian cells (Bockstahler and Lytle 1977, Rommelaere et al. 1981). Experiments investigating these mechanisms have used UV-C irradiation, and, in view of the shape of the average mammalian action spectrum in Figure 3.1, one would expect similar findings for UV-B radiation. The evidence for the existence of UV-induced repair is the ability of irradiated cells to reactivate UV-irradiated viruses that are used to infect the cells. Irradiation of cells before virus infection results in an enhanced survival of the infecting UV-irradiated viruses, and in some instances a higher frequency of mutations is observed in irradiated viruses infecting irradiated cells (Das Gupta and Summers 1978). The magnitudes of the observed effects are small, and the extrapolation of such data from effects on viruses to effects on the cells themselves has not been made.

RESEARCH RECOMMENDATIONS

The following list of unclarified issues is intended as a guide for future research. The list is not exhaustive. It has been limited to those issues that should receive attention first, but it is not organized according to priority.

1. It would be useful to know the shape of the action spectrum for affecting so-called DNA functions of mammalian cells at wavelengths greater than 313 nm.
2. An understanding is needed of why broad bands of UV (heterochromatic radiation) do not seem to act like a simple sum of monochromatic wavelengths in terms of their effects on DNA. Studies of synergistic effects between UV-A and UV-B (for example, in bacteria) are fundamental to understanding the mechanisms of cancer induction by sunlight.
3. The quantitative aspects of the immune response of mice to monochromatic wavelengths versus the response to broad bands of UV-B should be explored. The molecular and cellular mechanisms for immune system effects and wavelength dependence should be investigated.
4. An understanding of the mechanism of neoplastic transformation by UV in vitro is needed. In some rodent systems, the level of transformation is so high--close to 100 percent--that this transformation process looks suspiciously like a triggering mechanism that controls the regulation of cell growth rather than like an effect on a specific gene or genes (Kennedy et al. 1980).

5. The characteristics of UV-A-induced repair systems in bacteria need to be determined. The possibility of the existence of such systems in mammalian cells should be examined.

6. Data are needed on the rates of repair, in the dark and in the light, of UV-irradiated human skin cells as a function of UV dose. The differences, if any, between acute and chronic irradiations should be determined. One might be able to study the responses of individuals who are exposed to high levels of UV-B as part of therapy for psoriasis. The aim of such studies would be to determine whether the kinetics of dark repair of damage from dimers in human skin show two components, a slow and a fast one, as is true for human cells irradiated in vitro. (The fast component represents repair of DNA in the so-called linker regions of chromatin, and the slow reaction is the repair in the core regions of chromatin (Cleaver 1977, Smerdon et al. 1978). The latter is not as readily accessible to enzymes as is the former.) Equally important questions are, what other types of biologically important damages occur in skin, what are their lifetimes, and are any of them persistent?

Chapter 4

ECOSYSTEMS AND THEIR COMPONENTS

SUMMARY

In order to predict the effects of increased levels of UV-B on natural and cultivated ecosystems, the nature of the interactions of organisms with environmental variables and the adaptations of organisms to changes in nutrients, predation, climate, and light must first be understood. Only then can dose-response relationships for UV-B effects be established. Because these interactions are complex, mathematical models are often used to express them and to facilitate data handling. The usefulness of these models is limited, however, by the information available. Thus, although data are available on UV-B effects on specific organisms and on UV-B effects on life cycle stages of organisms, the relationship between the effects on the individual and the effects on the population is not clear.

It appears that the yield of food from domestic animals will not significantly decrease even with the most extreme projections of ozone depletion. Economically important cultivated crops may have reduced yields from increased UV-B levels. Further assessment is needed of the organismal and cellular properties and adaptations in plants and in most animals that modify the direct and indirect effects of UV-B. Such research must be conducted under carefully simulated field conditions. This capability exists in a number of laboratories; however, a subtropical facility would be valuable. Other food, fiber, and medicinal crops have received relatively little attention and the effects of enhanced UV-B on them cannot be assessed.

The potential impacts of increased UV-B on natural terrestrial ecosystems have received limited study. Any attempt at the present time to predict potential consequences would be subject to large uncertainties. Nevertheless, several physiological processes of plants (e.g., leaf growth and photosynthesis) have been shown to be adversely affected by UV-B. Most of these responses were determined in growth chambers or greenhouses where visible radiation, UV-A, and other environmental factors did not simulate ambient conditions. Since ambient visible radiation and UV-A ameliorate most, if not all, of the deleterious effects of UV-B, most of this research needs to be repeated under field conditions for verification.

Terrestrial faunal ecosystems have received almost no attention, and nothing can be said of their level of susceptibility. In part, such neglect may be justified because the few available studies indicate that in addition to possessing physiological and biochemical mechanisms to alleviate the effects of radiation impinging on the organism, some animals possess behavioral mechanisms that lessen exposure to presumptively damaging radiation.

Marine faunal ecosystems have received some attention. Both freshwater and marine systems have been examined in several U.S. laboratories. It is, however, difficult to assess the risk factor of increased UV-B directly. In part, the vertical mobility of aquatic creatures and the fact that the dose changes exponentially with depth in the sea make it difficult to measure the received dose as a function of time, although upper and lower bounds can be assigned. No assessment has been made of the consequences of the qualitative or quantitative changes in a natural aquatic ecosystem exposed to enhanced levels of UV-B to determine whether one or more members of that system are particularly sensitive to UV-B. Several aquatic organisms have been exposed to UV-B in laboratory situations, and most of these studies have shown deleterious effects on the organisms tested. However, as with plants, the effects appear to be modified by visible light and UV-A. The research needs to be repeated under field conditions for verification. If aquatic organisms are generally susceptible to UV-B damage in their natural setting, the effects of increased solar UV-B could have profound consequences on the stability of the food chains upon which the fish and shellfish used by humans depend. A reduction in primary food organisms (for example, algae) could drastically alter the protein supply for large numbers of people throughout the world and have obvious social and political consequences.

INTRODUCTION

Studies involving several aquatic organisms, insects, and terrestrial plants and animals exposed to various levels of enhanced UV-B have shown a number of detrimental effects on growth, reproduction, and physiological processes. The capacity to tolerate increased levels of UV-B through acclimation or repair of UV-induced damage appears to be limited and is, to some degree, species dependent. The value of most of these studies is in question, however, because they have evaluated UV-B doses either considerably beyond those expected to occur on the basis of currently projected estimates of atmospheric ozone reduction (approximately 7 percent) or under experimental conditions with limited extrapolation potential to natural conditions. The U.S. Environmental Protection Agency is currently supporting limited research programs on both terrestrial crop plants and aquatic organisms, but the preponderance of projects recommended by the NRC (1979a) to help resolve the great uncertainties regarding the biological effects of increased UV radiation have not been implemented. Therefore the uncertainties still exist.

EFFECTS ON PLANTS

Research Difficulties

The inherent complexity of simulating a reasonable representation of enhanced UV-B regimes under ambient field conditions has confined research almost exclusively to greenhouses and growth chambers. However, the levels of visible radiation and UV-A are considerably lower under these experimental conditions than under ambient field conditions. Recent evidence strongly suggests that ambient field levels of visible radiation can substantially reduce or even negate the damaging effects of UV-B (Sisson and Caldwell 1976, Teramura et al. 1980). Thus extrapolation of research conducted under low visible radiation regimes to effects under field conditions would be tenuous and would undoubtedly overestimate the potentially deleterious effects of enhanced UV-B on plants. Furthermore, the interactive effects of UV-B and other environmental stresses (e.g., water, temperature, air pollutants, and even UV-A) have not been adequately addressed in the context of reduced ozone concentrations.

Nevertheless, existing evidence suggests that an increase in terrestrial UV-B equivalent to that which would be caused by a 7 percent reduction in atmospheric ozone concentration could be potentially damaging to some higher plants (Biggs and Kossuth 1978). Credible predictions regarding the severity of damage to economically important crop plants or to natural plants are not yet feasible. This is even more true with regard to predictions of species displacement or perturbation within the world's natural ecosystems, because there is less information available.

Although relatively few plants have been evaluated, a wide range of sensitivity to UV-B radiation has been found in crop plants (Biggs and Kossuth 1978) and agriculturally derived varieties (cultivars) within single species (Krizek 1978). This range of sensitivities may be due in part to the differing capacity or efficiency of plants to repair UV-B damage. Plants also appear to differ in their capacity to attenuate UV-B before it is absorbed by the target molecules. Therefore information is needed regarding repair processes, the limits and rates of acclimation, and whether genetic control for acclimation is already present or must be developed (an evolutionary process) before the question of the effects of increased UV-B on plants can be addressed. This information is not currently available.

Advances in Knowledge

NRC (1979a) addressed the applicable research to 1979 and the limitations of existing light sources for simulating increases in UV-B corresponding to those that would be caused by reduced atmospheric ozone concentrations. Although the present discussion incorporates the results of some research done before 1979, advances made since 1979 are emphasized. Unfortunately, little applicable research has been done since then, because of the research difficulties described above and a lack of funding.

The variation among plant species in sensitivity to UV-B (Biggs et al. 1975, Biggs and Kossuth 1978, NRC 1979a) may be due, in part, to different acclimation potentials. An ability to tolerate even ambient UV-B levels appears to be induced by concomitant exposure to UV. For example, Bogenrieder and Klein (1977) found that *Rumex alpinus* seedlings grown in an environment free of

UV-B displayed severely depressed photosynthetic rates when exposed to ambient UV-B levels. A similar response was found in R. patientia again grown in a UV-B-free environment, but then exposed to enhanced levels of UV-B under ambient conditions (Sisson and Caldwell 1976). Ambient levels of UV-B even killed some R. alpinus plants after a 3-day exposure period, although all of the R. patientia exposed to enhanced UV-B survived. As discussed by Caldwell (1982), if acclimation to environments with high intensities of UV is a phenotypic response already available to the plant, the anticipated rate of atmospheric ozone depletion would not be of concern. If, however, acclimation involves genotypic changes that must occur over a long period of time, the rate of ozone depletion would be of considerable importance.

Several studies have recently addressed the question of acclimation by investigating leaf epidermal transmittance of UV-B. Wellman (1974) has shown that UV-B-absorbing pigments are synthesized in some species in response to UV levels equivalent to existing ambient fluxes. Leaf epidermal extracts (containing flavonoids and other related pigments) from several plants increased their absorbance at 380 nm after exposure to UV-B (Robberecht and Caldwell 1978). Extracts from squash (Cucurbita pepo) leaves exposed to three levels of UV-B increased their absorbance as dose increased (Sisson 1981). Although absorbance by these extracted pigments increased substantially with increasing UV-B dose rate, photosynthesis and leaf growth were repressed at the higher radiation level. Even though UV-B apparently induced a synthesis of leaf pigments, the attenuation appeared to be insufficient to protect leaf growth processes and the photosynthetic apparatus completely. Nevertheless, this response would be of significant value in alleviating the potentially deleterious effects of any increase in UV-B at the earth's surface.

The intensity of biologically effective UV-B (NRC 1979a) increases by a factor of more than 7 from the arctic (70°N) to the equator (Caldwell et al. 1980). The increase experienced in moving through this latitudinal gradient greatly exceeds the increase expected from atmospheric ozone reduction at temperate latitudes. The leaf epidermal transmittance of UV-B of several plants along this gradient was evaluated by Robberecht et al. (1980). The calculated mean effective UV-B dose transmitted by the epidermis to the physiologically active mesophyll cells was found to be similar along this gradient even though the ambient levels of UV increased substantially. Thus along this latitudinal gradient plants are apparently coping with ambient UV through acclimation processes that include a reduction in epidermal transmittance as intensity increases. Whether plants will be able to cope with additional UV-B along this gradient by increasing the attenuation properties of the cuticle, epidermis, or mesophyll cell structure and function is not known.

In attempting to evaluate the potential impacts on higher plants, it is important to know whether photosynthesis and other physiological processes have a reciprocal relationship (i.e., damage is cumulative and dose dependent) with UV-B damage. Reciprocity has been demonstrated in isolated chloroplasts (Jones and Kok 1966) and for photosynthesis in a sensitive plant exposed to UV-B over a 50-day period (Sisson and Caldwell 1977). Trocine et al. (1981) demonstrated that UV-B damage is cumulative in two of three seagrasses tested. They suggested that the differential degree of UV-B sensitivity within these species was a function of epidermal cell wall thickness and associated transmittance properties. Teramura et al. (1980) demonstrated that reciprocity also applies for photosynthesis in soybeans; the damaging effect of UV-B was shown to be more deleterious when visible radiation was low. In these studies, the apparent dependence of response on the total dose rather than on the dose rate suggests that repair of damage within the photosynthetic apparatus may not involve nucleic acid repair systems, which are dose rate dependent (Caldwell 1982). Although the reduction in photosynthesis is less pronounced at high visible radiation levels (Sisson and Caldwell 1976, Teramura et al. 1980), suggesting either photoprotection or photorepair (see Chapter 3) of the photosynthetic apparatus, the particular repair mechanism(s) involved has not been determined.

Caldwell et al. (1980) studied leaf inclination as an avoidance mechanism for reducing the solar UV radiation loads on plant leaves. However, at temperate latitudes 40 percent to 75 percent of solar UV-B is in the nondirect sunlight that reaches the leaves, which substantially reduces the effectiveness of leaf inclination as an avoidance mechanism. Their calculations indicated that even vertical foliage received at least 70 percent of the daily effective UV-B. Consequently, a breeding program directed at developing crop plants of economic importance with canopies to avoid direct-beam UV radiation may be of little use.

Several sites of inhibition within the photosynthetic apparatus have been determined (see review by Caldwell 1982). Inhibition of electron transport associated with photosystem II (Brandle et al. 1977, Yamashita and Butler 1968) disruption of thylakoid membranes and other structural components of the chloroplast (Brandle et al. 1977, Mantai et al. 1970), and inactivation of photosystem I (Okada et al. 1976) have been demonstrated after exposure to UV. Recently, Vu et al. (1981, 1982) demonstrated partial inhibition of carboxylating enzyme activity. Thus UV might be a nearly universal inhibitor of component reactions within the photosynthetic apparatus, as well as deleterious to its structural integrity. However, these studies were conducted in growth chambers or greenhouses with correspondingly low visible radiation levels. In order to predict real-life effects of increases in solar UV, the studies would need to be repeated under ambient conditions where UV-B can be adequately supplemented.

Since physiological responses of plants to UV are highly wavelength dependent, an appropriate action spectrum for weighting heterochromatic UV-B becomes necessary for expressing effective dose, determining threshold levels for damage, and developing predictive dose-response relationships. The inconsistency in conclusions that can be arrived at by using different action spectra has been thoroughly addressed elsewhere (Caldwell 1982, NRC 1979a). Although action spectra are useful for a prediction, their utility may be reduced because of apparent synergisms arising from interactions due to different wavelengths (Elkind et al. 1978, Elkind and Han 1978) (see Chapter 3). It is not known whether this synergistic effect is a general phenomenon within plant physiological processes. Attempts to develop new action spectra should therefore incorporate simultaneous investigations into potential synergistic action among the wavelengths involved. This will facilitate accuracy in dose-response relationships and make them more useful in predicting the consequences of increased intensities of UV-B at the earth's surface.

EFFECTS ON DOMESTIC ANIMALS

Among the domestic animals that are necessary for the maintenance of our food supply, only one breed of cattle appears to show deleterious effects from exposure to solar UV. The white-faced Hereford shows some susceptibility to injury and disease of the eye. Kopecky et al. (1979) have shown that UV-B is a probable causal factor in cancer eye (bovine ocular squamous cell carcinoma) and enhances the onset of infectious bovine keratoconjunctivitis (IBK), or pinkeye, in these cattle. It is questionable, however, whether the level of injury is serious enough to warrant extensive further investigation.

Ladds and Entwistle (1977) reported on squamous cell cancers occurring on the ears and nose of sheep. They found that incidence in tropical Queensland, Australia, was greater than that found in temperate areas in earlier studies (Lloyd 1961). Increasing incidence with advancing age was also demonstrated. The authors suggest that squamous cell cancer in sheep would provide a good model for studies of skin cancer in humans.

EFFECTS ON AQUATIC ORGANISMS

Research Difficulties

Measurement of the effects of enhanced UV-B on aquatic organisms presents difficult experimental problems. It is possible and desirable to assess effects of UV-B on individual organisms under controlled laboratory conditions. These experiments can at least establish which organisms may be sensitive to enhanced UV-B. It is in determining the relationship of the dose received by the organism in the laboratory to that received by the organism in its natural ecosystem that uncertainties occur, because of the many variables in natural systems. First, position of organisms in the water column is conditioned by UV-B, by light other than UV-B, by nutrient availability, by the nature of turbulent mixing of the waters by wind, and, in shallower regions, by tidal frictional forces along the bottom. Second, there are seasonal and annual variations in species compositions, in larval developmental stages, in predation success, and in the physical transport, turbidity, and pigment-absorptive characteristics of coastal and estuarine waters where major food chain productivity occurs. The large statistical variance associated with the reproductive success of organisms that produce large numbers of spawn (up to 10^8 per adult female) makes it difficult to assess the effects of small changes in environmental factors (e.g., a 10 percent change in UV-B levels) with any degree of statistical significance. This would be particularly difficult without precise base line data regarding the effects of present levels of UV-B. It is important to distinguish between assessing the possible effects of enhanced UV-B on individuals of a species, which can best be accomplished in the laboratory, and assessing the effects of damage to individuals on the populations of those individuals in the natural ecosystem, which requires extensive field studies.

Advances in Knowledge

The population ecology and dynamics of aquatic food chains--the anchovy, striped bass, herring, shellfish, crustaceans--are just beginning to be studied by interdisciplinary teams of physical hydrographers, phytoplankton-, zooplankton-, and fish-biologists, and biochemists and physiologists. This ecosystem research is expensive because it requires time on board ships and large numbers of personnel. But until the capability is developed for predictions of specific food chain effects under the range of present physical, chemical, and biological interactions, the inclusion of the UV-B variable (unless the effects of UV-B are catastrophic) will not lead to statistically significant conclusions.

There is strong experimental evidence that current levels of UV-B in surface waters depress near-surface productivity of organisms at the base of food chains (primary productivity) in marine waters (Calkins and Thordardottir 1980, Lorenzen 1979, Smith et al. 1980, Steemann Nielsen 1964). The measurement of the penetration of UV-B below the surface and the estimation of doses to aquatic organisms are very complex. The penetration of UV-B into waters with low transparency has not been as well documented as it has for clear waters (Smith and Baker 1981). Turbidity and blue-light-absorbing material (gelbstoff) in waters that have a high organic matter content severely limit UV-B penetration. This fact, the turbulent mixing of surface waters by wind, the ability of organisms to adjust their positions in the water column, and the 24-hour rhythms of vertical migration observed for many marine species all add to the difficulty of estimating the dose of UV-B to which aquatic organisms are exposed. Because of these difficulties, the possible effects of predicted increases in UV-B (due to ozone depletion) on phytoplankton populations within the entire photic zone are at present unknown.

Attempts have been made to maintain captured phytoplankton populations in tanks and to document species composition changes and changes in primary productivity resulting from enhanced UV-B (Worrest et al. 1978, 1981a,b). While these experiments show statistically significant effects of enhanced UV-B, the difficulties of mimicking natural mixing conditions, the visible and UV-A photic regimes, and the nutrient and predation conditions of the real world make extrapolation of these data to populations in the natural ecosystem difficult. Prediction of the effects on food chains for which phytoplankton serve as food sources is even more difficult. These experimental difficulties have not yet been overcome, and, except for the preliminary studies by Worrest et al. (1981a), have not been attempted on any concerted level.

Recently, it has been reported (Hunter et al. 1979) that the UV-B threshold for lesions and for retardation of growth in anchovy larvae can be reached after exposure during a 4-day period to UV-B intensities of 760 joules per square meter ($J\ m^{-2}$) (DNA effective dose). These intensities are equivalent to what would be expected at the surface of clear ocean water if stratospheric ozone concentrations were reduced by 25 percent. Current UV-B levels just below the sea surface were estimated to be 413 $J\ m^{-2}$ over a 4-day period. From Smith and Baker (1979) the absorption coefficient for UV-B can be calculated to be approximately 0.2 m^{-1}. This means, for example, that at 5 m below the surface the UV-B intensity is reduced to 37 percent of the surface intensity. These data indicate a marginal effect that depends strongly on the vertical distribution of larvae. The particular value of this study is that it attempts to relate in situ measurements of UV-B in seawater to physiological effects on a sensitive stage in the life cycle of a commercially important marine species. This physiological study, combined with sampling for the vertical distributions of larvae, physical hydrographic measurements of water mixing patterns, and further measurements of in situ UV-B doses, could serve as a model for investigating the effects of UV-B on specific food chains in the marine ecosystem. At the least, the research including photoreactivation effects would provide data from which predictions could be made about the significance of enhanced levels of UV-B on the anchovy populations.

A photorepair mechanism for UV-B lesions in anchovy larvae has been reported by Kaupp and Hunter (1981). The amount of light required to activate photorepair mechanisms fully was less than 10 percent of that available from the sun on a clear day even in March. Thus the authors concluded that even with increased UV-B, sufficient light exists in the sea to ensure photorepair of UV damage in anchovy larvae. This report again illustrates the difficulties inherent in studying the effects of enhanced UV-B in the laboratory where the effects of other wavelength regions of the solar spectrum cannot easily be considered.

A number of studies of UV effects on aquatic systems (Calkins and Thordardottir 1980, Karanas et al. 1981, Smith and Baker 1980, Smith et al. 1980, Thomson et al. 1980, Worrest et al. 1980) and on underwater penetration and characterization of UV (Green and Miller 1975; Green et al. 1980; Smith and Baker 1979, 1981) have been made recently. The phytoplankton studies of Worrest et al. (1978, 1981a,b) and the anchovy and physical optics studies of Hunter et al. (1979) have been used as examples.

RESEARCH RECOMMENDATIONS

Prediction of the possible effects of increased solar UV-B on economically important biological organisms, such as specific crop species, and on selected ecosystems is not currently possible for two reasons. First, two major areas of research that may have potential for improving predictive capability and that were proposed by NRC (1979a) have not been implemented. These are (1) studies of the effects of enhanced UV-B conducted in the presence of the natural photoperiodic intensities of the complete solar spectrum, and (2) the development of a low-latitude, UV-B-transmitting greenhouse facility where specific higher-latitude species of plants or specific aquatic organisms could be subjected to lower-latitude UV-B intensities while other conditions such as spectral intensities and temperatures were maintained close to higher-latitude ranges. Second, in the absence of meticulous attention to the need for careful simulation of environmental parameters implicit in the research areas described above, most of the research data reported appear to lack predictive capability. Our assessment therefore is essentially the same as that given in detail in previous reports (NRC 1979a, SRI 1980, 1981), and the research recommendations largely repeat those in the NRC report (1979a).

The following list of research recommendations is not exhaustive. It has been limited to those that should receive attention first, but it is not organized according to priority.

1. Most of the current experimental data on plants was gathered under greenhouse or growth chamber conditions where visible light and UV-A levels were considerably lower than plants would experience under ambient field conditions. The absence of ambient levels of visible light and UV-A appears to increase substantially the susceptibility of plants to damage by the UV levels tested. Thus economically important crop plants need to be evaluated in a field situation where enhanced levels of UV-B approximating those predicted to occur under reduced atmospheric ozone conditions are simulated.

These experimental conditions might best be obtained in a low-latitude (subtropical), minimal-cloud-cover, multiuser facility. Productivity, photobiological, and physiological studies should be conducted simultaneously. During these studies, UV levels incident upon plant surfaces should be carefully monitored. Before the data from such a facility could provide the most precise and unequivocal results, it would be necessary to develop instrumentation, data manipulation, and environmental regulating techniques that would enable the facility to simulate the changing ambient conditions of the higher-latitude location whose plants are under study. Without this capability, the data generated would have limited usefulness for predicting the effects of enhanced UV-B.

Priority should be given to screening for UV-B-sensitive species to look for adverse effects on productivity and to study the mechanisms of UV-B stress.

2. Sensitive phases of plant growth need to be determined. For example, is the reproductive stage especially sensitive to an increase in UV-B?

3. Dose-response curves and threshold levels for reduced crop yields are currently known for few plant species under ambient field conditions. Important crop plants, selected native plants, representative forage

plants, and economically and ecologically important forest species need to be similarly evaluated.

4. Virtually nothing is now known about the interactions of UV and other known stress factors. The interaction of UV with such factors as temperature, water stress, and air pollutants needs to be addressed.

5. Studies of the effects of UV on aquatic ecosystems must be approached on two levels: (1) the effects on individuals under controlled conditions where all environmental factors are reproduced and UV-B levels are varied, and (2) integration of laboratory dose-response data into ecosystem studies to establish possible effects among populations or specific food chains. Only in this way can it be determined whether a reduction in stratospheric ozone concentration will significantly affect aquatic populations. The interdisciplinary approach used in the Hunter et al. (1979) anchovy study (described earlier) to assess UV-B damage to food chains, together with the specific laboratory measurements, should serve as a model for future research proposals.

Chapter 5

DIRECT HUMAN HEALTH HAZARDS

SUMMARY

On the basis of current knowledge, we believe that ozone depletion and the resultant increase in UV would not result in new health hazards, but would increase existing ones as described in the following sections. The UV component of sunlight can cause direct damage to the skin, eyes, and immune system of humans. The UV wavelengths most affected by ozone concentration are essentially responsible for sunburn, an acute, inflammatory response of the skin. Although the exact targets and mechanisms for sunburn are not fully understood, enough is known about the doses required to predict the increased risk for any given increase in UV flux. For small increases in UV, simple sun avoidance measures would more than offset the increased risk of sunburn.

Much less is known about the long-term effects of sunlight on skin. Chronic exposure to sunlight leads to degenerative changes in skin. However, because the effective wavelengths and the relationships between UV dose and skin response are not known, the magnitude of the increased risk of degenerative changes that might accompany ozone depletion cannot be predicted. Epidemiological studies show that sunlight causes more than 90 percent of basal and squamous cell skin cancers and is a factor in melanoma. Experimental studies and theoretical considerations suggest that actually the wavelengths most affected by ozone (i.e., UV-B) cause basal and squamous cell tumors. Because the dosimetry for humans is uncertain, only crude estimates can be made for the increased risk of these cancers as the result of any given increase in UV flux. Techniques to measure individual UV exposures that either cause or prevent these cancers are still lacking.

Melanomas are undoubtedly related to sunlight, but the relationship is more complex and obscure than the relationship between sunlight and basal and squamous cell skin cancers. Consequently, the melanoma-sunlight relationship is more difficult to measure in epidemiological studies and to reproduce in animal experiments. The relationship of melanoma to UV-B is even less clear, and there is no animal model in which this relationship can be explored. Current epidemiological data suggest that individual sensitivity to sun damage, exposure to sunlight in childhood, the relationship of childhood nevi to melanoma, and the association between sunlight and specific histological types of melanoma should be explored.

In the eye, an acute painful irritation of the cornea, called photokeratitis, is caused by UV-B. The action spectrum for this effect is known, and the increased hazard for any given increase in UV flux is predictable. The symptoms are easily prevented by avoiding or reducing exposure to sunlight. There is some evidence that UV may be involved in the etiology of certain forms of cataracts, but the wavelengths most likely involved (UV-A) are not those affected by ozone.

Since the last NRC report (NRC 1979a), several new observations have heightened the awareness and broadened the understanding of the health hazards of human exposure to UV. These include a better understanding of the optical properties of skin and blood; the results of careful study of various exposure conditions that influence UV-induced skin cancer in laboratory animals; demonstration and quantification of two forms of DNA repair in vivo in human skin (see Chapter 3); and documentation that the immune system of animals and humans is affected by UV irradiation of skin. UV affects the immune system in a variety of potentially important ways. For example, systemic UV effects may well be a contributing factor to the efficacy of UV in inducing skin cancer.

ANATOMICAL AND OPTICAL PROPERTIES OF SKIN AND BLOOD

Humans, like most life forms, live in a complicated, dependent relationship with the sun. All life derives its energy from the sun; photosynthesis drives almost all food chains, and the sun is the major source of heat. UV photochemistry in the skin is an obligate step in vitamin

D synthesis, and visible light photochemistry within the retina allows vision. On the other hand, the UV component of sunlight can injure or kill cells, including intact living human tissue.

The organ most affected by UV is the skin. The optical properties of skin determine the amount of optical radiation reaching various depths in the tissue. Since the NRC (1976a, 1979a) reports, more accurate measurement techniques and useful optical models have made it possible to quantify, predict, and modify the optical properties of skin (Anderson and Parrish 1981, Wan et al. 1981). These advances may make it possible to localize important photobiologic chromophores (molecules or parts of molecules that absorb light), identify mechanisms of UV injury, and better quantify risks.

When light enters the skin, a portion is scattered back to the environment, some is absorbed as it reaches various layers, and part is transmitted inward to successive layers of cells, until all the energy of the incident beam has been dissipated (Figure 5.1). The epidermis is a 100-micrometer (μm) sheet of cells that can be viewed as an unpolished optical absorption filter. A 10-μm layer of dead cells, protein, and

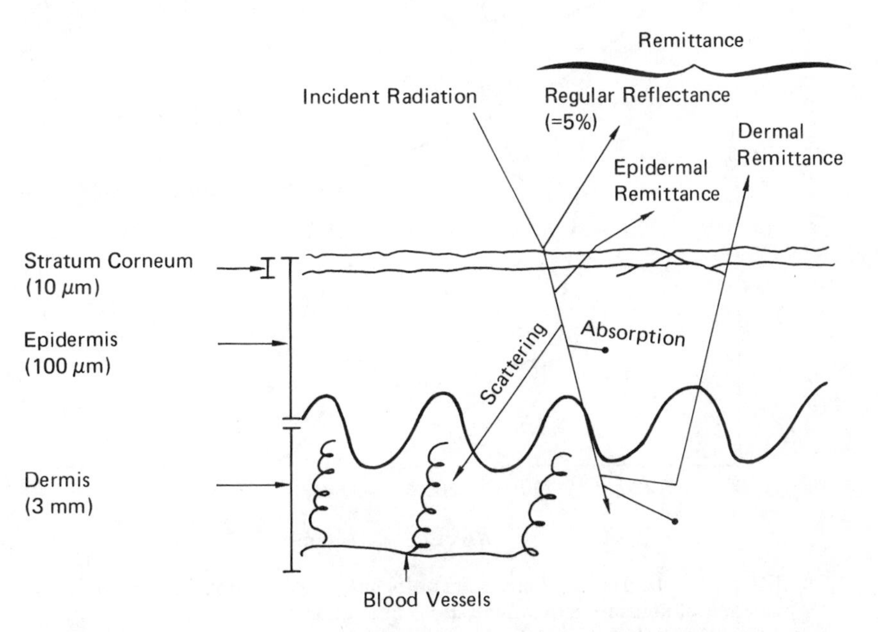

FIGURE 5.1 Optical interactions of skin layers with UV radiation (Parrish et al. 1978).

other biomolecules on the outermost surface of the epidermis is called the stratum corneum. Aromatic amino acids both free and in protein, urocanic acid, nucleic acids, and melanin are the major UV-absorbing chromophores in the epidermis (Figure 5.2). The dermis is a 1-millimeter (mm) to 4-mm layer of primarily collagenous connective tissue that provides much of the structural integrity of the skin. Optical scattering within the dermis largely determines the average pathlength and depth of penetration of various wavelengths of radiation. Dermal scattering is an inverse function of wavelength. The major pigments in the dermis include hemoglobin and bilirubin.

In considering the effects of possible changes in the terrestrial solar spectrum resulting from ozone depletion, it is important to know the depths to which optical radiation penetrates human skin (Table 5.1). UV-B is strongly absorbed by the stratum corneum and by many

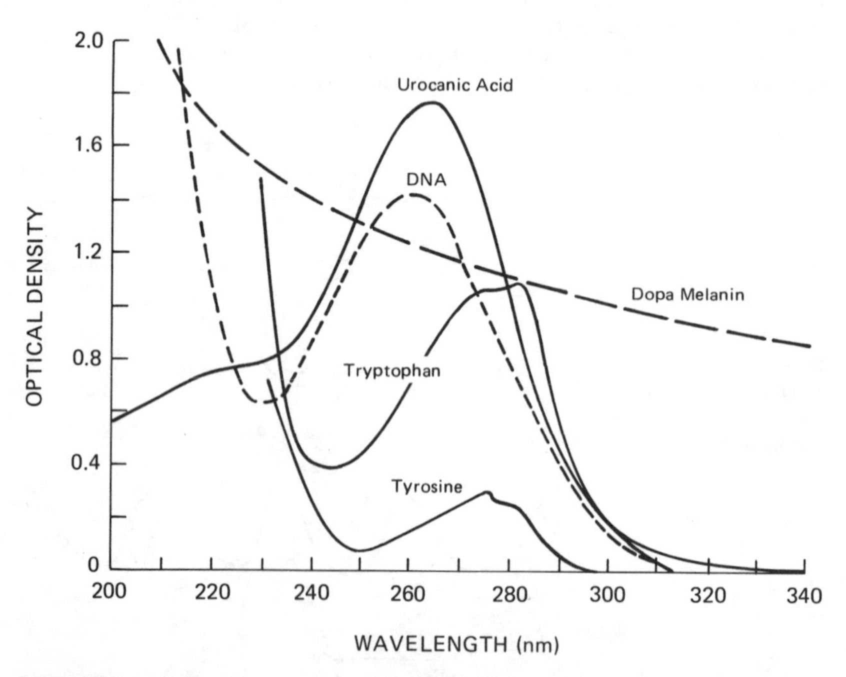

FIGURE 5.2 Optical absorption spectra of the major UV-absorbing chromophores in the epidermis of human skin (concentrations in aqueous solution: urocanic acid 100 μ molar, DNA 100 μg/ml, dopa melanin 15 μg/ml, tryptophan 200 μ molar, tyrosine 200 μ molar) (Anderson and Parrish 1982).

TABLE 5.1 Approximate Penetration Depths of Optical Radiation in Fair Caucasian Skin (μm)

	Depth to Which Following Percentages of Incident Energy Penetrate			
Wavelength (nm)	50%	37%	10%	1%
UV-C				
250	1.4	2	4.6	9.2
280	1	1.5	3.5	7.0
UV-B				
300	4	6	14	28
UV-A				
350	40	60	140	280
Visible				
400	60	90	200	400
450	100	150	350	690
500	160	230	530	1,100
600	380	550	1,300	2,500
700	520	750	1,700	3,500
Infrared				
800	830	1,200	2,800	5,500
1000	1,100	1,600	3,700	7,400
1200	1,500	2,200	5,100	10,000

SOURCE: Modified from Anderson and Parrish (1971).

molecules within living epidermal cells. About 50 percent of UV-A penetrates fair Caucasian epidermis to be largely attenuated within the first 50 μm of the papillary dermis. The longer visible wavelengths penetrate much further.

Biologically active UV reaches the level of cutaneous blood vessels. Endothelial cells and connective tissue elements may be directly affected by the radiation; blood cells, lymphatics, and humoral substances passing through the skin may be photochemically altered. The blood flow to the skin is 30 to 40 times greater than is necessary to supply nutrients and meet the metabolic needs of skin cells because it is primarily designed for heat regulation of the whole body. An equivalent of the entire blood volume may pass through the skin and be irradiated in a few minutes. The physiologic, pathologic, and possible therapeutic implications of this irradiation are just beginning to be understood.

EFFECTS OTHER THAN CANCER

Acute Responses of Normal Skin to UV: Whole Organ Inflammation

Many photochemical events are triggered by the absorption of UV and visible light by the variety of molecules within cells (see Chapter 3). Some of these alterations may have little consequence, whereas others may change cell function, cause cell death, or lead to the release of chemicals that affect adjacent cells or tissues. If there is sufficient damage to individual cells, the skin will react as a whole organ. Redness, swelling, heat, and pain appear after a latent period of several hours, and last for hours to days. The overall response of skin to UV is reparative and protective.

The tender redness or erythema (commonly called sunburn) is the manifestation of UV-induced inflammation that has received the most attention. The presence and degree of UV-induced delayed erythema depends on the exposure and the wavelengths in the irradiating UV. The reciprocal of the lowest exposure required to induce erythema plotted against wavelength is the action spectrum (Figure 2.1). The 250-nm to 290-nm portion of this curve is the most erythemogenic waveband. Ozone depletion would have little effect on this waveband. Erythemal effectiveness falls by a factor of more than 1000 from 290 nm to 320 nm, the UV-B range. Over a wide range of intensities, both high-intensity radiation for a short time and low-intensity radiation for a long time produce the same response (erythema) as long as the same total dose is given; thus reciprocity holds.

Because the action spectrum, dose-response curve, and intensity-time reciprocity relationship for sunburn are known, it is possible to calculate the decrease in time required to acquire a sunburn for any given ozone depletion. If UV-B increases by 10 percent, the decrease in time required to acquire a sunburn would be about 10 percent plus a small additional decrease in time because of the spectral shift to include shorter, more effective wavelengths.

Erythema, however, is only one component of a complex tissue response. Recent studies (Parrish et al. 1981) have revealed other important components in this response involving a variety of kinds of skin cells, blood vessels, and circulating factors, each having its own thresholds, dose-response curves, and action spectra. Cell injury and alterations of skin can occur without erythema. It has been shown that abnormal differentiation of keratinocytes, DNA injury, and pigment production can occur at suberythemogenic UV doses. The chromophore, molecular mechanisms, and complex cascade of mediators and events are poorly understood and may vary with wavelength.

Long-term Effects of UV on Skin: Solar Degeneration

Chronic exposure to the sun causes a complex of changes in skin called actinic or solar degeneration. The skin appears thick and furrowed but may also have zones of thinned epidermis. Hyperpigmentation and hypopigmentation, dilated blood vessels, and a leathery appearance are the other symptoms of the condition sometimes referred to as "sailor's skin" or "farmer's skin" because excessive occupational exposure often causes these changes, especially on the back of the neck. This condition has also been called "premature aging," but there is no convincing evidence that the cellular mechanisms and connective tissue alterations are the same as those that occur in natural aging. The action spectrum for solar degeneration is not known, and therefore the potential effects of ozone depletion are not known.

Effects of UV on the Immune System

The immune system is a complex and diverse collection of circulating and noncirculating cells in the body that provide protection against certain diseases and infections. The system recognizes foreign molecules or cells and initiates complex reactions to dilute, reject, or counteract them. Recently, it has been discovered that UV can alter the immune systems of animals and humans. For example, a mild sunburn results in the decreased viability and function of lymphocytes (circulating white blood cells) in humans for up to 24 hours (Morison et al. 1979), and in animals certain allergic reactions (Morison et al. 1981c), skin graft rejection (Morison et al. 1980), and other immune functions can be altered by giving otherwise tolerable doses of UV to intact skin. Most of what is now known about this topic, which is termed photoimmunology, has been learned since the NRC (1979a) report.

The immune system is well represented in the skin by Langerhans cells in the epidermis, mast cells and lymphocytes in the dermis, and other cellular elements percolating through the lymphatics and capillaries of the dermis. All these components of the immune system are therefore exposed to environmental light, and may be altered as a result of such exposure. The most detailed studies have been performed on experimental animals.

Skin cancers induced in mice by UV-B radiation are highly antigenic, and many are rejected by an immunologic reaction even when transplanted into genetically similar mice, i.e., mice from the same highly inbred strain. These tumors, however, grow in immunosuppressed mice. The tumors that survive in the primary host do so because the UV irradiation has induced systemic, immunologic alterations that suppress specific immune responses. The mechanism, in part, involves the generation of regulatory, thymus-derived (T) suppressor cells in the lymphoid tissue of UV-irradiated mice (Fisher and Kripke 1978, Spellman and Daynes 1978). Repeated exposure of mice to UV radiation induces a population of regulatory cells that prevent immunologic rejection of UV-induced tumors. This effect has been demonstrated by cell transfer and reconstitution experiments (Fisher and Kripke 1978). The suppression is specific for UV-induced tumors.

UV-irradiated mice also fail to respond to contact sensitizing antigens involved in allergic contact dermatitis and contact hypersensitivity. (The action spectrum for the inhibition of contact hypersensitivity is discussed in Chapter 3.) This represents a second systemic alteration in UV-irradiated mice. It is also associated with antigen-specific T suppressor cells and is thought to be caused by a UV-induced alteration in the cells (Langerhans cells or macrophages) that present antigen to lymphocytes in the initiation of an immune response (Greene et al. 1979; Kripke 1980, 1981; Noonan et al. 1981b).

Some evidence indicates that UV can alter antigen presentation in a way that activates the suppressor cell pathway (Sauder et al. 1980, Towes et al. 1980), and thus affect the immune response. It does this by direct interaction with antigen-processing cells. Suberythmogenic amounts of radiation are required for altering the function of Langerhans cells, and even the systemic alteration is produced by erythmogenic UV exposures. Some of the effects of UV on immunologic pathways could determine whether skin cancer develops or not (Fisher and Kripke 1981). Current efforts are directed toward understanding how UV alters antigen-presenting cells, defining the alteration in these cells that triggers suppressor cell production, and determining which immune responses are affected by the alteration.

Several lines of evidence suggest that at least some of the above observations may apply to species other than mice. UV-B exposure can suppress immune responses in guinea pigs and rabbits, and both local and systemic suppression have been reported (Haniszko and Suskind 1963; Morison et al. 1980, 1981c). UV irradiation of guinea pigs, for example, results in a suppression of delayed hypersensitivity responses to contact allergens and injected hapten-protein conjugates (Jessup et al. 1978, Morison et al. 1981c, Noonan et al. 1981b). In rabbits, the rejection of full-thickness skin grafts is delayed by treatment with oral methoxsalen and UV-A radiation (a photochemotherapy for psoriasis) (Morison et al. 1980). The role of T suppressors in these phenomena is being investigated. There are a few reports of local suppression of contact hypersensitivity in human subjects following UV exposure (Horowitz et al. 1974, O'Dell et al. 1980). There are increased UV-associated skin cancers in renal transplant patients (Marshall 1974). Although these cannot be ascribed to the immunosuppressive therapy with certainty, the observation is consistent with what would be expected if there were an immunological involvement in human photocarcinogenesis. The effects of UV-B radiation on Langerhans cells in human skin appear to be similar to those reported in rodents (Aberer et al. 1981).

Human studies in this new area of research are less advanced from the viewpoint of pathophysiologic mechanisms than animal studies, but it has been firmly established that exposure to UV radiation does affect the immune function in humans. In normal human volunteers, single exposures to sunburn-causing doses of UV-B radiation (Morison et al. 1979) or oral methoxsalen and UV-A radiation (Morison et al. 1981a) produce an alteration in the distribution and function of subpopulations of circulating lymphocytes. These effects are reversible within 48 to 72 hours. However, repeated exposure to such radiation may cause more long-lasting changes in lymphocyte viability and function (Morison et al. 1981b).

The quantitative implications of the above observations are not clear. It is possible that UV-induced alterations of immune function are important in mediating the beneficial effects of UV radiation in the treatment of skin disease and the harmful effects of such exposure, such as the development of skin cancer.

Effects of UV on the Eye

Because solar UV radiation is present during most of the daylight hours, the eye is exposed daily to some amount of solar UV radiation throughout life. UV-B is mostly absorbed within the cornea and conjunctiva. The acute effects of excessive exposure to these wavelengths are primarily conjunctivitis and a corneal inflammation reaction known as photokeratitis. After UV exposure, there is a period of latency varying more or less inversely with the amount of exposure. The latent period may be as short as 30 minutes or as long as 24 hours, but it is typically 6 to 16 hours. Photoconjunctivokeratitis causes the sensation of a foreign body or sand in the eye, varying degrees of excessive tearing, blinking, and intolerance of light. Corneal pain can be very severe, and the individual may be incapacitated for a period of time. These acute symptoms usually last from 6 to 24 hours, and almost all discomfort usually disappears within 48 hours. Rarely does exposure result in permanent damage. Unlike the skin, the ocular system does not develop tolerance or significant defenses against future UV exposures. Anatomic conformations protect human eyes from acute overexposure to the UV component of sunlight as do physiologic bright-light-avoidance responses when there is sufficient visible light to incite this protective response. There is epidemiological evidence that chronic exposure to sunlight may cause certain types of cataracts in humans and experimental evidence that UV-induced photochemical changes (with and without the addition of exogenous photosensitizers) in the lens can cause cataracts. The action spectrum for these changes, however, appears to be in the UV-A range and therefore would not be affected by ozone depletion.

CANCER EFFECTS

Basal and Squamous Cell Skin Cancers

What Was Known by 1979

Basal and squamous cell skin cancers constitute the most common malignancies in humans. These cancers are usually easily treatable but have a definite morbidity, cost, inconvenience, cosmetic liability, and mortality. The NRC Climatic Impact Committee (NRC 1975) and the Committee on Impacts of Stratospheric Change (NRC 1979a) were asked to predict whether increased exposure to UV-B would be likely to increase basal and squamous cell skin cancer incidence rates. To do so, they had to make judgments on the basis of limited evidence from epidemiological studies, backed up by clinical and pathological observations and the results of animal experiments, because human experimentation was out of the question. The same approach had to be used in 1964 to assess the relationship between cigarette smoking and lung cancer. The Surgeon General's report on smoking and health stated that determination of whether the confirmed association between an event and a disease is causal is a matter of judgment that goes far beyond any statement of statistical probability. It listed a number of criteria that must be used in assessing circumstantial evidence. These criteria included the consistency, strength, specificity, temporal relationship, and coherence of the association (Surgeon General 1964).

Over the years, collective evidence has confirmed the existence of an association between basal and squamous cell skin cancers and sunlight. Epidemiological surveys have consistently identified an overwhelming predominance of these cancers in Caucasians, increasing mortality and incidence rates with decreasing latitude, higher rates of disease in outdoor than in indoor workers, and rates of disease increasing more rapidly at older ages. These associations have always been relatively strong and have been accepted as biologically rational: the skin is indeed exposed to the sun, and increased incidence rates with advancing age favor a sequence in which disease does not precede but follows exposure.

The supporting clinical data have also been consistent in showing concentrations of these skin cancers among fair-complexioned individuals, particularly those with blue eyes who sunburn easily and have Celtic ancestors.

These clinical series have recorded another consistent finding, namely that most basal and squamous cell skin cancers occur on sites of the body habitually exposed to sunlight and often in the same tissue systems as sunburn. This overall picture was considerably strengthened by the finding that high rates of basal and squamous cell skin cancers are associated with defective DNA repair in patients with the inherited disease xeroderma pigmentosum (Kraemer 1980). This finding made a relationship between UV-B radiation and these cancers biologically plausible, on the basis of the knowledge that UV-B can damage DNA in skin cells (see Chapter 3).

There were very few hard data on the incidence of and mortality from skin cancers when the 1976 and 1979 reports of the Committee on Impacts of Stratospheric Change (NRC 1976a, 1979a) were prepared. The most solid evidence came from standardized measurements of skin cancer incidence (1971-1972 National Cancer Institute (NCI) survey) and of annual UV doses at four geographic locations in the United States with a range of UV exposures (NRC 1975, Parrish et al. 1978, Scott and Straf 1977, Scotto et al. 1974). The estimated annual UV dose at ground level was made in two ways: (1) by readings of Robertson-Berger (R-B) meters, and (2) by calculation from known solar fluxes, ozone concentrations, and estimated cloud cover (NRC 1975). The first method of estimating is supposed to measure the accumulated dose of wavelengths in the erythema action spectrum, although from Figure 2.2 it can be seen that the R-B meter measures more UV-A than is in the erythema spectrum. The second method makes a theoretical estimate of dose corresponding to wavelengths included in the DNA action spectrum. It is obvious from the plots of the R-B response to various wavelengths of UV and the action of UV on DNA as a function of wavelength (Figure 2.2) that the two dose measurements are not the same. A change in ozone concentration will change the annual "DNA dose" much more than the "R-B dose" (see NRC 1979a, Figure D.5 and Table D.1). A 1 percent change in ozone will produce, at 40° north latitude, an approximate 2.3 percent change in the DNA-damaging dose, but only an approximate 0.8 percent change in the R-B dose (NRC 1979a).

Both measures of UV dose are strong functions of latitude, increasing markedly as latitude decreases. The measurements of incidence and UV exposure mentioned above were used to show that the reported statistical correlation between latitude and skin cancer incidence could, in fact, be a correlation between the occurrence of skin cancer and the local annual exposure to UV-B. The measurements were used to make crude estimates of the increases in skin cancer incidence rate to be expected from various percentage decreases in ozone (NRC 1975, Scott and Straf 1977) or from the percentage increases in units of exposure as measured by an R-B meter (Scotto et al. 1974).

With such limited, epidemiological findings, more weight than usual was given to clinical observations from both published and unpublished sources. The panels preparing the human health sections of the NRC (1976a) and (1979a) reports believed that death rates from skin cancer other than melanoma, based on this weighted evidence, were too low to have national significance, even though the cost of treatment and the disfigurement and morbidity resulting from these types of cancers were considerable.

NRC (1979a) provided preliminary observations from ongoing NCI programs to relate UV (and by inference UV-B) to the incidence of basal and squamous cell skin cancers. The report provided a brief summary of the survey methods that are described in detail in NRC (1975). The information from the preliminary results available in 1979 confirmed earlier observations that Caucasians living in areas of high insolation do have higher rates of basal and squamous cell skin cancer than those living in areas of low insolation and that fair-skinned Caucasians, particularly those who sunburn easily or have limited ability to tan, are at measurably increased risk. The report concluded that in the United States, most basal and squamous cell cancer is found in these people. It also acknowledged that there are signs of an upward trend in the incidence rate of these cancers.

Advances in Knowledge

<u>Experimental Photocarcinogenesis</u>. Since the NRC (1979a) report, new information from animal experiments has shed more light on the dose rate and the quantitative relationship between skin cancers other than melanoma and the biologically effective UV wavelengths. Skin cancer has been induced in experimental laboratory animals by exposing them to wavelengths shorter than 320 nm. The UV wavebands most affected by alterations in ozone concentrations (UV-B) are carcinogenic in animal studies. Using

a xenon arc to represent extraterrestrial sunlight and a series of filters to simulate various atmospheric ozone concentrations, Forbes et al. (1980) showed that incremental additions of shorter wavelength UV increased carcinogenic effects in hairless mice.

Other factors have been studied. Freeman and Knox (1964) found that increased temperature at the time of UV exposure accelerated tumor production. Using environmental chambers to irradiate experimental animals, Owens et al. (1974) found that animals exposed to UV and wind developed more tumors than animals receiving the same dose of UV alone. In other animal groups, Owens et al. (1975) noted that animals maintained at high humidity developed tumors more rapidly than those maintained at low humidity.

Mice are not perfect models for human photocarcinogenesis. The susceptibility to UV-induced cancer varies with the strain. All known strains have poor excision repair of DNA compared with people (see Chapter 3), and the optical properties of mouse skin differ from those of human skin. But many important models and concepts have resulted from the decades of work accumulated by using mouse models. Recent and ongoing studies have shown that the relationship between cumulative dose and cancer production is not simple. Under certain experimental conditions, the tumor yield can be increased by alteration of the exposure regime. The same total dose given at lower irradiance produces more tumors. Intermittent exposure may be more photocarcinogenic than the same dose given continuously, although the opposite holds for rats (Strickland et al. 1979). The susceptibility to UV-induced cancer seems to vary with age of the animal. UV also alters the immune system of animals (see the discussion earlier in this chapter), which markedly influences the susceptibility to UV-induced skin cancer. All of these observations may eventually help us to better understand, predict, and make models of human photocarcinogenesis.

Chemicals and UV can interact in a variety of ways to affect tumor yield in skin. Either enhancement or inhibition of photocarcinogenesis may occur depending on the chemical carcinogen and the wavelength and dose of radiation used. Certain chemicals can promote but cannot initiate tumors. Antioxidants have been shown to either increase or decrease tumor yield under certain conditions (Forbes et al. 1981). Skin cancer can result from the interacting effects of a chemical and UV irradiation at doses at which neither agent alone is a primary carcinogen (Black et al. 1978). The combination of certain psoralens and UV-A is an example. Not all chemicals that enhance photobiologic effects on cells or tissue are photocarcinogenic agents, but nevertheless as chemical pollution of our environment grows, chemical enhancement of photocarcinogenesis may be an increasing concern. (The state of knowledge about UV interactions with other environmental stresses on ecosystems is discussed in Chapter 4.)

Epidemiology. This committee was given access to new epidemiological data, as well as to extended analysis of updated existing data. We were also provided with revised estimates of predicted increases in the incidence rates of skin cancer for a range of possible percentage reductions in stratospheric ozone concentration and/or changes in R-B meter units. The new information made available to us by the NCI came from three sets of data. The first consisted of counts of cases of newly diagnosed (nonrecurring) skin cancers from June 1, 1977, through May 31, 1978, among residents of Atlanta (Standard Metropolitan Statistical Area, SMSA), Georgia; Detroit (SMSA), Michigan; Minneapolis/St. Paul (SMSA), Minnesota; New Mexico; New Orleans (metropolitan area), Louisiana; San Francisco/Oakland (SMSA), California; King County, Washington (Seattle); and Utah. These eight geographic locations were chosen because they receive various intensities of solar radiation and most are participants in the Surveillance, Epidemiology, and End Results (SEER) program of the NCI. The second set of new data consisted of preliminary R-B meter measurements from meters installed in 1978 (Berger and Urbach 1982), which were used to estimate the UV dose accumulated over one year in five geographic areas (1978 in Figure 5.3). When taken together with measurements from the meters installed in 1974, estimates of the annual accumulated UV dose for all of the eight locations with new incidence data were available. The third set of data came from a telephone interview survey of patients with skin cancer and of general population controls at the same eight locations. The questionnaire was designed to obtain information on several host and environmental factors that may affect the risk of developing skin cancer. Descriptions of the questionnaire and the survey itself are in the literature (Scotto and Fraumeni, Jr. 1982). However, in the analyses by Scotto and co-workers at NCI of the relation between the incidence data and UV accumulated dose (as

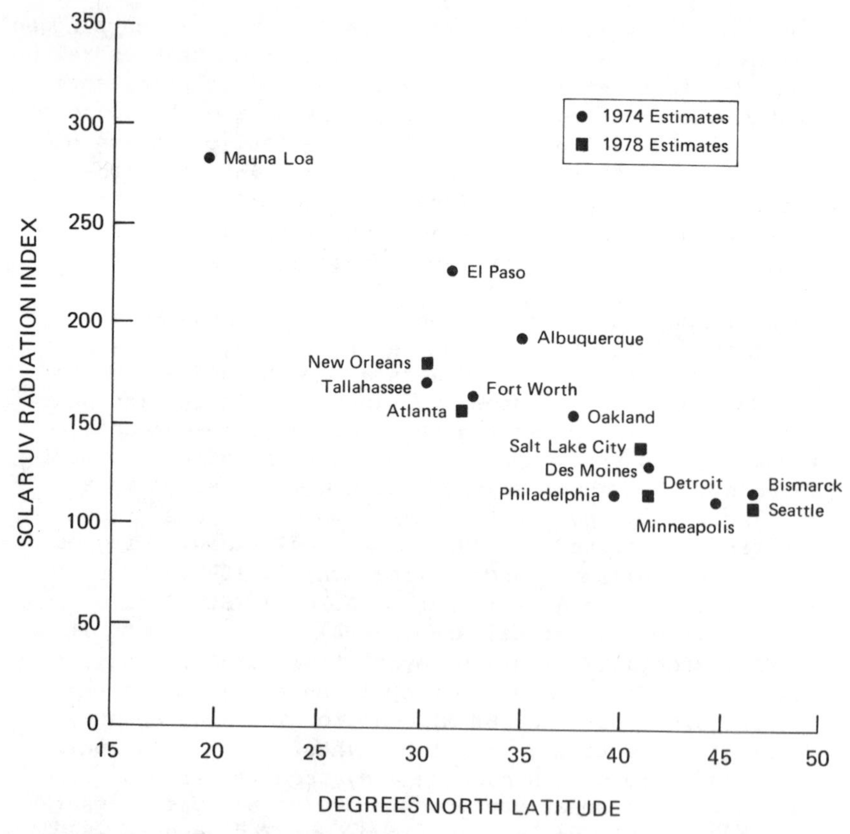

FIGURE 5.3 Annual UV measurements by latitude, 1974 and 1978. The UV radiation index is total Robertson-Berger meter counts over a one-year period multiplied by 10^{-4} (preliminary monthly averages provided by Daniel Berger of Temple University for the 1978 estimates). The meters read UV-B between 290 nm and 320 nm, as well as some UV-A. (Modified from Scotto et al. (1982).)

measured by an R-B meter), the incidence rates were not corrected for any confounding factors other than age and sex.

The incidence data from the eight geographic locations provided age- and sex-specific rates for basal and squamous cell cancers, as well as rates of occurrence on different sites of the body. The UV accumulated dose (as measured by an R-B meter) clearly correlated with latitude (Scotto et al. 1982) (Figure 5.3). When age-adjusted rates were plotted against latitude, the incidence of skin cancer was inversely associated with latitude, whereas the incidence for all other cancers combined was not (Scotto et al. 1982) (Figure 5.4). In the southern part of the United States, the annual rates of skin cancer other than melanoma far exceeded the total annual rate for all other cancers. More detailed analysis showed that basal and squamous cell skin cancers were reported at earlier ages in the South. When the new annual sex- and age-adjusted incidence rates of these skin cancers were plotted against R-B meter measurements of UV, all incidence rates were found to be lower in

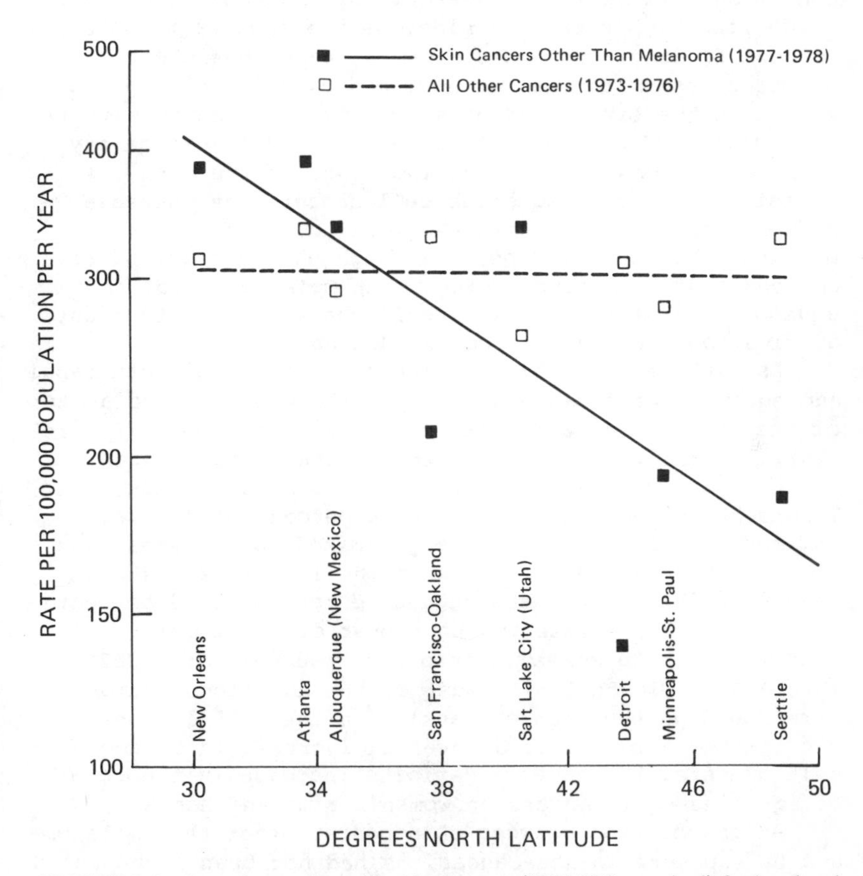

FIGURE 5.4 Annual age-adjusted incidence rates (1970 U.S. standard) for basal and squamous cell skin cancer (1977-1978) and all other cancers (1973-1976) by latitude in the U.S. white population (Scotto et al. 1982).

locations with the lowest annual UV dose (Scotto et al. 1981) (Figure 5.5). This was true for both basal and squamous cell cancers. The slope representing the correlation between accumulated UV dose and incidence was steeper for squamous cell than for basal cell cancers for both sexes (Figure 5.5).

The 1977-1978 incidence rates of both basal and squamous cell cancers increased steadily from younger to older age groups for both sexes, except for some leveling-off at extreme old age. These age-specific rates confirm repeated past observations that older people get more skin cancers than younger people. The higher frequency of incidence for older people is true in cohort as well as in cross-sectional analyses. In other words, the higher rate in older people appears to be a correct finding. The existence of these consistently higher rates of disease at advanced ages is believed to mean that the likelihood of skin cancer increases with accumulated UV exposure (Fears et al. 1977). Mortality data have suggested that the incidence of the more invasive and lethal squamous cell cancers may increase with age more rapidly than the more common and less malignant basal cell type. Although only one out of five new cases of skin cancer other than melanoma is of the squamous cell type, squamous cell cancers cause four out of five nonmelanoma skin cancer deaths.

As expected, the highest incidence rates of both basal and squamous cell cancers were for those on exposed areas of the face, head, and neck in both men and women (Figure 5.6). Overall, about 80 percent of the cancers began on the head and neck, 10 percent on the arms and hands, 6 to 7 percent on the trunk, and 2 to 4 percent on the legs and feet (Table 5.2). The very limited comparisons that the NCI staff could make between the 1971-1972 and the 1977-1978 incidence rates suggested to them a 15 percent to 20 percent increase in the number of basal cell cancers for both sexes (Scotto and Fraumeni, Jr. 1982) (Table 5.3, Figure 5.5). Most of the additional tumors were found on the trunks (back) of males. While there was limited evidence of an overall increase in squamous cell cancers, there was a definite increase in the rate of squamous cell cancers on women's arms and hands.

An association between skin cancer other than melanoma and UV exposure in non-Caucasians had not been found until recently. The 1977-1978 survey found 68 black patients with basal cell or squamous cell skin cancers. In spite

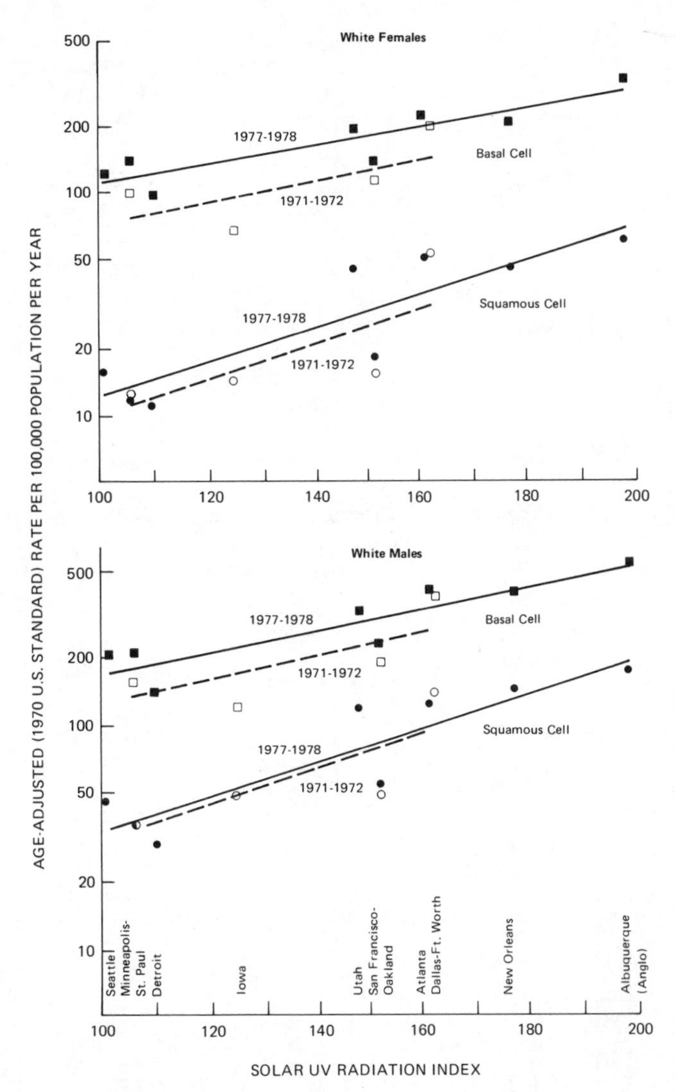

NOTE: For each sex and cell type, the model used to fit the data is log incidence = $\alpha + \beta F$, where F is the flux in annual R-B meter counts x 10^{-4}.

FIGURE 5.5 Annual age-adjusted incidence rates for basal and squamous cell cancers among white females and males for two surveys, 1977-1978 (closed symbols) and 1971-1972 (open symbols), according to one year's UV measurements at selected areas of the United States. The UV radiation index is the total Robertson-Berger meter counts over a one year period multiplied by 10^{-4}. The meters read UV-B between 290 nm and 320 nm, as well as some UV-A. (Adapted from Scotto et al. (1981).)

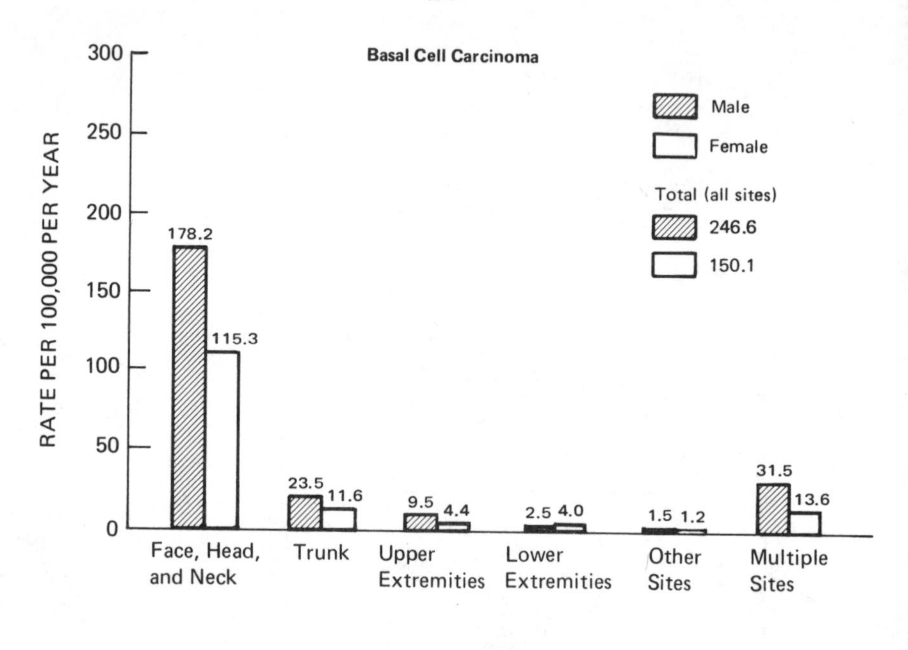

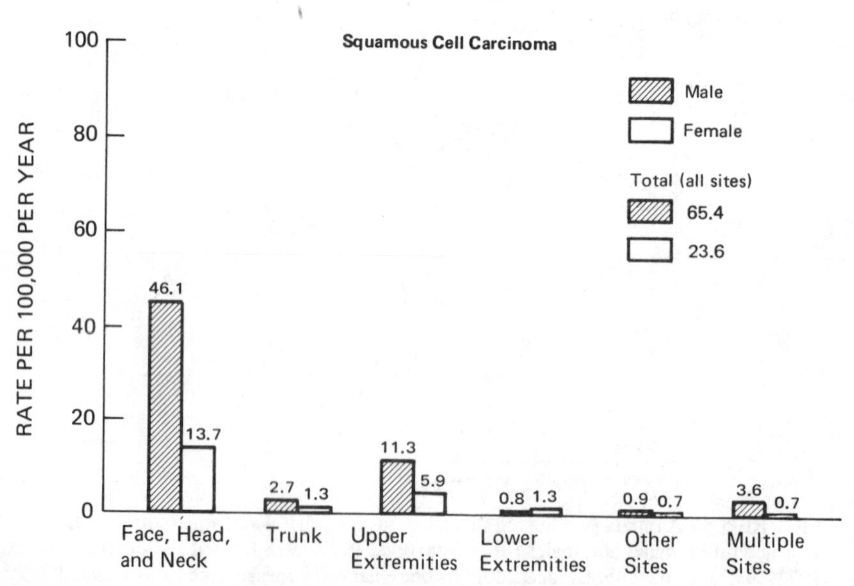

FIGURE 5.6 Annual age-adjusted incidence rates (1970 U.S. Standard) of basal and squamous cell cancers according to anatomic site and sex (U.S. white population, 1977-1978) (Scotto and Fraumeni, Jr. 1982).

TABLE 5.2 Percentages of Basal and Squamous Cell Skin Cancers by Sex and Anatomic Site Among U.S. Whites, 1977-1978

	Males	Females
Face, head, and neck	80	81
Upper extremities	11	9
Trunk	7	6
Lower extremities[a]	2	4
Total	100	100

[a]Includes a small number of genital and unspecified tumors.

SOURCE: NCI survey 1977-1978.

TABLE 5.3 Annual Age-Adjusted Incidence Rates (per 100,000) for Basal and Squamous Cell Skin Cancers Among U.S. Whites by Cell Type and Sex, 1971-1972 and 1977-1978

	1971-1972 NCI Survey	1977-1978 NCI Survey
All Survey Areas		
Basal cell cancers		
Male	202.1	246.6
Female	115.8	150.1
Squamous cell cancers		
Male	65.5	65.4
Female	21.8	23.6
San Francisco-Oakland		
Basal cell cancers		
Male	197.9	239.0
Female	117.2	145.1
Squamous cell cancers		
Male	51.7	56.3
Female	15.8	18.4
Minneapolis-St. Paul		
Basal cell cancers		
Male	165.0	213.1
Female	102.8	144.0
Squamous cell cancers		
Male	36.5	36.6
Female	12.3	11.8

NOTE: The 1971-1972 survey included four areas, and the 1977-1978 survey included eight. However, only two locations, San Francisco-Oakland and Minneapolis-St. Paul, were common to both surveys.

SOURCE: Scotto and Fraumeni, Jr. (1982).

of the small number of cases, the Scotto and Fraumeni, Jr. (1982) report a suggested latitude gradient among blacks for both cell types. Squamous cell cancers were more common than basal cell.

In summary, the new information provided by analysis of the NCI population-based incidence data (corrected only for age and sex) confirmed and strengthened existing evidence in favor of a causal relationship between basal and squamous cell skin cancers and UV. It also gave new insight into the magnitude of this problem. Basal and squamous cell skin cancers from UV radiation have become so common that although the fatality rate is one death in every hundred cases, the overall national mortality figures actually resemble those for melanoma (Mason et al. 1975). The long-assumed protective effect of skin pigmentation was confirmed by the very small number of cases reported in blacks, and the suggested association (latitude gradient) of those cases with UV exposure may be an important piece of confirmatory evidence.

It was hoped that the data from the questionnaire survey could be used to assess the relative personal risk to individuals of developing basal and squamous cell cancers from their reported susceptibility characteristics, e.g., skin and eye color and ancestry. It was further hoped that, in this way, enough information could be gained for estimates to be made of the total amount of disease in a given community that could be attributed to individual susceptibility. This work has not yet been done.

Early results of the questionnaire survey confirmed previous clinical observations that among people with skin cancer there is a higher occurrence of fair complexions, blue eyes, red or blond hair, and Scottish or Irish ancestors than among the general population. They also confirmed suspicions that differences in the frequencies of these particular characteristics in case-and-comparison groups were sufficiently variable across geographic areas to affect overall measures of the correlations between incidence and UV exposure. However, this survey points out that at least three out of every ten individuals diagnosed as having basal or squamous cell skin cancers do not have fair complexions and that fewer than 50 percent have blue eyes. In other words, basal and squamous cell skin cancers are not confined to the fair-skinned, blue-eyed descendents of Scottish-Irish immigrants. This is important to remember in directing research into causes of skin cancer and in planning measures for its control. For example, the most important marker may be a measure of sunburn response and the ability to tan.

There are no studies collecting and measuring the total UV dose that an individual is naturally exposed to in a lifetime and correlating it with the subsequent incidence of skin cancer. Prospective studies needed to detect initiators of skin cancer in humans may be too unwieldy to consider. Thousands of participants would have to be monitored for decades. It may, however, be feasible to design studies of the promoting effects of UV in human skin cancer other than melanoma.

Studies of cohorts (groups born in a specific time interval and followed through life) of patients with psoriasis undergoing two different types of UV phototherapy are contributing indirect but valuable information about skin cancer mechanisms. Groups of patients with psoriasis are being treated with phototherapy using high-intensity UV-B sources, mainly in the 290-nm to 300-nm part of the spectrum. This therapy would be expected to damage DNA in the same way as solar radiation of the same wavelengths. Although this form of phototherapy has been widely used for decades, it has only recently been used in high doses. Consequently, prospective studies of the long-term toxicity or skin cancer risk are just beginning. One retrospective study suggests that psoriatic patients who have had massive cumulative doses of UV-B and topical crude coal tar have an increased incidence of skin cancer (Stern et al. 1982). Information is also becoming available from a relatively large cohort of psoriasis patients in 16 medical centers treated with photochemotherapy (oral methoxsalen and UV-A). Although this treatment damages DNA in a different way, the study raises some important questions about dose-response relationships, synergism, and cocarcinogenesis. For example, after four years of follow-up, rates of basal and squamous cell skin cancer in 1373 psoriasis patients treated with photochemotherapy using UV-A (PUVA) were 3 times higher than were expected on the basis of age, sex, and geographic-location incidence rate data. The proportion of squamous cell cancers was much higher than average, and the additional cancers were mostly on parts of the body that are not normally exposed to sunlight but were exposed to photochemotherapy. A history of exposure to ionizing radiation greatly increased the risk of developing (particularly) squamous cell cancer (Stern et al. 1981). When measured in treatment units, much more UV-A exposure was needed to initiate squamous cell than basal cell cancers.

Data from the four locations in the 1972 incidence survey were added to data from the eight 1978 survey locations (which gave 12 measurements from 10 locations). Scotto et al. (1981) plotted the age-adjusted incidence rates for basal and squamous cell cancers on a logarithmic scale against the annual UV dose measured by an R-B meter (Figure 5.5). The incidence data are plotted on logarithmic scales because of the observation that the 1971-1972 incidence data (and the more recent data) gave similar slopes on such a graph for all age groups and for males and females even though the absolute value of the incidence was appreciably greater for males than for females and higher age groups had a much greater incidence than lower age groups (Fears et al. 1976, NRC 1975). Such an exponential model implies that over the dose range considered there is no threshold for the population.

The model was applied to the 1971-1972 data to obtain estimates of the relative increases in skin cancer incidence accompanying a relative decrease in ozone (calculated as a DNA dose) or as a relative increase in the R-B meter reading (NRC 1979a). A similar method of analysis was applied to the 1977-1978 data. For a 1 percent increase in the R-B meter reading, Scotto et al. (1982) predicted the incidence of skin cancer other than melanoma to increase from a low of 1.9 percent at high latitudes to a high of 2.9 percent at low latitudes. The overall estimate is an approximate 2.5 percent increase in skin cancer incidence for a 1 percent increase in the R-B meter reading (Scotto et al. 1982). Since a 1 percent decrease in ozone concentration corresponds to an approximate 0.8 percent increase in the R-B meter reading (NRC 1979a), these data imply that a 1 percent decrease in ozone concentration would result in an approximate 2 percent (0.8 x 2.5) increase in skin cancer other than melanoma among the white population of the United States. A correction needs to be made for statistical bias and for the variability in dose received among individuals of a particular population (Scott and Straf 1977). The estimated increase in squamous cell cancers due to increased UV exposures is predicted to be greater than the increase in basal cell cancers (see Figure 5.5). For females the squamous cell values would be at least twice those for basal cell cancer, and for males at least 1.5 times (Scotto et al. 1981).

Percentage increases in various measures of skin cancer corresponding to various decreases in stratospheric ozone concentration have been predicted by the Panel to Review Statistics on Skin Cancer of the NRC Committee on National Statistics (NRC 1975, Appendix C). Using the simple model that the logarithm of the incidence is proportional to annual UV flux (NRC 1975, Appendix C), the panel has also estimated increases in skin cancer incidence from age- and sex-specific data from the 1977-1978 NCI survey (Scott 1981). The panel has made a number of refinements in the measures of UV used in their 1975 work. Theoretical computations have been done on more recent ozone data, with more wavelengths, and for more areas. The panel's estimates of annual UV dose reflect weights corresponding to the action spectrum for DNA damage (Figure 2.2) and are corrected for cloud cover and transmission through the skin.

The relationship between the panel's values of UV and those reported from readings of the R-B meter is very nearly linear. Thus either measure of UV will exhibit the approximately linear relationship between log incidence and UV, and, using the simple model, estimates of changes in log incidence corresponding to equivalent changes in the two measures of UV are found to be approximately the same. However, the estimates of percentage increases in skin cancer incidence that correspond to a given percentage decrease in ozone depend on the resulting increase in the magnitude of the measure of UV used in the model, and thus depend on the locality, being greater at lower latitudes. Some predictions by the panel of percentage increases in skin cancer incidence in two localities for 5 percent and 10 percent reductions in stratospheric ozone are shown in Table 5.4. Those values are appreciably greater than those computed using R-B meter readings (Table 5.5).

The differences between the two methods of calculation reviewed above are not the result of different epidemiological data, since they both use the same incidence rates, but of different ways of estimating the change in UV dose per unit change in ozone concentration. In addition, the NRC panel made corrections for variations in doses among individuals and in time (E. L. Scott, University of California, Berkeley, personal communication, 1982).

All measures of annual light flux at the surface of the earth are functions of latitude. Visible light, UV-A, and UV-B increase with decreasing latitude. Assume

TABLE 5.4 Estimates, Derived from Calculations of the UV Flux Corresponding to a DNA Action Spectrum, of Percentage Increases in Skin Cancer Incidence in the U.S. White Population Due to Reduction of Stratospheric Ozone Concentration (with 90 Percent Confidence Bounds) for Two Localities, by Sex and Cell Type

	Percentage Increase in Skin Cancer for Ozone Reduction of:			
	5%		10%	
	Not Corrected	Bias Corrected	Not Corrected	Bias Corrected
Basal cell				
Minneapolis-St. Paul				
Male	7.7 (2-14)	13.0 (10-15)	17.0 (11-25)	29 (22-38)
Female	5.8 (4-9)	9.8 (8-12)	12.3 (8-20)	21 (16-28)
Dallas-Ft. Worth				
Male	16.7 (10-25)	28 (20-35)	34 (21-56)	67 (49-91)
Female	11.3 (8-20)	20 (15-31)	26 (18-42)	47 (35-64)
Squamous cell				
Minneapolis-St. Paul				
Male	12.7 (8-19)	24 (17-29)	29 (19-42)	54 (39-70)
Female	12.0 (6-19)	21 (16-27)	27 (18-40)	50 (37-65)
Dallas-Ft. Worth				
Male	26 (18-44)	49 (36-63)	65 (38-96)	136 (95-190)
Female	25 (16-41)	46 (34-60)	60 (37-90)	126 (89-174)

NOTE: The relation, log incidence = $\alpha + \beta F$ + error, is assumed, where F is the annual flux weighted according to the DNA action spectrum in Figure 2.2. The values in the "Bias Corrected" columns are corrected for variations in doses among individuals and in time (Scott and Straff 1977). The correction is to multiply the estimate of β by 1.7.

SOURCE: E. L. Scott, University of California, Berkeley, personal communication, 1982.

for simplicity that they are all linear functions of latitude whose coefficients of proportionality depend on the wavelength region. In this special case, a 10 percent change in any wavelength region, such as the visible, as a result of a change in location in the United States, would correspond to a 10 percent change in

TABLE 5.5 Estimates, Derived from Measurements of UV Flux by a Robertson-Berger Meter, of Percentage Increases in Skin Cancer Incidence in the U.S. White Population Due to Reduction of Stratospheric Ozone Concentration (with 95 Percent Confidence Bounds) for Two Localities, by Sex and Cell Type

	Percent Increase in Skin Cancer for Ozone Reduction of:	
	5%	10%
Basal cell		
Minneapolis-St. Paul		
Male	5.6 (3-8)	11.2 (5-17)
Female	4.4 (2-7)	8.9 (4-14)
Dallas-Ft. Worth		
Male	8.4 (4-13)	16.9 (8-26)
Female	6.8 (3-11)	13.6 (6-21)
Squamous cell		
Minneapolis-St. Paul		
Male	8.7 (5-13)	17.4 (9-26)
Female	9.2 (4-14)	18.4 (9-28)
Dallas-Ft. Worth		
Male	13.3 (7-20)	27 (14-40)
Female	14 (7-22)	28 (13-43)

NOTE: The model used to fit the data is log incidence = $\alpha + \beta F$ + error, where F is the flux in annual Robertson-Berger meter counts times 10^{-4}. The values in the table were calculated using Scotto et al. (1981) estimates of percentage increases in incidence per 1 percent increase in R-B counts. These estimates were multiplied by 5 (or 10) percent increase in R-B counts times a radiation amplification factor of 0.8 (NRC 1979a), which is approximately the percent increase in R-B counts per 1 percent decrease in ozone concentration. (The actual radiation amplification factor will be slightly larger as the reduction in ozone concentration gets larger.) No bias corrections were made as in Table 5.4.

SOURCE: Derived from Scotto et al. (1981) and NRC (1979a).

UV-B. But a change in stratospheric ozone concentration would change the UV-B component without changing the visible component of sunlight. Hence changes in flux computed from latitude dependencies cannot be used to predict increases in skin cancer, unless the wavelength region used for prediction corresponds to that for skin cancer induction. The epidemiological data, by themselves, do not enable one to determine which wavelength region or regions are important in skin cancer incidence. Thus it is understandable that estimates derived from calculations of the UV flux corresponding to a DNA action spectrum and those from measurements of UV flux by an R-B meter give different estimates of the predicted increase in skin cancer per unit decrease in ozone. To calculate such a prediction, an action spectrum must be assumed, and although the R-B meter is supposed to measure incident UV in the erythema action spectrum, in fact, as was noted earlier, it measures substantially more UV-A (Figure 2.2), which is not sensitive to changes in ozone concentration. Moreover, at present, the best action spectrum to use for such a calculation is the DNA one, even though there are reservations about the roles played by combinations of wavelengths and immunological effects (see Chapter 3). Hence theoretically the best estimates are obtained by the calculations used in Table 5.4.

A completely different way of estimating the impacts of ozone depletion on skin cancer other than melanoma is to assume a knowledge of the shape of the dose-response curves from animal data (de Gruijl and Van der Leun 1980, Rundel and Nachtwey 1978). The increase in skin cancer as a function of age at a particular location is assumed to follow such a dose-response curve, the dose increasing proportionately with age. Calculations based on such theories (de Gruijl and Van der Leun 1980) predict an approximate 5.5-fold increase in skin cancer per 1 percent decrease in ozone concentration, if one assumes that the 1 percent decrease in ozone corresponds to an approximate 2.3 percent increase in the UV that would damage DNA (NRC 1979a).

Predictions by the NRC panel (Scott 1981) estimate the overall increase in basal cell skin cancer incidence per 1 percent decrease in ozone concentration to be between 2 percent and 5 percent depending on latitude. For squamous cell cancer the values are approximately double. The estimates by Rundel and Nachtwey (1978) and de Gruijl and Van der Leun (1980) are in agreement with the panel's. The uncertainties, at present, in such values are approximately the same as the uncertainties in the predictions of expected decrease in ozone concentration.

Experts agree that exposure to sunlight causes more than 90 percent of basal and squamous cell cancers in the United States. These estimates can be crudely tested in at least two ways. The first is to assume (a) that the incidence rate of basal and squamous cell cancers is the same in blacks and whites for all causes except sunlight, and (b) that in blacks these cancers are not caused by sunlight. For these crude calculations it is not necessary to take into account the probable differences in the relationship of basal versus squamous cell cancer to sunlight (Brodkin et al. 1969, Urbach et al. 1972). If these gross assumptions are accepted, the difference between white and black incidence rates would be an estimate of the rate of basal and squamous cell cancer in whites that is due to sunlight. For example, the annual age-adjusted incidence rate of both types of cancer in blacks is 3.4 per 100,000, compared with 232.6 in whites (Scotto et al. 1981). These figures indicate that 99 percent are related to sunlight. Even if it was assumed that all the cancers in blacks were squamous cell--since in blacks this type is relatively more common than basal cell--92 percent of squamous cell cancers in whites would still be related to sunlight. However, given that the incidence rates in blacks have some correlation with latitude, as was suggested earlier, these values would be overestimates.

The second way to get a crude estimate of the percentage of cancers caused by exposure to sunlight is to assume that in whites the incidence rates on body sites virtually never exposed to sunlight (genital areas) are base line measures of incidence from all other causes. For example, one tenth of 1 percent of basal cell cancer was found on the genital areas of men's bodies and 0.3 percent on women's (Scotto et al. 1981). The comparable figures were 0.8 percent and 2.8 percent for squamous cell cancers. With these base lines, more than 99 percent of basal cell cancers and more than 90 percent of squamous cell cancers would be attributed to UV irradiation.

Melanomas

What Was Known by 1979

Melanoma is less common and more dangerous than other types of skin cancer. Data on mortality rates are shown in Table 5.6. In the NRC (1975) report, Environmental Impact of Stratospheric Flight, melanoma was accepted as a disease associated with exposure to sunlight and found on parts of the body exposed to sunlight although not concentrated on the sites that receive the highest intensity. There was a higher incidence at lower latitudes and among fair-skinned people. The reasons in favor of translating this association into one between melanoma and the UV-B component of sunlight were summarized in NRC (1975) and repeated in NRC (1976a, 1979a). The argument was seen at that time to be clearly less substantial than that supporting the inferred relationship between UV-B exposure and basal and squamous cell skin cancer.

TABLE 5.6 Mortality of Melanoma by Country and Latitude for White Populations

Country or Geographical Area	Latitude of Center of Population	Deaths from Melanoma per Million per Year
South Island, New Zealand	45° S	8
North Island, New Zealand	39° S	12
Victoria	38° S	8
New South Wales	34° S	15
Cape of Good Hope	32° S	13
Natal	29° S	18
Transvaal and Orange Free State	27° S	10
Queensland	27° S	23
California	37° N	12
Northeast United States	42° N	9
Italy	43° N	2
Switzerland	47° N	6
France	48° N	1
Canada	50° N	5
Netherlands	52° N	5
England	53° N	6
Eire	53° N	3
Scotland	56° N	4
Sweden	59° N	9
Norway	61° N	10

SOURCE: Lancaster (1956).

The argument was built on the facts that sunburn and melanoma are often found in the same tissue, melanomalike lesions can be induced by irradiation of chemically induced benign pigmented lesions in experimental animals, and individuals with xerodema pigmentosum have an extraordinarily high prevalence of melanoma (Kraemer 1980, Takebe et al. 1977). An additional argument was that one could read a pathogenetic relevance in the similarity of the erythema and DNA-damaging action spectra. As discussed in the next section, since 1976, the case for an association between UV-B and melanoma has been weakened rather than strengthened by the results of additional clinical, pathological, and epidemiological studies. Furthermore (with the exception of a single animal), it has not been possible to use UV-B alone to induce melanomas in experimental animals.

The only statistical association that has been repeatedly found nationwide and worldwide is the one between melanoma incidence or mortality rates and latitude. Although widespread, the association is not totally consistent. There is still no clear evidence, although cohort analysis shows an increased incidence with age, that the latitude association is a dose-related relationship. There seems to be no doubt that Western countries have been living through a rapid increase in melanoma incidence (Houghton et al. 1980, Lee et al. 1979). Each successive cohort studied has had higher incidence rates. The epidemic has been affecting populations at many different latitudes with varying background incidence levels. The epidemiological picture of higher incidence rates in each successive birth cohort is reminiscent of the earlier lung cancer epidemic in those same countries, which resulted from the progressive adoption of the habit of cigarette smoking by more and more members of each younger generation.

In spite of grave reservations about the nature of the observed statistical association, NRC (1975, 1976a, 1979a) used the existing statistical association between either latitude or R-B meter readings and melanoma incidence or mortality to make predictions about the likely increase in melanoma incidence, given future increases in UV-B exposure. It was clearly recognized that this decision was made without knowledge of the percentage of skin melanomas in the United States likely to be caused either wholly or in part by UV-B exposure and without good evidence pinpointing other factors that would be more powerful determinants of the future incidence rates. It was argued that the predictions would be useful even if the association between latitude and melanoma incidence turned out to be indirect or an extremely remote index of the true causative factor. Lacking any epidemiological clues to major etiological factors other than sunlight, NRC (1976a, 1979a) tried to provide a behavioral explanation (changes in exposure patterns) for epidemiological inconsistencies on the basis of variability in personal susceptibility recognized in series of clinical observations. Following through on this line of reasoning the earlier studies had emphasized the need for new and more extensive data that would permit, for individuals with varying levels of innate susceptibility, analysis of measurements of exposure to sunlight.

Advances in Knowledge

Much new information about melanoma has been collected and published since 1979. Some of this information confirms the association between melanoma and latitude and levels of UV intensity. Much of it underscores and extends the inconsistent and sometimes paradoxical findings from past epidemiological studies, and some of it provides interesting new avenues for exploration.

When the 1973-1976 incidence data from the NCI SEER program are plotted against the 1977-1978 NCI R-B meter measurements of accumulated dose in eight geographic locations, the results are consistent with those of earlier analyses and show a definite relationship between melanoma and measurements of annual solar UV flux (Scotto et al. 1982) (Figure 5.7). The slope resulting from this statistical analysis is similar to that obtained for basal cell cancers (Figure 5.5) and is virtually the same as the slopes previously developed from other bodies of data (Scott and Straf 1977, Scotto et al. 1982). Other newly published studies of incidence data again consistently report higher rates of melanoma at lower latitudes (Crombie 1979, Jensen and Bolander 1980, Malec and Eklund 1978). Several reported studies of trends in incidence have confirmed the continuation of the worldwide increase.

Although the overall increase in the incidence of melanoma is virtually universal, the incidence on specific body sites has increased at various rates (Scotto et al., 1982). A report that melanoma of the eye has not

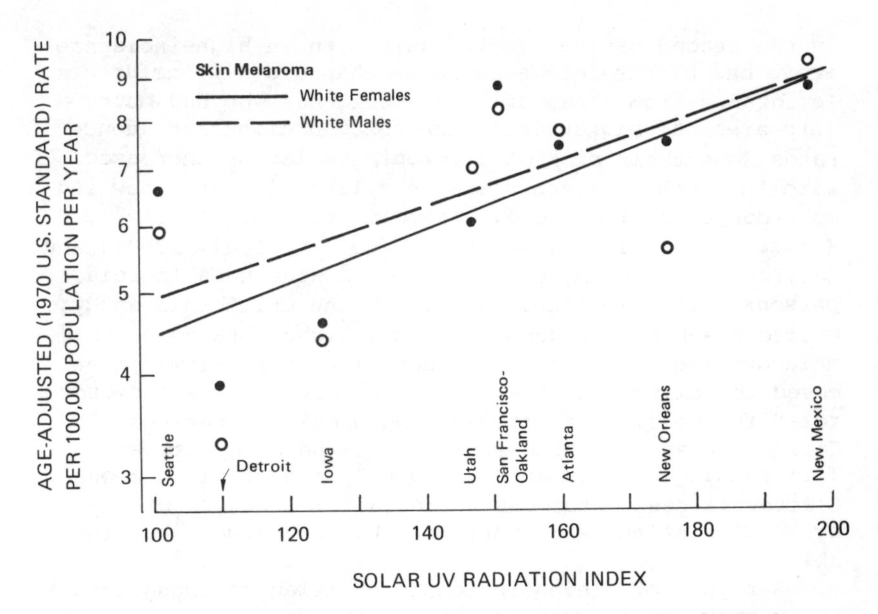

FIGURE 5.7 Annual age-adjusted incidence rates for skin melanoma (SEER data, 1973-1976) among white females (open symbols) and males (closed symbols), according to one year's UV measurements at selected areas of the United States. The UV radiation index is the total Robertson-Berger meter counts over a one-year period multiplied by 10^{-4}. The meters read UV-B between 290 nm and 320 nm, as well as some UV-A. (J. Scotto, National Cancer Institute, personal communication, 1981.)

increased in incidence has some importance, because this tumor occurs in the back of the eye, where UV does not penetrate, and is therefore unlikely to be associated with exposure to UV (Strickland and Lee 1981). A new series of reports on the occupational incidence of melanoma has provided very consistent information (Lee 1981), as has a second series of studies of the incidence of melanoma in immigrant and indigenous residents of Israel (Movshovitz and Modan 1973). Each series, however, provides information that has to be reconciled with that of the other. In the occupational series, four studies from different parts of the world and from very different latitudes failed to demonstrate any excess incidence of or mortality from melanoma among outdoor as compared with indoor workers of similar status. All four studies confirmed earlier reports of the correlation of increasing incidence with higher socioeconomic status.

In the second series, individuals born in high-incidence areas had higher incidence rates than did all immigrants. Immigrants from areas of lower incidence who had moved into areas of higher incidence assumed higher incidence rates, and their risk of developing melanoma increased with the number of years they had lived in their new and more dangerous locations. This is true for Israel (Anaise et al. 1978), Australia (Holman et al. 1980), and California (T. Mack, University of Southern California, personal communication, 1981). In the California study, California-born residents have the highest rates of melanoma incidence and immigrant midwesterners who have moved to California have the lowest rates. The incidence rates for California residents born halfway between California and the Midwest fall somewhere in between. This ranking of incidence rates by place of birth among California residents does not hold for melanoma of the eye or for melanoma in parts of the body other than the skin.

A report of a high incidence of melanoma among workers at the Lawrence Livermore National Laboratory may provide a unique opportunity to identify contributory, if not causative, etiological factors (Austin et al. 1981).

Most other recent information concerns individual susceptibility and the etiology of specific histological types of melanoma, and precancerous conditions.

Individual Susceptibility. Although Scandinavian populations have unusually high incidence rates of skin melanoma for their latitude of residence, it is now known that they also have high incidence rates of melanoma of the eye, which cannot readily be related to either sunlight or UV-B exposure (Strickland and Lee 1981). This combination suggests an underlying susceptibility to melanomas in general.

There are a number of reports of higher incidence rates of melanomas in women during the later years of reproductive life in populations with different base rates of melanoma incidence and among different ethnic groups (Jensen and Bolander 1980, Lee and Storer 1980). The possibility of a specific hormonal component in the etiology of a certain proportion of these tumors is now being considered. Some early analysis of data collected at the University of Sydney suggests that in the women with higher incidence rates the ratio of superficial spreading melanoma to nodular and other histological types (Table 5.7) is higher than average. This increased

TABLE 5.7 Primary Melanoma of Skin (in White Persons)

Type of Melanoma	Median Age (years)	Specific Sites[a]	Rate of Development	Appearance
Lentigomaligna	70	Face, neck, and hands	Slow: 5-20 years	Predominantly flat spot 2-20 cm in size, with irregular borders and pigment pattern. Raised areas indicate areas of invasive tumor.
Superficial spreading melanoma	47	Face, neck, upper trunk, and (in females) lower legs	Moderately slow: 1-7 years	Predominantly a slightly raised lesion with a raised distinct irregular border that may be notched. The brown and black color is admixed with blue, white, red, or their half tones gray and pink.
Nodular	50	–	Rapid: months	Isolated small (3.0 cm) nodule with smooth borders; color uniform blue-black.
Arcolentiginous[b]	–	Palms, soles, nail beds, mucous membrane	–	Predominantly flat spot with irregular borders. Raised areas indicate invasive tumor.

[a]The first three types occur either on the exposed parts of the face, neck, and hands or on the relatively exposed areas of the chest, back, and legs. Only a few lesions are seen on covered areas such as the breasts of females, bathing trunk areas of males, and bathing suit areas of females.
[b]This type predominates in blacks and orientals.

SOURCE: NRC (1979a).

ratio has also been found in other groups, for example, among men in the highest socioeconomic classes. It is also found on body sites with the highest rates of increasing incidence, namely, women's lower legs and men's trunks (backs) (McCarthy et al. 1980).

These early findings could explain a number of other recent reports that note that the proportion of small and thin newly diagnosed melanomas is steadily increasing in populations with both very high and moderate incidence rates. One interpretation would be that the superficial spreading melanoma is the major cause of the increased rate in the groups identified above; another is that lesions are being detected at an earlier stage in those groups and are being treated while they are small and thin. Clinicians have always associated the lentigo melanoma with excessive exposure to UV. Some clinicians believe that melanomas of this type are undercounted, particularly in the lower latitudes, where they are more common. It is the view of these clinicians that these cancers usually progress so slowly that they rarely reach a serious point during life, and thus remain undiagnosed. Without good diagnosis and reliable reporting, there can be no valid assessment of the distribution of these cancers over the body. Clinical observation would lead us to be believe that they virtually always appear on exposed areas of the body. If they are relatively rarely diagnosed and are not lethal, the published incidence rates of 6 percent to 10 percent among all melanomas must be taken as uncertain. If, for example, superficial spreading and nodular melanomas were not associated with exposure to UV and all incidence cases could be accurately counted, the proportion of lentigo melanomas among all melanomas would be higher in high-incidence areas, such as Texas or Australia, than in low-incidence areas. There are no available data with which to test this hypothesis (Lee and Strickland 1981). This is one example of the growing interest in specific histological types of melanoma. Much more detailed histological descriptions at the time of diagnosis are needed to provide the basis for pursuing this potentially fruitful research.

Other preliminary data suggest that individuals with melanomas sunbathe less and use more sunscreens than do control subjects. Furthermore, they may have less residential, occupational, and recreational exposure to the sun. These findings seem to apply to both susceptible and less susceptible individuals (S. Graham, State

University of New York at Buffalo, personal communication, 1981). There is considerable evidence of familial concentration of melanoma (S. Graham, State University of New York at Buffalo, personal communication, 1981). In a recent study of 214 patients with melanomas, an appropriate group of controls, and family members of both groups, it was found that family members of patients with melanomas have high relative risks that are of the order of eightfold for all first-degree relatives (parents, offspring, and siblings) and twelvefold for parent-offspring pairs (Duggleby et al. 1981). These very high risks in family members could be consistent with other studies reported below. It is perhaps important to mention that this high relative risk found in relatives of the individuals first identified as having melanoma (which may be of great importance in clarifying the etiology of melanoma) occurs in relatively few instances and can account for only a few among all cases of melanoma.

Precursor Lesions. There have been a number of recent reports of precursor dysplastic nevi (Clark et al. 1978; Elder et al. 1980, 1981; Reimer et al. 1978; Wiskemann 1977). Dysplastic nevi are usually large irregular moles on the skin that exhibit evidence of abnormal histological development (dysplasia). It was first believed that these lesions were always part of a familial condition called B-K Mole Syndrome (Clark et al. 1978; Green et al. 1978, 1980). However, it is now believed that there are both familial and sporadic dysplastic nevus syndromes and that the progression from a typical (i.e., histologically normal) nevus to a melanoma is analogous to the progression in the cervix from normal endothelial cells to squamous cell carcinoma in situ. It is also believed that on the skin, as on the cervix or in the bronchi, dysplasia is likely to occur in multiple sites. A body of histopathological, clinical, and biochemical evidence is being accumulated to explore this hypothesis, and some tentative results from laboratory experiments suggest that the fibroblast cells of patients with dysplastic nevi and hereditary cutaneous melanoma are peculiarly sensitive to UV radiation (Smith et al., in press).

During the past 5 years, there has been an increased number of laboratory and case control studies of individual human beings and population-based incidence studies. Preliminary results are available from very few. Other investigators expect that their results are likely to strengthen evidence favoring individual types of susceptibility and to emphasize the need to analyze melanoma incidence rates by histological type (Sober et al. 1979). There does not seem to be any reason to expect strengthening of evidence in support of a hypothesis that lengthy accumulation of exposures to UV radiation per se is the overriding or even one of the most important causes of melanoma other than lentigo maligna melanoma. There is an increasing number of individually inconclusive reports that all suggest that a history of acute exposures such as sunburn or marked skin sensitivity to sun exposure may be particularly important (Beitner et al. 1981, Jung et al. 1981, Paffenbarger et al. 1978, Sober et al. 1979). In light of the inconsistent and inconclusive state of knowledge about a possible dose-response relationship between melanoma and UV, we are unwilling to make quantitative estimates of the effects of reduced concentrations of atmospheric ozone on the incidence of melanoma.

PROTECTION AGAINST DAMAGE FROM SUNLIGHT

Most of the direct human health hazards predicted to result from a depletion of stratospheric ozone concentration, and a consequent increase in solar UV, stem from exposure of the skin--increased incidence of sunburn, solar degeneration, skin cancer, and immune system effects. All skin is not equally susceptible to UV damage, however. There are two principal intrinsic barriers to UV. One is the stratum corneum on the outermost surface of the skin, which absorbs the most biologically active wavelengths of UV. This layer is approximately the same in all individuals and can be thickened as a reparative response to UV injury of skin. The other physiologic, chemical, and optical protector against UV is a pigment called melanin, which is produced by cells in the epidermis called melanocytes. This pigment gives skin its brownish color. The production of this pigment is increased after sun exposure (tanning). The base line amount of melanin and the capacity to increase melanin production are genetically determined. White persons have much less melanin than blacks. Caucasians have different levels of melanin in their skin. In general, those with the least base line pigment have the least capacity for tanning. These individuals are the most susceptible to sun damage of all kinds. The base line pigmentation of very dark skinned races protects against UV-B radiation 30 times better, and that of moderately dark-skinned races 3 to 5 times better, than that of fair Caucasians.

The range of base line pigmentation, and the capacity for tanning (i.e., for increasing melanin production), in fair Caucasians has been arbitrarily divided into four categories (see NRC 1979a, Appendix H), depending on the person's assessment of his or her own propensity to sunburn (relative absence of base line melanin) and ability to tan. Information is obtained by asking a standardized question about response to sun exposure. This method, called "skin typing," has proved to be a useful shorthand for categorizing persons in terms of responses to phototherapy, sunscreen testing, and clinical surveys. It is, however, not quantitative and is subject to cohort and interviewer bias. It simply predicts photobiologic response on the basis of the subject's memory of past photobiologic response.

Two additional excellent barriers against UV are available. There are now excellent sunscreens--chemicals that when applied to the skin absorb UV before it reaches viable cells. They provide a wide range of added protection that can reach a factor of more than 10. This means that if it normally requires 25 minutes of sun exposure at noon in June to cause minimal sunburn in a fair person, a sunscreen with a protection factor of 10 would change the requirement to 250 minutes. This large amount of protection is more than enough to cover the UV increases likely to result from possible ozone depletion. Screening provided by protein, melanin, and topically applied sunscreens is most likely additive (Hawk and Parrish 1982).

The other means of protection is the most effective. Avoidance of sunlight between 11:00 a.m. and 2:00 p.m. greatly reduces the exposure of skin to UV-B. Even modest changes in human behavior can decrease solar UV exposure by factors that are much greater than the least conservative factors estimated for ozone-related increases in UV.

Finally, the possible anticarcinogenic effects of β-carotene and synthetic retinols are being explored, but the roles of these compounds are complicated and controversial at this time.

RESEARCH RECOMMENDATIONS

The following list of research recommendations is not exhaustive but has been limited to those issues that should receive attention first. Two of the several direct human health hazards that might be expected to result from an increase in the intensity of solar UV radiation should be emphasized in future research: immune system effects and skin cancer. The list is not organized according to priority.

1. Photoimmunology is a new and important area of research. It appears that erythmogenic (sunburn-causing) UV exposures can cause systemic alterations in the immune systems of animals and humans. The implications of these findings for understanding the pathogenesis of skin cancer and certain other diseases must be investigated. The identification of common mechanisms would be an important contribution. As an initial step, studies to determine the magnitude of UV-B effects on the human immune system, the dose-response relationships, and the effective wavelengths should be vigorously pursued.
2. The use of animal models to study UV-induced skin cancer (experimental photocarcinogenesis) has proved valuable in understanding the role of UV in the development of human skin cancer other than melanoma. Further animal studies are needed to understand interactions among parameters such as intermittent exposures, different wavelengths, dose rates, and agents that modify cellular responses to UV irradiation.
3. An animal model for light-induced melanoma must be discovered before it will be possible to determine if a reduction in stratospheric ozone concentration will cause an increased incidence of melanoma in humans. Dose-response relationships and effective wavelengths should be determined.
4. Prospective studies of patients undergoing various forms of phototherapy and photochemotherapy could be helpful in obtaining quantitative information about the relationship of certain UV wavebands to human skin cancer.
5. Epidemiological studies of skin cancer incidence and mortality rates have supplied valuable evidence confirming the existence of an association between basal and squamous cell skin cancers and sunlight. As basal and squamous cell skin cancers are not routinely reported to cancer registries, it will be necessary to maintain routine surveillance by periodic surveys during the next

50 years. These incidence surveys should be at intervals no longer than 10 years and should collect data that can be subjected to cohort as well as cross-sectional analysis.

6. In addition to (5) above, epidemiological research on skin cancer other than melanoma should concentrate on retrospective and prospective studies of individual human beings. The latter will need some simple measures of effective individual exposure to UV-B to correlate with incidence and/or documentation of complete protection from UV exposure to correlate with prevention of skin cancer.

7. Information obtained since 1979 makes it clear that the etiology of malignant melanoma is even more complex than previously believed. A number of risk factors are involved, and, in addition, there are various subtypes of melanoma. In order to determine the association between UV and melanoma, it is essential to determine the incidence of and latitude dependence of the various melanoma subtypes. To do this, careful epidemiological studies based on reliable clinical and much more detailed histological descriptions at the time of diagnosis are needed.

8. Epidemiological studies of individual human beings and their effective exposures are essential in learning more about the etiology of melanoma. These studies should include some that focus on the experience of children, some that explore associations between the development of nevi and the sensitivity to sunlight exposure, and some that explore the protective aspects of exposure to wavelengths other than UV-B.

REFERENCES

Aberer, W., G. Shuler, G. Stinge, H. Honigsmann, and K. Wolff (1981) Ultraviolet light depletes surface markers of Langerhans cells. Journal of Investigative Dermatology 76:202-210.

Anaise, D., R. Steinitz, and N. Ben Hur (1978) Solar radiation: A possible etiological factor in malignant melanoma in Israel: A retrospective study (1960-72). Cancer 42:299-304.

Ananthaswamy, H.N. and M.S. Fisher (1980) Photoreactivation of ultraviolet radiation-induced pyrimidine dimers in neonatal BALB/c mouse skin. Cancer Research 41:1829-1833.

Anderson, R.R. and J.A. Parrish (1971) Microvasculature can be selectively damaged using dye lasers: A basic theory and experimental evidence in human skin. Lasers in Surgery and Medicine 1:263-276.

Anderson, R.R. and J.A. Parrish (1981) The optics of human skin. Journal of Investigative Dermatology 77:13-19.

Anderson, R.R. and J.A. Parrish (1982) Optical properties of human skin. Chapter 26, Photomedicine. New York, N.Y.: Plenum Press. (In press)

Austin, D.F., P.J. Reynolds, M.A. Snyder, M.W. Biggs, and H.A. Stubbs (1981) Malignant melanoma among employees of Lawrence Livermore National Laboratory. Lancet II:712-716.

Beitner, H., U. Ringborg, G. Wennersten, and B. Lagerlot (1981) Further evidence for increased light sensitivity in patients with malignant melanoma. British Journal of Dermatology 104(3):289-294.

Berenblum, I. and V. Armuth (1981) Two independent aspects of tumor promotion. Biochimica et Biophysica Acta 651:51-63.

Berger, D.S. and F. Urbach (1982) A climatology of sunburning ultraviolet radiation. Photochemistry and Photobiology 35:187-192.

Berger, D., D.F. Robertson, and R.E. Davies (1975) Field measurements of biologically effective UV radiation. Impacts of Climatic Change on the Biosphere, CIAP Monograph 5, Part I: Ultraviolet Radiation Effects, edited by D.S. Nachtwey, M.M. Caldwell, and R.H. Biggs. Washington, D.C.: U.S. Department of Transportation.

Biggs, R.H. and S.V. Kossuth (1978) Impact of solar UV-B radiation on crop productivity. Final report of UV-B biological and climate effects research (BACER). Terrestrial FY 77. Gainesville, Fla.: University of Florida. (Unpublished)

Biggs, R.H., W.B. Sisson, and M.M. Caldwell (1975) Response of higher terrestrial plants to elevated UV-B irradiance. Pages 4-34 to 4-50, Impacts of Climatic Change on the Biosphere, CIAP Monograph 5, Part I: Ultraviolet Radiation Effects, edited by D.S. Nachtwey, M.M. Caldwell, and R.H. Biggs. Washington, D.C.: U.S. Department of Transportation.

Black, H.S., J.T. Chan, and G.E. Brown (1978) Effects of dietary constraints on ultraviolet light mediated carcinogenesis. Cancer Research 38:1384-1387.

Blum, H.F. (1959) Carcinogenesis by Ultraviolet Light. Princeton, New Jersey: Princeton University Press.

Bockstahler, L.E. and C.D. Lytle (1977) Radiation enhanced reactivation of nuclear replicating mammalian viruses. Photochemistry and Photobiology 25:477-482.

Bogenrieder, A. and R. Klein (1977) Die Rolle des UV-Lichtes beim sog. Auspflanzungsschock von Gewachshaussetzlingen. Angewandte Botanik 51:99-107.

Brandle, J.R., W.F. Campbell, W.B. Sisson, and M.M. Caldwell (1977) Net photosynthesis, electron transport capacity, and ultrastructure of Pisum sativum L. exposed to ultraviolet-B radiation. Plant Physiology 60:165-169.

Brodkin, R.H., A.W. Kopf, and R. Andrade (1969) Basal cell epithelioma and elastosis: A comparison of distribution. Pages 581-618, The Biological Effects of Ultraviolet Radiation, edited by F. Urbach. Oxford, England: Pergamon Press.

Cabrera-Juarez, E., J.K. Setlow, P.A. Swenson, and M.J. Peak (1976) Oxygen-independent inactivation of Haemophilus influenzae transforming DNA by monochromatic radiation: Action spectrum, effect of histidine and repair. Photochemistry and Photobiology 23:309-313.

Caldwell, M.M. (1971) Solar UV irradiation and the growth and development of higher plants. Photophysiology 6:131-177.

Caldwell, M.M. (1982) Plant response to solar ultraviolet radiation. In Encyclopedia of Plant Physiology, Vol. 12A, Interactions of Plants with the Physical Environment, edited by O.L. Lang, P.S. Nobel, O.B. Osmond, and H. Ziegler. Berlin: Springer-Verlag. (In press)

Caldwell, M.M., R. Robberecht, and W.D. Billings (1980) A steep latitudinal gradient of solar ultraviolet-B radiation in the arctic-alpine life zone. Ecology 61:600-611.

Calkins, J. and T. Thordardottir (1980) The ecological significance of solar UV radiation on aquatic organisms. Nature 282:563-566.

Chan, G.L. and J.B. Little (1976) Induction of oncogenic transformation in vitro by ultraviolet light. Nature 264:442-444.

Cicerone, R.J. (1981) Halogens in the atmosphere. Reviews of Geophysics and Space Physics 19:123-139.

Clark, W.H., Jr., R.R. Reimer, M. Greene, A.M. Ainsworth, and M.J. Mastrangelo (1978) Origin of familial malignant melanomas from heritable melanocytic lesions. Archives of Dermatology 114:732-738.

Cleaver, J.E. (1977) Nucleosome structure controls rates of excision repair in DNA of human cells. Nature 270:451-453.

Crombie, I.K. (1979) Variation of melanoma incidence with latitude in North America and Europe. British Journal of Cancer 40:774-781.

D'Ambrosio, S.M., L. Slazinski, J.W. Whetstone, and E. Lowney (1981a) Excision repair of UV-induced pyrimidine dimers in human skin in vivo. Journal of Investigative Dermatology 77:311-313.

D'Ambrosio, S.M., J.W. Whetstone, L. Slazinski, and E. Lowney (1981b) Photorepair of pyrimidine dimers in human skin in vivo. Photochemistry and Photobiology 34:461-464.

Das Gupta, U.B. and W.C. Summers (1978) Ultraviolet reactivation of herpes simplex virus is mutagenic and inducible in mammalian cells. Proceedings of the National Academy of Sciences USA 75:2378-2381.

DeFabo, E.C. and M.L. Kripke (1979) Dose-response characteristics of immunological unresponsiveness to

UV-induced tumors produced by UV irradiation of mice. Photochemistry and Photobiology 30:385-390.

DeFabo, E.C. and F.P. Noonan (1980) Preliminary studies on an in vivo action spectrum for UV-induced suppression of contact sensitivity in mice. Page 345, Book of Abstracts, Eighth International Congress on Photobiology, July 20-25, Strasbourg.

DeGruijl, F.R. and J.C. Van der Leun (1980) A dose-response model for skin cancer induction by chronic UV exposure of a human population. Journal of Theoretical Biology 83:487-504.

DiPaolo, J.A. and P.J. Donovan (1976) In vitro morphologic transformation of Syrian hamster cells by UV-irradiation is enhanced by x-irradiation and unaffected by chemical carcinogens. International Journal of Radiation Biology 30:41-53.

Doniger, J., E.D. Jacobson, K. Krell, and J.A. DiPaolo (1981) Ultraviolet light action spectra for neoplastic transformation and lethality of Syrian hamster embryo cells correlate with spectrum for pyrimidine dimer formation in cellular DNA. Proceedings of the National Academy of Sciences USA 78:2378-2382.

Duggleby, W.F., H. Stoll, R.L. Priore, P. Greenwald, and S. Graham (1981) A genetic analysis of melanoma-polygenic inheritance as a threshold trait. American Journal of Epidemiology 114(1):63-72.

Elder, D.E., L.I. Goldman, S.C. Goldman, M.H. Greene, and W.H. Clark (1980) Dysplastic nevus syndrome: A phenotypic association of sporadic cutaneous melanoma. Cancer 46:1787-1794.

Elder, D.E., M.H. Greene, E.E. Bondi, and W. H. Clark, Jr. (1981) Acquired melanocytic nevi and melanoma: The dysplastic nevus syndrome. Pathology of Malignant Melanoma, edited by A.B. Ackerman. New York, N.Y.: Masson. (In press)

Elkind, M.M. and A. Han (1978) DNA single-strand lesions due to (sunlight) and UV light: A comparison of their induction in Chinese hamster and human cells, and their fate in Chinese hamster cells. Photochemistry and Photobiology 27:717-724.

Elkind, M.M., A. Han, and C.M. Chiang-Liu (1978) "Sunlight"-induced mammalian cell killing: A comparative study of ultraviolet and near-ultraviolet inactivation. Photochemistry and Photobiology 27:709-715.

Ellison, J.J. and J.D. Childs (1981) Pyrimidine dimers induced in Escherichia coli DNA by ultraviolet radiation present in sunlight. Photochemistry and Photobiology 34:465-469.

Erickson, L.C., M.O. Bradley, and K.W. Kohn (1980) Mechanisms for the production of DNA damage in cultured human and hamster cells irradiated with light from fluorescent lamps, sunlamps, and the sun. Biochimica et Biophysica Acta 610:105.

Fears, T.R., J. Scotto, and M.A. Schneiderman (1976) Skin cancer, melanoma and sunlight. American Journal of Public Health 66:461-464.

Fears, T.R., J. Scotto, and M.A. Schneiderman (1977) Mathematical models of age and ultraviolet effects on the incidence of skin cancer among whites in the United States. American Journal of Epidemiology 105:420-427.

Fels, S.B., J.D. Mahlman, M.D. Schwarzkopf, and R.W. Sinclair (1980) Stratospheric sensitivity to perturbations in ozone and carbon dioxide: Radiative and dynamical response. Journal of Atmospheric Science 37:2265-2297.

Fisher, M.S. and M.L. Kripke (1977) Systemic alteration induced in mice by ultraviolet light irradiation and its relationship to ultraviolet carcinogenesis. Proceedings of the National Academy of Sciences USA 74:1688-1692.

Fisher, M.S. and M.L. Kripke (1978) Further studies on the tumor-specific suppressor cells induced by ultraviolet radiation. Journal of Immunology 121:1139-1144.

Fisher, M.S. and M.L. Kripke (1981) The role of UV-induced suppressor cells in the development of primary skin cancers in UV-irradiated mice. Page 178, Proceedings of the American Society for Photobiology, 9th Annual Meeting. Williamsburg, Va., June 14-18, 1981. Bethesda, Md.: American Society for Photobiology.

Forbes, P.D., R.E. Davies, D. Berger, and F. Urbach (1980) A Study on Photocarcinogenesis in Hairless Mice. Final Report, Contract No. NO-1 CP 43271. Bethesda, Md.: National Cancer Institute.

Forbes, P.D., H.F. Blum, and R.E. Davies (1981) Photocarcinogenesis in hairless mice: Dose-response and the influence of dose-delivery. Photochemistry and Photobiology 34:361-365.

Freeman, R.G. and J.M. Knox (1964) Influence of temperature on ultraviolet injury. Archives of Dermatology 89:858-864.

Friedberg, E.C., U.K. Ehmann, and J.I. Williams (1979) Human diseases associated with defective DNA repair. Advances in Radiation Biology 8:85-174.

Green, A.E.S. and J.H. Miller (1975) Measures of biologically effective radiation in the 280-340 nm region. Pages 2-60 to 2-70, Impacts of Climatic Change on the Biosphere, Part I: Ultraviolet Radiation Effects, edited by D.S. Nachtwey et al. DOT-TST-75-55. Washington, D.C.: U.S. Department of Transportation.

Green, A.E.S., K.R. Cross, and L.A. Smith (1980) Improved analytic characterization of ultraviolet skylight. Photochemistry and Photobiology 31:59-65.

Greene, M.H., R.R. Reimer, W.H. Clark, Jr., and M.J. Mastrangelo (1978) Precursor lesions in familial melanoma. Seminars in Oncology 5:85-87.

Greene, M.I., M.S. Sy, M.L. Kripke, and B. Benacerraf (1979) Impairment of antigen-presenting cell function by ultraviolet radiation. Proceedings of the National Academy of Sciences USA 76:6592-6595.

Greene, M.H., et al. (1980) Precursor naevi in cutaneous malignant melanoma: A proposed nomenclature. Lancet II:1024.

Griggs, H.G. and M.A. Bender (1973) Photoreactivation of ultraviolet-induced chromosomal aberrations. Science 179:86-88.

Hanawalt, P.C., P.K. Cooper, A.K. Ganesan, and C.A. Smith (1979) DNA repair in bacteria and mammalian cells. Annual Review of Biochemistry 48:783-836.

Haniszko, J. and R.R. Suskind (1963) The effect of ultraviolet radiation on experimental cutaneous sensitization in guinea pigs. Journal of Investigative Dermatology 40:183-191.

Hariharan, P.V. and P.A. Cerutti (1977) Formation of products of the 5, 6-dihydroxydihydrothymine type by ultraviolet light in HeLa cells. Biochemistry 16:2791.

Harm, H. (1980a) Damage and repair in mammalian cells after exposure to non-ionizing radiation. Mutation Research 69:157-165 and 167-176.

Harm, W. (1980b) Biological Effects of Ultraviolet Radiation. Cambridge, Mass.: Cambridge University Press.

Hawk, J.L.M. and J.A. Parrish (1982) Responses of normal skin to ultraviolet radiation. Photomedicine. New York, N.Y.: Plenum Press. (In press)

Holman, O.D.J., C.D. Mulroney, and B.K. Armstrong (1980) Epidemiology of pre-invasive and invasive malignant melanoma in Western Australia. International Journal of Cancer 25:317-323.

Horowitz, S., D. Cripps, and R. Hong (1974) Selective T cell killing of human lymphocytes by ultraviolet radiation. Cellular Immunology 14:80-86.

Houghton, A., J. Flannery, and M.V. Viola (1980) Malignant Melanoma in Connecticut and Denmark. International Journal of Cancer 25:95-104.

Hudson, R.D. and E.I. Reed (1979) The Stratosphere: Present and Future. NASA R.P. 1049. Greenbelt, Md.: National Aeronautics and Space Administration.

Hudson, R.D. et al., eds. (1982) The Stratosphere 1981: Theory and Measurements. WMO Global Research and Monitoring Project Report No. 11. Geneva: World Meteorological Organization. (Available from National Aeronautics and Space Administration, Code 963, Greenbelt, Md. 20771.)

Hunter, J.R., J.H. Taylor, and H.G. Moser (1979) Effect of ultraviolet irradiation on eggs and larvae of the northern anchovy Engraulis morolax, and the Pacific mackerel, Scoaker japonicus, during the embryonic stage. Photochemistry and Photobiology 29:325-338.

Jacobson, E.D., K. Krell, and M.J. Dempsey (1981) The wavelength dependence of ultraviolet light-induced cell killing and mutagenesis in h5178Y mouse lymphoma cells. Photochemistry and Photobiology 33:257-260.

Jensen, O.M. and A.M. Bolander (1980) Trends in malignant melanoma of the skin. World Health Statistics 33:2-26.

Jessup, J.M., N. Hanna, E. Palaszynski, and M.L. Kripke (1978) Mechanisms of depressed reactivity to dinitrochlorobenzene and ultraviolet-induced tumors during ultraviolet carcinogenesis in BALB/c mice. Cellular Immunology 38:105-115.

Johnson, F.S., T. Mo, and A.E.S. Green (1976) Average latitudinal variation in ultraviolet radiation at the earth's surface. Photochemistry and Photobiology 23:179-188.

Jones, L.W. and B. Kok (1966) Photoinhibition of chloroplast reactions. I. Kinetics and action spectra. Plant Physiology 41:1037-1043.

Jung, E.G., K. Gunthart, R.F.G. Metzger, and E. Bohnert (1981) Risk factors of the cutaneous melanoma phenotype. Archives of Dermatology Research 270:33-36.

Kantor, G.J., J.C. Sutherland, and R.B. Setlow (1980) Action spectra for killing non-dividing normal human and xeroderma pigmentosum cells. Photochemistry and Photobiology 31:459-460.

Karanas, J.J., R.C. Worrest, and H. Van Dyke (1981) Impact of UV-B radiation (290-320 nm) on the fecundity of Acartia clausii (Copepoda). Marine Biology 65:125-133.

Kaupp, S.E. and J.R. Hunter (1981) Photorepair in larval anchovy Engraulis mordax. Photochemistry and Photobiology 33:253-256.

Kennedy, A.R., M. Fox, G. Murphy, and J.B. Little (1980) Relationship between x-ray exposure and malignant transformation in C_3H 10T cells. Proceedings of the National Academy of Sciences USA 77:7262-7266.

Kolb, C.E. and J.B. Elgin (1976) Gas phase chemical kinetics of sodium present in the upper atmosphere. Nature 263:488-490.

Kopecky, K.E., G.W. Pugh, Jr., D.E. Hughes, G.D. Booth, and N.F. Cheville (1979) Biological effect of ultraviolet radiation on cattle: Bovine ocular squamous cell carcinoma. American Journal of Veterinary Research 40(12):1783-1788.

Kraemer, K.H. (1980) Xeroderma pigmentosum. Unit 19-7, Clinical Dermatology, Vol. 4, edited by D.J. Demis, R.L. Dobson, and J. McGuire. New York, N.Y.: Harper & Row.

Kripke, M.L. (1980) Immunologic effects of UV radiation and their role in photocarcinogenesis. Pages 257-292, Photochemical and Photobiological Reviews, Vol. 5, edited by K. Smith. New York, N.Y.: Plenum Publishing Corp.

Kripke, M.L. (1981) Immunologic mechanisms in UV radiation carcinogenesis. Advances in Cancer Research 34:69-106.

Krizek, D.T. (1978) Differential sensitivity of two cultivars of cucumber (Cucumis sativus L.) to increased UV-B irradiance. I. Dose-response studies. Final report of UV-B biological and climate effects research (BACER). Terrestrial FY 77. Beltsville, Md.: Agricultural Research Center. (Unpublished)

Ladds, P.W. and K.W. Entwistle (1977) Observations on squamous cell carcinomas of sheep in Queensland, Australia. British Journal of Cancer 35:110-114.

Lancaster, H.O. (1956) Some geographical aspects of the mortality from melanoma in Europeans. Medical Journal of Australia 1:1082-1087.

Lee, J.A.H. (1981) Melanoma Epidemiology Since the Academy Reports of 1979: With particular reference to the relationship between incidence and exposure to ultraviolet light. Notes for NRC Workshop on Biological Effects of Increased Solar Ultraviolet Radiation, July 30-31, 1981, Environmental Studies

Board, National Academy of Sciences, Washington, D.C. (Unpublished)

Lee, J.A.H., G.R. Petersen, R.G. Stovens, and K. Vesanen (1979) The influence of age, year of birth, and data on mortality from malignant melanoma in the populations of England and Wales, Canada, and the white population of the United States. American Journal of Epidemiology 110:734-739.

Lee, J.A.H. and B.E. Storer (1980) Excess of malignant melanomas in women in the British Isles. Lancet II:1337-1339.

Ley, R.D., B.A. Sedita, D.D. Grube, and R.J.M. Fry (1977) Induction and persistence of pyrimidine dimers in the epidermal DNA of two strains of hairless mice. Cancer Research 37:3243-3248.

Ley, R.D., B.A. Sedita, and D.D. Grube (1978) Absence of photoreactivation of pyrimidine dimers in the epidermis of hairless mice following exposure of ultraviolet light. Photochemistry and Photobiology 27:483-485.

Lloyd, L.C. (1961) Epithelial tumors of the skin of sheep. British Journal of Cancer 15:780.

Lorenzen, C.J. (1979) Ultraviolet radiation and phytoplankton photosynthesis. Limnology and Oceanography 24:1117-1120.

McCarthy, W.H., A.L. Black, and G.W. Milton (1980) Melanoma in New South Wales. Cancer 46:427-432.

Maher, V.M. and J.J. McCormick (1976) Effect of DNA repair on the cytotoxicity and mutagenicity of UV irradiation and of chemical carcinogens in normal and xeroderma pigmentosum cells. Pages 129-145, Biology of Radiation Carcinogenesis, edited by J.M. Yuhas, R.W. Tennant, and J.D. Regan. New York, N.Y.: Raven Press.

Malec, E. and G. Eklund (1978) The changing incidence of malignant melanoma of the skin in Sweden, 1959-68. Scandinavian Journal of Plastic Reconstructive Surgery 12:19-27.

Mantai, K.E., J. Wong, and N.I. Bishop (1970) Comparison studies on the effects of ultraviolet irradiation on photosynthesis. Biochimica et Biophysica Acta 197:257-266.

Marshall, V. (1974) Premalignant and malignant skin tumours in immuno-suppressed patients. Transplantation 17:272-275.

Mason, T.J., F.W. McKay, R. Hoover, W.J. Blot, and J.F. Fraumeni, Jr. (1975) Atlas of Cancer Mortality for U.S. Counties: 1950-1969. DHEW Publication No. (NIH)

75-780. Washington, D.C.: U.S. Government Printing Office.
Milo, G.E., S.A. Weisbrode, R. Zimmerman, and J.A. McCloskey (1981) Ultraviolet radiation-induced neoplastic transformation of normal human cells, in vitro. Chemico-Biological Interactions 36(1):45-59.
Morison, W.L., J.A. Parrish, K.J. Bloch, and J.I. Krugler (1979) In vivo effect of UV-B on lymphocyte function. British Journal of Dermatology 101:513-519.
Morison, W.L., J.A. Parrish, M.E. Woehler, and K.J. Bloch (1980) The influence of PUVA and UVB radiation on skin-graft survival in rabbits. Journal of Investigative Dermatology 75:331-333.
Morison, W.L., J.A. Parrish, K.J. Bloch, and J.I. Krugler (1981a) In vivo effects of PUVA on lymphocyte function. British Journal of Dermatology 104:405-413.
Morison, W.L., J.A. Parrish, R. Moscicki, and K.J. Bloch (1981b) Abnormal lymphocyte function following long-term PUVA therapy for psoriasis. Journal of Investigative Dermatology 76(4):303.
Morison, W.L., J.A. Parrish, M.E. Woehler, J.I. Krugler, and K.J. Bloch (1981c) Influence of PUVA and UVB radiation on delayed hypersensitivity in the guinea pig. Journal of Investigative Dermatology 76:484-488.
Movshovitz, M. and B. Modan (1973) Role of sun exposure in the etiology of malignant melanoma: Epidemiological inference. Journal of the National Cancer Institute 51:777-779.
Murad, E., W. Swider, and S.W. Benson (1981) Possible role for metals in stratospheric chlorine chemistry. Nature 289:273-274.
Natarajan, A.T., A.A. van Zealand, E.A.M. Verdegaal-Immerzed, and A.R. Filon (1980) Studies on the influence of photoreactivation on the frequencies of UV-induced chromosomal aberrations, sister-chromatid exchanges and pyrimidine dimers in chicken embryonic fibroblasts. Mutation Research 69:307-317.
National Research Council (1975) Environmental Impact of Stratospheric Flight: Biological and Climatic Effects of Aircraft Emissions in the Stratosphere. Climatic Impact Committee. Washington, D.C.: National Academy of Sciences.
National Research Council (1976a) Halocarbons: Environmental Effects of Chlorofluoromethane Release. Committee on Impacts of Stratospheric Change, Assembly of Mathematical and Physical Sciences. Washington, D.C.: National Academy of Sciences.

National Research Council (1976b) Halocarbons: Effects on Stratospheric Ozone. Panel on Atmospheric Chemistry, Committee on Impacts of Stratospheric Change, Assembly of Mathematical and Physical Sciences. Washington, D.C.: National Academy of Sciences.

National Research Council (1977) Response to the Ozone Protection Sections of the Clean Air Act Amendments of 1977: An Interim Report. Washington, D.C.: National Academy of Sciences.

National Research Council (1978) Nitrates: An Environmental Assessment. Panel on Nitrates, Committee for Scientific and Technical Assessments of Environmental Pollutants, Environmental Studies Board, Commission on Natural Resources. Washington, D.C.: National Academy of Sciences.

National Research Council (1979a) Protection Against Depletion of Stratospheric Ozone by Chlorofluorocarbons. Committee on Impacts of Stratospheric Change, Assembly of Mathematical and Physical Sciences and Committee on Alternatives for the Reduction of Chlorofluorocarbon Emissions, Commission on Sociotechnical Systems. Washington, D.C.: National Academy of Sciences.

National Research Council (1979b) Stratospheric Ozone Depletion by Halocarbons: Chemistry and Transport. Panel on Chemistry and Transport, Committee on Impacts of Stratospheric Change, Assembly of Mathematical and Physical Sciences. Washington, D.C.: National Academy of Sciences.

National Research Council (1981) Testing for Effects of Chemicals on Ecosystems, Committee to Review Methods for Ecotoxicology, Environmental Studies Board, Commission on Natural Resources. Washington, D.C.: National Academy of Sciences.

Neely, W.B. and J.H. Plonka (1978) Estimation of the time-averaged hydroxyl radical concentration in the troposphere. Environmental Science and Technology 8:317-321.

Noonan, F.P., E.C. DeFabo, and M.L. Kripke (1981a) Suppression of contact hypersensitivity by UV radiation and its relationship to UV-induced suppression of tumor immunity. Photochemistry and Photobiology 34:683-690.

Noonan, F.P., M.L. Kripke, G.M. Pedersen, and M.I. Greene (1981b) Suppression of contact hypersensitivity in mice by ultraviolet radiation is associated with defective antigen presentation. Immunology 43:527-533.

O'Dell, B.L., R.T. Jessen, L.E. Becker, R.T. Jackson, and E.B. Smith (1980) Diminished immune response in sun-damaged skin. Archives of Dermatology 116:559.

Okada, M., M. Kitajima, and W.L. Butter (1976) Inhibition of photosystem I and photosystem II in chloroplasts by UV-irradiation and heat treatment. Plant Physiology 43:2037-2040.

Owens, D.W., J.M. Knox, H.T. Hudson, and D. Troll (1974) Influence of wind on ultraviolet injury. Archives of Dermatology 109:200-201.

Owens, D.W., J.M. Knox, H.T. Hudson, and D. Troll (1975) Influence of humidity on ultraviolet injury. Journal of Investigative Dermatology 64:250-252.

Paffenbarger, R.S., A.L. Wing, and R.T. Hyde (1978) Characteristics in youth predictive of adult-onset malignant lymphomous melanomas, and leukemias: brief communication. Journal of the National Cancer Institute 60(1):89-92.

Parrish, J.A., R.R. Anderson, F. Urbach, and D. Pitts (1978) UV-A: Biological Effects of Ultraviolet Radiation with Emphasis on Human Responses to Longwave Ultraviolet. New York, N.Y.: Plenum Press.

Parrish, J.A., S. Zaynoun, and R.R. Anderson (1981) Cumulative effects of repeated subthreshold doses of ultraviolet radiation. Journal of Investigative Dermatology 76:356-359.

Peak, M.J. and R.W. Tuveson (1979) Revised spectra for the inactivation of Haemophilus influenzae transforming DNA by monochromatic ultraviolet light: Effect of histidine. Photochemistry and Photobiology 29:855-856.

Peters, J. and J. Jagger (1981) Inducible repair of near-UV-radiation lethal damage in E. coli. Nature 289:194-195.

Rasmussen, R.A. and M.A.K. Khalil (1981) Atmospheric methane: Trends and seasonal cycles. Journal of Geophysical Research 86:9826-9832.

Reimer, R.R., W.H. Clark, Jr., M.H. Greene, A.M. Ainsworth, and J.F. Fraumeni, Jr. (1978) Precursor lesions in familial melanoma: A new genetic preneoplastic syndrome. Journal of the American Medical Association 239:744-746.

Reynolds, R.J., A.J. Natarajaw, and P.H.M. Lohman (1979) Micrococcus luteus UV-endonuclease-sensitive sites and sister-chromated exchanges in Chinese hamster ovary cells. Mutation Research 64:353-356.

Robberecht, R. and M.M. Caldwell (1978) Leaf epidermal transmittence of ultraviolet radiation and its implications for plant sensitivity to ultra-violet-radiation induced injury. Oecologia 32:277-289.

Robberecht, R., M.M. Caldwell, and W.D. Billings (1980) Leaf ultraviolet optical properties along a latitudinal gradient in the arctic-alpine life zone. Ecology 61:612-619.

Rommelaere, J., J.M. Vos, J.J. Cornelis, and D.C. Ward (1981) UV-enhanced reactivation of minute-virus-of-mice: Stimulation of a late step in the viral life cycle. Photochemistry and Photobiology 33:845-854.

Rosenstein, B.S. and R.B. Setlow (1980) Photoreactivation of ICR 2A frog cells after exposure to monochromatic ultraviolet radiation in the 252-313 nm range. Photochemistry and Photobiology 32:361-366.

Rothman, R.H. and R.B. Setlow (1979) An action spectrum for cell killing and pyrimidine dimer formation in Chinese hamster V-79 cells. Photochemistry and Photobiology 29:57-61.

Rundel, R.D. and D.S. Nachtwey (1978) Skin cancer and ultraviolet radiation. Photochemistry and Photobiology 28:345-356.

Sauder, D.N., K. Tamaki, H. Fujiwara, A.N. Moshell, and S.I. Katz (1980) Induction of immunologic hyporesponsiveness using ultraviolet light irradiated haptenated epidermal cells. Clinical Research 28:1.

Scott, E.L. (1981) Summary of Studies Made Since December, 1979: Estimating the Increase in Skin Cancer That Will Result from the Depletion of Stratospheric Ozone. University of California, Berkeley. (Unpublished, transmitted to J.R. Hibbs, U.S. Environmental Protection Agency, Washington, D.C.)

Scott, E.L., and M.L. Straf (1977) Ultraviolet radiation as a cause of cancer. Pages 529-546, Origins of Human Cancer, Book A, Incidence of Cancer in Humans, edited by H.H. Hiatt, J.D. Watson, and J.A. Winsten, Vol. 4 of The Cold Spring Harbor Conferences on Cell Proliferation. New York, N.Y.: The Cold Spring Harbor Laboratory.

Scotto, J. and J.F. Fraumeni, Jr. (1982) Skin (other than melanoma). Pages 996-1011, Cancer Epidemiology and Prevention, edited by D. Schottenfeld and J.F. Fraumeni, Jr., Philadelphia, Pa.: W.B. Saunders Company.

Scotto, J., A.W. Kopf, and F. Urbach (1974) Nonmelanoma skin cancer among Caucasians in four areas of the United States. Cancer 34:1333-1338.

Scotto, J., T.R. Fears, and J.F. Fraumeni, Jr. (1981) Incidence of Non-Melanoma Skin Cancer in the United States. DHHS Publication No. (NIH) 82-2433. Bethesda, Md.: National Cancer Institute.

Scotto, J., T.R. Fears, and J.F. Fraumeni, Jr. (1982) Solar radiation. Pages 254-276, Cancer Epidemiology and Prevention, edited by D. Schottenfeld and J.F. Fraumeni, Jr. Philadelphia, Pa.: W.B. Saunders Company.

Setlow, R.B. (1974) The wavelengths in sunlight effective in producing skin cancer: A theoretical analysis. Proceedings of the National Academy of Sciences USA 71:3363-3366.

Setlow, R.B. (1978) Repair deficient human disorders and cancer. Nature 271:713-717.

Setlow, R.B. and J.K. Setlow (1972) Effects of radiation on polynucleotides. Annual Review of Biophysics and Bioengineering 1:293-346.

Sisson, W.B. (1981) Photosynthesis, growth, and ultraviolet irradiance absorbance of Cucurbita pepo L. leaves exposed to ultraviolet-B radiation (280-315 nm). Plant Physiology 67:120-124.

Sisson, W.B. and M.M. Caldwell (1976) Photosynthesis, dark respiration, and growth of Rumex patientia L. exposed to ultraviolet irradiance (288-315 nm) simulating a reduced atmospheric ozone column. Plant Physiology 58:563-568.

Sisson, W.B. and M.M. Caldwell (1977) Atmospheric ozone depletion: Reduction of photosynthesis and growth of a sensitive higher plant exposed to enhanced UV-B radiation. Journal of Experimental Botany 28:691-705.

Smerdon, M.J., T.D. Tlsty, and M.W. Lieberman (1978) Distribution of ultraviolet-induced DNA repair synthesis in nuclease sensitive and resistant regions of human chromatin. Biochemistry 17:2377-2386.

Smith, P.J. and M.C. Paterson (1981) Abnormal responses to mid-ultraviolet light of cultured fibroblasts from patients with disorders featuring sunlight sensitivity. Cancer Research 41:511.

Smith, P.J. M.H. Greene, D.A. Devlin, E.A. McKeen, and M.C. Paterson (In press) Abnormal sensitivity to UV-radiation in cultured skin fibroblasts from patients with hereditary cutaneous malignant melanoma and dysplastic nevus syndrome. (Submitted to Cancer Research)

Smith, R.C. and K.S. Baker (1979) Penetration of UV-B and biologically effective dose-rates in natural waters. Photochemistry and Photobiology 29:311-323.

Smith, R.C. and K.S. Baker (1980) Stratospheric ozone, middle ultraviolet radiation, and carbon-14 measurements of marine productivity. Science 208:592-593.

Smith, R.C. and K.S. Baker (1981) Optical properties of the cleanest natural waters (200-800 nm). Applied Optics 20:177-184.

Smith, R.C., K.S. Baker, O. Holm-Hansen, and R. Olson (1980) Photoinhibition of photosynthesis in natural waters. Photochemistry and Photobiology 31:585-592.

Sober, A.J., R.A. Lew, T.B. Fitzpatrick, and R. Marvell (1979) Solar exposure patterns in patients with cutaneous melanoma: A case control series. Clinical Research 27:5361.

Spellman, C.W. and R.A. Daynes (1978) Properties of ultraviolet light-induced suppressor lymphocytes within a syngeneic tumor system. Cell Immunology 36:383-387.

Stanford Research Institute International (1980) Effects of Stratospheric Ozone Depletion. Contract No. 68-02-3939 for the U.S. Environmental Protection Agency, Palo Alto, Calif.

Stanford Research Institute International (1981) Research Needs in Photobiology as Related to Stratospheric Ozone Depletion. Contract No. 68-02-3632 for the U.S. Environmental Protection Agency, Palo Alto, Calif.

Steemann Neilsen, E. (1964) On a complication in marine productivity work due to the influence of ultraviolet light. Journal du Conseil, Conseil International pour Exploration de la Mer 29:130-135.

Stern, R.S., L.A. Thibodeau, R.A. Kleinerman, J.A. Parrish, T.B. Fitzpatrick, and H.L. Bleich (1981) Effect of methoxsalen photochemotherapy on cost of treatment for psoriasis: An example of technologic assessment. Journal of the American Medical Association 245:1913-1918.

Stern, R.S., S. Zierler, and J.A. Parrish (1982) Psoriasis and the risk of cancer. Journal of Investigative Dermatology. (In press)

Strickland, D. and J.A.H Lee (1981) Melanomas of eye: Stability of rates. American Journal of Epidemiology 113:700-702.

Strickland, P.T., F.J. Burns, and R.E. Albert (1979) Induction of skin tumors in the rat by single exposure to ultraviolet radiation. Photochemistry and Photobiology 30:683-688.

Surgeon General's Advisory Committee on Smoking and Health (1964) Smoking and Health: Report of the Advisory Committee to the Surgeon General of the Public Health Service. Public Health Service Publication No. 1103, Washington, D.C.

Sutherland, J.C. and K.P. Griffin (1981) Absorption spectrum of DNA for wavelengths greater than 300 nm. Radiation Research 86:399-410.

Sutherland, B.M., P. Runge, and J.C. Sutherland (1974) DNA photoreactivating enzyme from placental mammals: Origin and characteristics. Biochemistry 13:4710-4715.

Sutherland, B.M., J.S. Cimino, N. Delihas, A.G. Shih, and R.P. Oliver (1980a) Ultraviolet light-induced transformation of human cells to anchorage-independent growth. Cancer Research 40:1934-1939.

Sutherland, B.M., L.C. Harber, and I.E. Kochevar (1980b) Pyrimidine dimer formation and repair in human skin. Cancer Research 40:3181-3185.

Sutherland, B.M., N.C. Delihas, R.P. Oliver, and J.C. Sutherland (1981) Action spectra for ultraviolet light-induced transformation of human cells to anchorage-independent growth. Cancer Research 41:2211-2214.

Suzuki, F., A. Han, G.R. Lankas, H. Utsumi, and M.M. Elkind (1981) Spectral dependencies of killing, mutation, and transformation in mammalian cells and their relevance to hazards caused by solar ultraviolet radiation. Cancer Research 41(12):4916-4924.

Taber, H., J. Pomerantz, and G.N. Halfenger (1978) Near ultraviolet induction of growth delay studied in a menaguinone-deficient mutant of Bacillus subtillis. Photochemistry and Photobiology 28:191-196.

Takebe, Hiraku, et al. (1977) DNA repair characteristics and skin cancers of xeroderma pigmentosum patients in Japan. Cancer Research 37:490-495.

Teramura, A.H., R.H. Biggs, and S.V. Kossuth (1980) Effects of ultraviolet-B irradiance on soybean. II. Interaction between ultraviolet-B and photosynthetically active radiation on net photosynthesis, dark respiration, and transpiration. Plant Physiology 65:483-488.

Thomson, B.E., R.C. Worrest, and H. Van Dyke (1980) The growth response of an estuarine diatom (Melosira nummuloides [Dillw.] Ag.) to UV-B (290-320 nm) radiation. Estuaries 3:69-72.

Towes, G.B., P.R. Bergstresser, and J.W. Streilein (1980) Epidermal Langerhans cell density determines whether contact hypersensitivity or unresponsiveness follows skin painting with DNFB. Journal of Immunology 124:445-453.

Trocine, R.P., J.D. Rice, and G.N. Wells (1981) Inhibition of seagrass photosynthesis by ultraviolet-B radiation. Plant Physiology 68:74-81.

Tsai, S.C. and J. Jagger (1981) The roles of the rel gene and of 4-thiouridine in killing and photoprotection of E. coli by near-ultraviolet radiation. Photochemistry and Photobiology 33:825-834.

Turner, M.A. and R.B. Webb (1981) Comparative mutagenesis and interaction between near-ultraviolet (313 to 405 nm) and far-ultraviolet (254 nm) radiation in Escherichia coli strains with differing repair capabilities. Journal of Bacteriology 147:410-417.

Tyrrell, R.M. (1978) Mutagenic interaction between near-(365 nm) and far-(254 nm) ultraviolet radiation in repair-proficient and excision-deficient strains of Escherichia coli. Mutation Research 52:25-35.

Tyrrell, R.M. (1980) Mutation induction by and mutational interaction between monochromatic wavelength radiations in the near-ultraviolet and visible ranges. Photochemistry and Photobiology 31:37-46.

Urbach, F., D.B. Rose, and M. Bonnem (1972) Genetic and environmental interactions in skin carcinogenesis. Pages 355-371, Environment and Cancer. Baltimore, Md.: Williams and Wilkins Co.

U.S. Congress, Senate (1975) Stratospheric Ozone Depletion Hearings before the Subcommittee on the Upper Atmosphere, Committee on Aeronautical and Space Sciences. Part 1, Appendix 1. Sept. 8, 9, 16, and 17, 94th Congress.

Vu, C.D., L.H. Allen, Jr., and L.A. Garrard (1981) Effects of supplemental UV-B radiation on growth and leaf photosynthetic reactions of soybean (Glycine max L.) Physiologia Plantarum 52:353-362.

Vu, C.D., L.H. Allen, Jr., and L.A. Garrard (1982) Effects of supplemental UV-B radiation on primary photosynthetic carboxylating enzymes and soluble proteins and leaves of C_3 and C_4 crop plants. Physiologia Plantarum. (In press)

Wan, S., R.R. Anderson, and J.A. Parrish (1981) Analytical modeling for the optical properties of the skin with in vivo and in vitro applications. Photochemistry and Photobiology 34:493-499.

Webb, R.B. (1977) Lethal and mutagenic effects of near-ultraviolet radiation. Pages 169-261, Photochemical and Photobiological Reviews, Vol. 2, edited by K.C. Smith. New York, N.Y.: Plenum Press.

Webb, R.B., M.S. Brown, and R.M. Tyrrell (1978) Synergism between 365 and 254 nm radiations for inactivation of Escherichia coli. Radiation Research 74:298-311.

Weiss, R.F. (1981) The temporal and spatial distribution of tropospheric nitrous oxide. Journal of Geophysical Research 86:7185-7196.

Weiss, R.F. and H. Craig (1976) Production of atmospheric nitrous oxide by combustion. Geophysical Research Letters 3:751-753.

Wellman, E. (1974) Regulation der Flavonoidbiosynthese durch ultraviolettes Licht und Phytochrom in Zellkulturen und Keimlingen von Petersilie (Petroselinum hortense Hoffm.). Berichte der Deutschen Botanischen Gesellschaft 87:267-273.

Wiskemann, A. (1977) Sunlight and melanomas. Pages 479-484. Research in Photobiology, edited by A. Castellani. New York, N.Y.: Plenum Publishing Corp.

Watkin, E.M. (1976) Ultraviolet mutagenesis and inducible DNA repair in Escherichia coli. Bacteriological Reviews 40:869-907.

Worrest, R.C., H. Van Dyke, and B.E. Thomson (1978) Impact of enhanced simulated solar ultraviolet radiation upon a marine community. Photochemistry and Photobiology 27:471-478.

Worrest, R.C., D.L. Brooker, and H. Van Dyke (1980) Results of a primary productivity study as affected by the type of glass in the culture bottles. Limnology and Oceanography 25:360-364.

Worrest, R.C., B.E. Thomson, and H. Van Dyke (1981a) Impact of UV-B radiation upon estuarine microcosms. Photochemistry and Photobiology 33:861-867.

Worrest, R.C., K.U. Wolniakowski, J.D. Scott, D.L. Brooker, B.E. Thomson, and H. Van Dyke (1981b) Sensitivity of marine phytoplankton to UV-B radiation: Impact upon a model ecosystem. Photochemistry and Photobiology 33:223-227.

Yamashita, T. and W.L. Butler (1968) Inhibition of chloroplasts by UV-irradiation and heat treatment. Plant Physiology 43:2037-2040.

Zelle, B., R.J. Reynolds, M.J. Kottenhagen, A. Schuite, and P.H.M. Lohman (1980) The influence of the wavelength of ultraviolet radiation on survival, mutation induction and DNA repair in irradiated Chinese hamster cells. Mutation Research 72:491-507.

GLOSSARY

Actinic or solar degeneration: A complex of degenerative changes in skin caused by chronic exposure to sunlight. The skin appears thick and furrowed but may also have zones of thinned epidermis.

Action spectrum: A graph or mathematical expression indicating the relative effectiveness of radiation at different wavelengths for producing a photochemical or photobiological effect.

Antigen: Any substance that, when introduced into the body, stimulates the production of an antibody, a protein that acts to neutralize the antigen and thus produce immunity.

Backscatter ultraviolet (BUV): Solar radiation in the ultraviolet region that is scattered from the atmosphere back into space.

Basal cell skin cancer: A relatively common type of skin cancer that can result from exposure to sunlight. Its tendency to metastasize is small. This cancer arises in the basal cell layer of the epidermis where cells continually divide and replace dead cells in the epidermis.

Carcinogenesis: The production and development of cancer. The process of carcinogenesis may be divided into at least two parts. The first part, initiation, involves the interaction of a physical or chemical carcinogen with cells, resulting in altered cells that are potentially cancerous, or precancerous. Such an altered cell may remain quiescent for a long time before subsequent cell proliferation and the expression of a tumor. The second part, promotion, involves the subsequent proliferation of the altered cells. Substances called promoters, administered after, even long after, the initiating event, may result in observed tumors. But, if administered before the initiating event, promoters will not enhance the observed carcinogenic effects of initiators. Initiators are thought to act via reactions with cellular macromolecules--in many instances DNA. The molecular mechanisms of promotion are not well understood, but are hypothesized to affect the regulatory activities of cells or cell-cell interactions (Berenblum and Armuth 1981).

Catalysis: A means by which the rate of a chemical reaction is enhanced through the action of a catalyst (a substance that itself remains chemically unaltered).

Catalytic cycle: A set of chemical reactions wherein one or more reactive species are alternately consumed and generated. The net effect is to cause a reaction between the partners of the reactive species to yield the products. The simplest example of a catalytic cycle involving ozone is

$$O_3 + NO \rightarrow NO_2 + O_2$$
$$O + NO_2 \rightarrow NO + O_2$$
$$\overline{O_3 + O \rightarrow O_2 + O_2 \quad \text{(net)}.}$$

In these reactions the molecules NO and NO_2 act as catalysts for the combination of O-atoms with O_3 to produce O_2. The direct reaction can occur, but the presence of NO and NO_2 (causing the same net change) increases the rate by means of the two-reaction pathway.

Chlorocarbon: A hydrocarbon in which one or more chlorine atoms are substituted for hydrogen atoms.

Chlorofluorocarbon: A hydrocarbon containing chlorine and fluorine as substituents for hydrogen atoms.

Chlorofluoromethane: A methane derivative containing chlorine and fluorine as substituents.

Chromatin: A complex of highly polymerized DNA with basic proteins (histone or protamine) that stains intensely with basic dyes; regarded as the physical carrier of genes.

Chromophores: Molecules or parts of molecules that absorb light.

Cohort: A group born in a specific time interval, e.g., one calendar year.

Cohort analysis: The study of a cohort from its inception to its final dissolution, e.g., a study of all the people born in one calendar year followed until the last person dies.

Contact hypersensitivity: An immune system response at the cellular level to an antigen applied to the skin, to which the animal has been previously sensitized. The response is specific to the antigen due to specific cell-surface antigen receptors.

Core regions of chromatin: Mammalian DNA may be visualized as a strand with aggregates of protein occurring at intervals along the strand. The protein aggregates are called nucleosomes, and are known as core regions, and the intervening areas are the linker regions.

Cross-sectional analysis: A type of empirical analysis, i.e., analysis concerned with the establishment of quantitative or qualitative relations between observable variables, using cross-section data. Cross-section data are observations on variables at a point in time, as opposed to time-series data.

Dermis: A 1-mm to 4-mm layer of primarily collagenous connective tissue that provides much of the structural integrity of the skin. It is located beneath the epidermis.

Ecosystem: A dynamic, integrated assemblage of plants, animals, and microorganisms that is definable by the interactions among the living and nonliving components of the functional unit.

Eddy diffusion: A process whereby, through the action of random eddies in a turbulent fluid, heat and/or matter is transported along a gradient.

Epidermis: The outermost layer of the skin, approximately 100 μm thick, separated from the next layer (dermis) by a basement membrane. The epidermis consists of five layers: (1) the outermost protective stratum corneum (no nuclei); (2) several layers of transparent nucleated cells; (3) the granular layer; (4) the Malpighian layer, consisting of multiple squamous or prickle cells; and (5) the basal layer, composed of germinative cells. Less than 10 percent of incident UV-B may pass through the epidermis.

Erythema: A reddening of the skin due to a dilation of the blood capillaries. It is one of the components of the syndrome commonly known as sunburn.

Excision repair: A cellular repair mechanism that eliminates photoproducts in DNA, thereby ameliorating UV damage to DNA. In this process, products of UV irradiation are removed from one strand of a DNA double helix by specific enzymes that work in the dark. The opposite unaltered strand is then used as a template on which a new complementary strand is built.

Fibroblast: A connective tissue cell, flat and elongated. It performs the function of supporting and binding tissues of all sorts in higher organisms.

Freon: The U.S. trade name for several chlorofluorocarbons.

F-11, F-12: The halocarbon F-11 is trichlorofluoromethane $CFCl_3$ and F-12 is dichlorodifluoromethane CF_2Cl_2. These are the two most-used chlorofluorocarbons and they constitute most of the threat to ozone by this class of compounds.

Halocarbon: A hydrocarbon in which one or more halogen atoms are substituted for hydrogen atoms.

Heterochromatic light or radiation: Light or radiation consisting of a range of (more than a single) wavelengths. <u>See also</u> monochromatic light or radiation.

Hydrocarbon: A compound of hydrogen and carbon.

Immunosuppression: The suppression of a natural immune response of an organism to a foreign agent.

Initiator: <u>See</u> carcinogenesis.

Langerhans cells: In the skin, dendritic cells in the epidermis that function as part of the immune system.

Lentigo malignant melanoma: A subtype of melanoma occurring almost exclusively on the exposed parts of the face, neck, and hands; characterized by the presence of brownish pigmented spots on the skin (lentigines or freckles) that increase in size and darken in color. The spots are predominantly flat, 2 cm to 20 cm in diameter with irregular borders and pigment pattern. Raised areas indicate invasive tumor.

Linker regions of chromatin: <u>See</u> core regions of chromatin.

Lymphocytes: Circulating white blood cells that are part of the immune system.

Mast cell: A connective tissue cell whose physiologic function remains partly unknown; after a variety of insults or stimuli the cell releases chemicals that are mediators of inflammation.

Melanin: A dark pigment found in skin (epidermis), hair, and various tumors. The epidermal melanin unit is composed of melanocytes and associated Malpighian cells. <u>See also</u> epidermis, melanocytes.

Melanocytes: Cells located in the basal layer of the epidermis with dendritic (armlike) projections that extend into the Malpighian layer of the epidermis. These cells form the pigment melanin, which they pass into the Malpighian cells.

Melanoma: A tumor made up of melanin-pigmented cells. Melanoma is a serious, sometimes fatal form of skin cancer, usually developing from a nevus and consisting of black masses of cells with a tendency to metastasize.

Mixing ratio: The ratio of the concentration of a gaseous species to the total concentration of the gas.

Model: In the context of Part I of this report a model is a mathematical representation of the transport and chemical behavior of species in the atmosphere. In principle, with suitable specification of initial and boundary conditions, the distributions of any and all (relevant) chemical species in space and time can be computed by means of the model.

Monochromatic light or radiation: Light or radiation consisting of a single wavelength. See also heterochromatic light or radiation.

Nevus: Any congenital growth or mark on the skin, such as a birthmark.

Odd-hydrogen compound: Specifically one of the radical species OH (hydroxyl) and HO_2 (hydroperoxyl) that contain a single H-atom each. The term odd-hydrogen is used more in analogy with odd-nitrogen and odd oxygen (see below).

Odd-nitrogen compound: One of the species containing a single nitrogen atom such as NO, NO_2, HNO_3, $ClONO_2$, HONO, and $HOONO_2$.

Odd-oxygen species: Specifically O-atoms and O_3 (as opposed to the "even oxygen" species O_2). Since O-atoms are rapidly converted mainly to O_3, reactions which remove them are considered as effectively removing O_3. The set of reactions that remove both odd-oxygen species thus constitutes the means by which ozone abundance becomes reduced.

Ozone (O_3): An allotrope of oxygen containing three atoms. It is a reactive, toxic, acrid smelling, colorless gas under atmospheric conditions. It is created naturally in the stratosphere where its abundance is the largest and where it exists as a permanent layer. Ozone is created in the troposphere by the so-called smog reactions involving the oxides of nitrogen and hydrocarbons.

Ozonesonde: (a) One of several devices that are carried aloft through the atmosphere by balloons to measure the vertical concentration distribution of ozone. (b) The plotted record of the vertical concentration distribution of ozone obtained by such a device.

Photochemical reaction: Any reaction in which one or more of the reactants or their reactive precursors are created by the interaction of light with a molecule or atom. Generally the term refers to free radical reactions wherein the radicals or their radical precursors were created by photolysis of a stable molecule.

Photobiology: That branch of biology that deals with the effects of light on living organisms.

Photochemistry: That branch of chemistry that deals with the chemical effects of light.

Photoimmunology: That branch of immunology that deals with the effects of light on the immune system. Most of what is now known about the effects of UV on the immune systems of animals and humans has been learned within the past five years.

Photokeratitis: An acute, painful irritation of the cornea of the eye caused by exposure to UV-B or UV-C.

Photon: A "particle" of light. A photon is the smallest unit (quantum) of light that exists; its energy depends on the wavelength of the light.

Photoproducts: Specific changes in molecular structure that result from the absorption by molecules of photons (in this report, photons in the UV band specifically).

Photoprotection: A protective cellular mechanism whereby a preceding illumination with UV-A may decrease the damage to DNA caused by UV-B. Photoprotection involves the induction by UV-A of a delay in growth, allowing for more time after UV-B irradiation is completed for error-free dark-repair systems to repair the damaged DNA.

Photoreactivation: A cellular repair mechanism that eliminates photoproducts in DNA, thereby ameliorating UV damage to DNA. In this process, an enzyme binds to a DNA molecule containing pyrimidine dimers. The complex of enzyme and damaged DNA can absorb UV-A or visible light, which causes the dimer to split, thereby repairing the damage.

Planetary waves: Longitudinally ranging motions of the atmosphere organized on a scale of the magnitude of the distance around the earth.

Promoter: See carcinogenesis.

Pyrimidine dimers: Biologically damaging products of UV irradiation, formed when two adjacent pyrimidine residues within one DNA strand bond to each other.

Radical (also called free radical): Any chemical species (atom or molecule or ion) that contains one or more unpaired electrons. In this report the term refers to reactive species such as OH, HO_2, Cl, and ClO, each of which contains a single unpaired electron. The ease of combining with other molecules to form bonds having paired electron spins is understood as the seat of their reactivity.

Rate-limiting reaction (or process): In a sequence of reactions, that reaction which is the slowest and thus limits the rate at which the initial reactants ultimately become converted to products.

Robertson-Berger (R-B) meter: A meter that records, after each 30-minute interval, a measure of the cumulative amount of UV that passes through its filters and is absorbed by its photosensors. Half-hourly recordings may range from 0 to slightly over 1000 depending on geographical location and prevailing meteorological conditions. The meters are designed to measure UV effective in producing skin erythema (sunburn), but in fact measure some longer UV wavelengths as well (see Figure 2.2). A count of about 400 in a half hour is estimated to produce skin erythema on the typical Caucasian skin.

Solar backscatter UV (SBUV): See backscatter ultraviolet.

Solar degeneration: See actinic or solar degeneration.

Spectrophotometer: An electro-optical device that measures the intensity of light distributed over a spectral range (of wavelength or frequency). The Dobson meter is an example of a specialized spectrophotometer that measures and intercompares the relative intensities of sunlight at four different wavelengths, two of which correspond to absorption peaks in the ozone spectrum. From this measurement, the column abundance of ozone can be calculated.

Squamous cell skin cancer: A relatively common type of skin cancer that can result from exposure to sunlight. Its tendency to metastasize is small. This cancer arises in the Malpighian layer of the epidermis.

Stratosphere: The region of the atmosphere above the tropopause (altitude range 6 km to 17 km) and below the stratopause (altitude about 55 km). The principal characterizing feature of the stratosphere is its thermal stability. That is, the temperature increases with increasing height in the stratosphere.

Stratum corneum: A 10-μm layer of dead cells, protein, and other molecules on the outermost surface of the epidermis. UV-B is strongly absorbed by the stratum corneum.

T suppressor cells: A certain class of lymphocytes that suppresses cellular immune response.

Transformation (in vitro): An inheritable change wherein cells in culture are altered such that they do not stop growing when they encounter similar cells. Colonies of UV-transformed rodent cells are often tumorigenic when injected into certain animals, but no tumorigenicity has been shown for the UV-transformed human cells described in the experimental results shown in Figure 3.1.

Tropopause: The surface that is the boundary between the troposphere and the stratosphere. The height of this surface varies in the range 6 km to 17 km depending upon latitude and season.

Troposphere: The main layer of the atmosphere between the surface and the tropopause.

Ultraviolet radiation (UV): Light in the range of wavelengths less than 400 nm. The lower limit is a matter of definition to distinguish between UV and x-rays at even shorter wavelengths.

UV-A: Ultraviolet radiation in the wavelength region from 320 nm to 400 nm; near-UV.

UV-B: Ultraviolet radiation in the wavelength region from 290 nm to 320 nm; mid-UV.

UV-C: Ultraviolet radiation in the wavelength region from 190 nm to 290 nm; far-UV.

Umkehr method: A mathematical manipulation of the data from the Dobson spectrophotometer that creates a rough representation of the vertical profile of ozone. The data come from measurements of the zenith sky as the sun rises or sets.

Xeroderma pigmentosum: A genetically inherited, sunlight-sensitive, cancer-prone disease; rare and often fatal if the individuals are not protected from sun exposure; characterized by brown spots and ulcers of the skin. Cells from individuals with this disease are almost always defective in DNA repair, and the high prevalence of skin cancer in such individuals is ascribed to this defect.

LIST OF CHEMICAL SYMBOLS

CCl_4	carbon tetrachloride
$CFCl_3$	a chlorofluorocarbon (F-11)
CF_2Cl_2	a chlorofluorocarbon (F-12)
CH_4	methane
CH_3Cl	methyl chloride
CH_3CCl_3	methyl chloroform
Cl	atomic chlorine
$ClNO_3$	chlorine nitrate
ClO	chlorine monoxide
ClO_x	the oxides of chlorine
CO_2	carbon dioxide
H	atomic hydrogen
H_2	molecular hydrogen
H_2O	water
HCl	hydrochloric acid
HO_x	the oxides of hydrogen
N	atomic nitrogen
NO	nitric oxide
NO_2	nitrogen dioxide
NO_x	the oxides of nitrogen
N_2O	nitrous oxide
O	atomic oxygen
O_2	diatomic molecular oxygen
O_3	triatomic molecular oxygen, ozone
OH	hydroxyl radical

APPENDIXES

Appendix A

PERTURBATIONS OF THE STRATOSPHERE AND OZONE DEPLETION

Ralph J. Cicerone
National Center for Atmospheric Research
Boulder, Colorado

HISTORICAL BACKGROUND

Central to the concern that man's activities can modify the atmospheric ozone layer is the concept of chemical catalysis. A substance is a catalyst if it serves to drive or enhance a process or reaction--without itself being consumed in the process. In the earth's stratosphere, nitric oxide, NO, can catalyze the destruction of ozone through the cyclic chain reactions:

$$\begin{array}{lcl} NO + O_3 & \rightarrow & NO_2 + O_2 \\ O_3 + h\nu & \rightarrow & O_2 + O \\ NO_2 + O & \rightarrow & NO + O_2 \\ \hline O_3 + O_3 + h\nu & \rightarrow & 3O_2 \quad \text{(net).} \end{array}$$

The corresponding catalytic chain reaction involving chlorine atoms proceeds through:

$$\begin{array}{lcl} Cl + O_3 & \rightarrow & ClO + O_2 \\ O_3 + h\nu & \rightarrow & O_2 + O \\ ClO + O & \rightarrow & Cl + O_2 \\ \hline O_3 + O_3 + h\nu & \rightarrow & 3O_2 \quad \text{(net).} \end{array}$$

In these reactions, NO and Cl are not consumed as they destroy O_3 because they are regenerated in the last reaction of the cycle. Chemical catalysis can be an extremely efficient process; some industrial catalysts mediate millions of cyclical reactions before they themselves require regeneration. The number of times that the catalytic cycle proceeds is called the chain length. With a chain length of 10,000 one can see how a substance present at part per billion concentrations is capable of chemically controlling another substance present at 10 parts per million.

The realizations that (a) efficient reactions like these were occurring in the natural stratosphere and (b) certain pollutants could mimic nature's ozone-destroying catalytic cycles were nearly simultaneous and have helped measurably to improve our understanding of the natural atmosphere and of man's potential for perturbing it. The research of Crutzen (1970, 1971) and Johnston (1971) showed that natural nitrogen oxides and aircraft-injected NO could have important roles in counterbalancing natural ozone production and providing extra, artificial ozone-destroying capacity, respectively. Earlier the need for identifying unspecified natural loss processes for stratospheric ozone had been noted by Hampson (1964) and Hunt (1966), who based their work on earlier theory from Bates and Nicolet. The proposal (Berkner and Marshall 1967) that the evolution of life on the exposed earth surface began with the formation of the ultraviolet light-absorbing ozone screen, coupled with the realization that extant or planned human activities could destroy some of the ozone and a generally growing environmental awareness caused scientists to respond seriously to suggestions of stratospheric chemical perturbations. The ongoing release of synthetic chlorofluorocarbons, shown by Molina and Rowland (1974) to be capable of delivering chemically effective amounts of ozone-destroying chlorine atoms to the stratosphere, remains in 1981 the largest and most plausible threat. The biological UV-shield function of atmospheric ozone has focused attention on chemical pollutants capable of reducing the total amount of ozone in a vertical column of the atmosphere. Proposed fleets of stratospheric supersonic aircraft (releasing NO and H_2O), space shuttle rockets (releasing HCl), the use of bromine-containing chemicals, the surface release of N_2O from agricultural nitrogen-fertilizer usage and from some types of fuel combustion, and the emissions to the air of certain chlorinated solvents have been proposed as possible ozone reducers. Research in the United States and elsewhere (as documented in earlier NRC and NASA reports) has focused too narrowly on possible reductions in the total vertical column of ozone in the stratosphere--probably because of the UV shield that ozone provides to life on earth. Too little emphasis has been placed on inquiring whether ozone spatial redistributions (in altitude and latitude) can result from man's activities. Climatic effects could ensue from large redistributions. Further, tropospheric ozone (about 10 percent of total atmospheric ozone) has been relatively unstudied. Until recently, the prevailing view has been that the only interesting tropospheric photochemistry involving O_3 takes place in highly polluted urban areas. Better understanding of tropospheric chemistry and more complete photochemical kinetic data bases are changing this view, and consequently a fuller concept of man's impact on atmospheric ozone is emerging.

NATURAL PERTURBATIONS TO ATMOSPHERIC OZONE

Both the chemical and the dynamical forces that control the atmospheric ozone distribution are subject to natural perturbations and variations. It is important to understand the consequent natural ozone responses and variability; cause-effect relations must be fathomed if we are to be able to predict ozone changes. The very existence of natural variations affects our ability to detect secular trends in ozone.

On human time scales the most pertinent natural perturbations to atmospheric ozone appear to arise from: solar proton events, relativistic electron bombardments, quasi-biennial oscillation (and temperature change) effects, temperature changes on other time scales, and, possibly, 11-year solar UV irradiance changes and volcanic chemical inputs.

In August of 1972, a burst of solar protons entered the high-latitude atmosphere. High-latitude ozone amounts were observed to decrease almost simultaneously by the Nimbus 4 BUV instrument (Heath et al. 1977, Reagan et al. 1981); these decreases persisted for several weeks. The first theoretical analysis of the effects of this solar proton event (Heath et al. 1977) found good qualitative agreement between observed O_3 decreases and those calculated in a 2-dimensional parameterized transport model including only the direct chemistry, i.e., ion-pair production by proton impact, dissociative recombination to yield NO, then NO-catalyzed O_3 depletion. Recently, Solomon and Crutzen (1981) have expanded the computational model's chemistry to include chlorine chemistry and the production of hydrogen oxides (HO_x) by the arriving protons. They also included the expected temperature-decrease feedback in their model. Their calculated O_3 decreases due to the solar protons agreed very well with the measured decreases except above 50 km. A further analysis has been performed by McPeters et al. (1981), who have provided certain corrections to the Heath et al. (1977) BUV ozone data. The revised August 1972 data on the ozone perturbation evidently agree more closely with Solomon and Crutzen's calculated ozone reductions. McPeters and coworkers also analyzed two earlier (1971) smaller solar-proton events after which high-altitude ozone was observed to be depleted; the measured O_3 depletions were larger than McPeters and co-workers calculated with their photochemical model. All of these investigations, when combined with earlier studies of a 1969 solar proton event and observed ozone reductions (Weeks et al. 1972, Swider and Keneshea 1973), have utilized a natural NO_x-injection event to establish that NO_x catalytic cycles do reduce ozone in the middle and upper stratosphere.

Natural variations in the solar UV output may have influenced stratospheric ozone during the recent past when Dobson instruments and satellite instruments have measured ozone. While there is no argument in principle that UV irradiance changes would modulate ozone amounts, there is disagreement over the reality of solar cycle variations in UV irradiance. Recently, Brasseur and Simon (1981) have expressed this concern, reviewed earlier calculations and presented new calculations of altitude, latitude, and temporal problems to be expected from solar-cycle-related UV changes. A more empirical approach has been taken by Keating et al. (1981) and Reber and Huang (1982). From the monthly global average ozone amounts derived from Nimbus IV BUV measurements (1970-1977), Tolson (1981) and Keating et al. (1981) sought an empirical relationship between solar UV output (as indicated by the 10.7-cm radiowave flux) and global total ozone. They found a very high correlation between the two; this strong correlation suggested a causal relationship. An independent analysis by Reber and Huang (1982) shows that much of this correlation is due to a secular trend in both. Further, the remaining correlation maximizes for a zero time lag or for ozone changes one month preceding the 10.7-cm flux change. Coupled with the uncertainty (several references cited by Reber and Huang) in the stability of the BUV instrument for total ozone measurements over this seven-year time period, firm conclusions about relationships between total ozone and solar UV seem impossible at this time (Reber and Huang 1982). Thus, while photochemical theory calls for such a relationship, it has been difficult to observe.

Perturbations to stratospheric ozone can also be caused by altered circulation patterns or temperature fields. Episodic phenomena such as sudden stratospheric warmings should affect ozone, but more interesting for our present purpose are those large-scale temperature changes that can be sustained for a year or more. Recently, Angell has extended earlier analyses that have found a significant cooling of the upper stratosphere (cited in Hudson et al. (1982)). In the 46- to 55-km region there has been about a 5°K cooling between 1971 and 1980; a less pronounced cooling is evident down to 36 km. Such a cooling should have led to slightly larger local ozone concentrations. These were not observed by Heath with the NIMBUS IV BUV instrument. Instead, he reported ozone decreases of nearly 9 percent at the 40-km level from 1971 to 1977 (NASA/WMO Stratosphere Workshop, Hampton, Virginia, May 18-22, 1981; see also Science, September 4, 1981, pp. 1088-1089).

It is also important to recognize the possibility of a large natural change in tropospheric ozone, both because the ozone column would be affected and because of possible climatic effects that could ensue. Data suggest that northern hemispheric tropospheric ozone has increased substantially in the last decade; this is discussed below.

Finally, although explosive volcanoes can in principle affect stratospheric ozone by direct injections of water and chlorine, there are no indications of measurable effects due to volcanoes during the life of the Dobson instrument network. A related question involves the ability of stratospheric dust to confound the Dobson measurement technique (Dave et al. 1981, De Luisi et al. 1975). One would feel more confident with a complete absorption spectrum rather than discrete wavelength pairs at which absorption is measured.

MAN'S IMPACT: ASSESSMENT AND UNCERTAINTY IN 1981

Of all the potential anthropogenic influences on atmospheric ozone the continued release of chlorofluorocarbons 11 and 12 and of trichloroethane remains in 1981 that of most immediate and apparently largest concern. The anticipated magnitude of the effect continues to change as the laboratory photochemical data base grows. It is worth noting that there have been few, if any, results reported from coupled meteorological/chemical model calculations. Our estimates of man's impact on ozone due to fluorocarbon release continue to be based on models that have not changed conceptually since before the fluorocarbon problem was identified. Further, even the most elegant and difficult field measurements such as the Cl and ClO profile measurements of Anderson et al. (1980) and those of total chlorine in the lower stratosphere by Berg et al. (1980) have not altered the initial view of this environmental problem. Field measurements in general have substantiated all elements of the original Molina-Rowland hypothesis; quantitative adjustments to the size of the ozone perturbation have arisen frequently from new or changed laboratory kinetic data.

If we focus on the expected reduction in total ozone due to continued release of CF_2Cl_2 and $CFCl_3$ at their 1975 annual rates (see Figure A.1), we see that major changes have resulted from altered chemical reaction rates and from the inclusion of previously omitted reactions and species (e.g., $ClNO_3$ and HNO_4). The calculations that produced Figure A.1 were performed with 1-dimensional photochemical models with parameterized transport. Since 1980 there have also been similar 2-dimensional models that have been able to include as

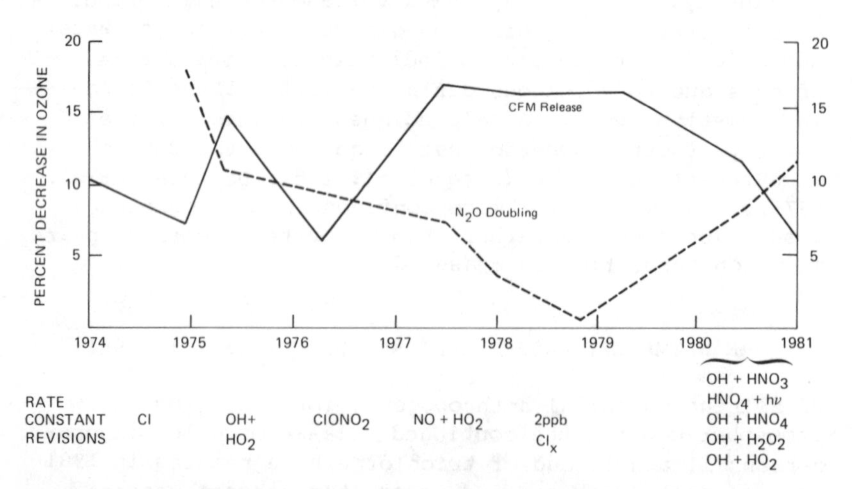

FIGURE A.1 Brief schematic history of the estimates of the steady state column ozone reduction due to (a) continued release of CF_2Cl_2 and $CFCl_3$ at 1975 annual rates, and (b) doubling of N_2O (from 300 to 600 ppb). More detail on reasons for changes between 1979 and 1981 is in Hudson et al. (1982, Chapter 3).

many chemical processes as the preexisting 1-dimensional models. The change in globally averaged total ozone amounts due to CF_2Cl_2 and $CFCl_3$ from the available 2-dimensional models is very close to that from 1-dimensional models (see Chapter 3 of Hudson et al. (1982)). As Figure A.1 indicates, since 1979 there has been a considerable downward revision of the predicted steady state ozone depletion. With currently accepted chemical reaction rates one calculates steady state, globally averaged ozone reductions of perhaps 6 percent.

The principal chemical data changes since 1979 are: (i) a faster rate for $OH + HNO_3 \rightarrow H_2O + NO_3$ especially at low temperatures, (ii) a faster rate for $OH + H_2O_2 \rightarrow H_2O + HO_2$, (iii) slower photolysis of HNO_4 than previously measured, (iv) faster reaction of $OH + HNO_4 \rightarrow$ products, presumably $H_2O + NO_2 + O_2$ than originally estimated, and (v) upward revision of the $OH + HO_2$ rate constants. Key references for these recent changes are: (i) Wine et al. (1981); (ii) Keyser (1981) and Kaufman (1980); (iii) Molina and Molina (1981); (iv) and (v) NASA/JPL Kinetics Panel (1981). While these changes have the effect of lowering calculated OH (and ClO) concentrations below 30 km and thus permitting more favorable comparison with Anderson's ClO measurements below 30 km than before (Cicerone and Walters 1980, Duewer and Wuebbles 1980, Sze and Ko 1981), they are not the final word (see next paragraph). It would not be surprising if the best estimates of column ozone changes due to CF_2Cl_2 and $CFCl_3$ and N_2O increases continue to oscillate as on Figure A.1.

More detail on the chemical reaction rates that have changed since 1979 and the effect each change has had on ozone-depletion predictions can be found in Chapter 3 of Hudson et al. (1982). The effects of each of the processes mentioned above as well as discussion of recent findings on the reactions $HO_2 + ClO \rightarrow$ products and $ClO + NO_2 \rightarrow$ products (isomers of $ClNO_3$) are spelled out in that report. The remaining uncertainties in every one of these processes except $OH + H_2O_2 \rightarrow H_2O + HO_2$ are considerable. Unfortunately, most of these processes involve working with notoriously difficult laboratory systems, e.g., any study of HNO_4 properties and the reactions of the radicals like $OH + HO_2$.

Besides the laboratory kinetic uncertainties one must also note that measurements of most of these apparently important polyatomic species in the atmosphere have not yet been achieved: there has been no positive detection of $ClONO_2$ (chlorine nitrate), H_2O_2, HOCl, HNO_4, or N_2O_5. In this regard, one must state that there is considerable remaining uncertainty in ozone-depletion estimates. As with most scientific questions, in this case there is uncertainty on the high and low sides, i.e., if certain predicted species do not actually exist or certain reactions proceed at predicted rates, the curves on Figure A.1 could go in either direction. To make matters worse, these key polyatomic species are predicted to exist (and to mediate the critical chemistry) in the lower to middle stratosphere, precisely where physical transport in dynamical meteorological systems is important and simple photochemistry is not the controlling factor. Accordingly, it appears that the resolution of present uncertainties involving chlorine nitrate, HOCl, HNO_4, H_2O_5, H_2O_2, etc., will require not only difficult laboratory measurements but much more complete coupling of chemical and dynamical meteorology models. Phrased in the terminology of chemical catalysis, we must be able to calculate the catalytic chain lengths of the chlorine and nitrogen oxide chains and the effectiveness of the methane-oxidation ozone production reactions. Such a calculation must accurately account for (a) processes that can interrupt catalytic chains that form, for example, HNO_4 or HOCl or $ClONO_2$, and (b) meteorological motions that can rapidly move the reacting chemicals to locations with different pressures and temperatures.

The possibility of a separate anthropogenic effect on atmospheric ozone has been raised by Liu et al. (1980). In a research report concerned with the natural origins of tropospheric ozone they found evidence that ozone is produced photochemically in the upper troposphere where subsiding stratospheric NO_x encounters rising hydrocarbons. If so, then the NO_x emitted by commercial and military subsonic aircraft should lead to ozone production near the 10 km (flight altitude) level. Liu et al. (1980) calculated that increased subsonic air traffic could have increased northern hemispheric tropospheric ozone by about 15 percent from 1970 to 1980. Such an increase, while important in its own right, would also amount to a 1.5 percent increase in total overhead ozone. This increase could mask a 1.5 percent decrease in the stratospheric ozone column. At eight of nine northern hemisphere stations where tropospheric ozone profiles are measured regularly there was a measured increase of about the predicted amount (Liu et al. 1980). This apparent increase in northern hemispheric tropospheric ozone has also been discussed by Pittock (see Appendix F) and in the 1981 NASA/WMO Stratosphere Report (Hudson et al. 1982). From all these sources it appears clear that an increase might have occurred but the present state of measurement capability for tropospheric ozone above the surface is poor enough to be unable to settle this question.

The changes in our photochemical reaction schemes and data of the past two years have had relatively little effect on our view of the upper stratosphere. One still expects ozone at the 40-km level to be strongly attacked by chlorine compounds. No serious doubt at all exists on this point; one must state that a stong perturbation of the upper stratosphere is under way. However, because of the large uncertainties in the region below 30 km, mostly with the polyatomic species mentioned above, one cannot say with much confidence what the total column ozone depletion will be eventually. It is possible that as the upper stratospheric ozone decreases and N_2O and CO_2 increase, there could be extra ozone production below about 25 km so that the vertical column of ozone could be changed only slightly. In this event there would probably be a significant redistribution of ozone in latitude and altitude, leading to concern over climatic effects.

Two other human activities need updating. First, in the case of atmospheric N_2O, Weiss (1981) has shown through measurements that N_2O has increased by about 0.2 percent per year since 1976 and most likely at a similar rate since 1963. The relative contributions of combustion-produced N_2O and fertilizer-produced N_2O are not yet clear although Weiss's data can be explained roughly by the former. Second, atmospheric detonation of nuclear explosives has been examined once again, and it appears as before that there are significant uncertainties in estimating the NO yields (and thus the chemical effects) of such explosions (McGhan et al. 1981).

RECOMMENDATIONS FOR RESEARCH

Although the exact size of the effect has proven difficult to predict, the hypothesis that continued chlorofluorocarbon release will have a significant global impact on atmospheric ozone appears correct--it has withstood over seven years of reexamination. Because of the need for industry and government to make decisions on production and regulation and because other global anthropogenic pollutants (e.g., NO_x and N_2O) remain items of concern, further atmospheric chemistry research is indicated. The goals of the research areas listed below are (i) to understand relevant atmospheric chemistry and meteorology well enough to permit better prognostic mathematical models, (ii) through measurements, to better characterize the atmosphere's chemical behavior, and (iii) to obtain more accurate and precise measurements of atmospheric ozone to permit variations and trends to be detected earlier and more clearly.

1. Continue the operation, maintenance, and data analysis of the WMO/NOAA Dobson instrument ozone measurements.

2. Develop improved ground-based instrumentation for measurements of the ozone column. It should be feasible to take entire absorption spectra for ozone determinations rather than the isolated measurements at a few discrete wavelengths. We are fortunate to have the Dobson instruments, but one suspects that it is possible to improve accuracy and precision with modern techniques (the Dobson instrument was invented in 1927).

3. Develop improved methods for measurements of tropospheric ozone. Both lidar and stable chemical sensors seem like good prospects.

4. Continue and expand, if possible, in situ measurements of key chemical species and the ratios of key reactive species in spatial regions where the reactants are important and where photochemical time constants are smaller than those for transport.

5. Accelerate the development of mathematical models of atmospheric chemistry with coupled atmospheric dynamical fluid motions.

6. Encourage extant models to focus on more complicated scenarios, e.g., increasing CFMs and increasing CO_2 and increasing N_2O, CH_4, and CH_3CCl_3.

7. Accelerate research on climatological effects of redistribution of atmospheric ozone and of trace gas increases.

8. Continued monitoring of background concentrations of CF_2Cl_2, $CFCl_3$, N_2O, and CH_4. While no evidence exists for tropospheric removal processes of CF_2Cl_2, $CFCl_3$, or N_2O, it is very important to obtain a continuous record of their concentrations. The

preparation and stable maintenance of calibrated standards for each of these gases is still an important research problem deserving commitment of government and industrial resources and scientific talent.

9. Improved satellite sensors and continued data-reduction effort focused toward stratospheric ozone and trace-constitutent monitoring are needed.

10. Expanded high-altitude (upper stratospheric) whole-air sampling is needed to obtain vertical profiles of H_2O, CH_4, and N_2O and other stable trace gases. These are needed to provide ground truth values for overflights of satellite sensors and to begin to acquire a climatology of the upper stratosphere for multi-dimensional models to employ in validation tests.

REFERENCES

Anderson, J.G., H.J. Grassl, R.E. Shetter, and J.J. Margitan (1980) Stratospheric free chlorine measured by balloon-borne in situ resonance fluorescence. Journal of Geophysical Research 85:2869-2887.

Berg, W.W., P.J. Crutzen, F.E. Grahek, S.M. Gitlin, and W.A. Sedlacek (1980) First measurements of total chlorine and bromine in the lower stratosphere. Geophysical Research Letters 7:937-940.

Berkner L.W. and L.L. Marshall (1967) The rise of oxygen in the earth's atmosphere with notes on the martian atmosphere. Advances in Geophysics 12:309-331.

Brasseur, G. and P.C. Simon (1981) Stratospheric chemical and thermal response to long-term variability in solar UV irradiance. Journal of Geophysical Research 86:7343-7362.

Cicerone, R.J. and S. Walters (1980) NO_2-Catalyzed Removal of Stratospheric HO_x. Paper presented at Fourteenth Informal Conference on Photochemistry, March 31. (Unpublished)

Crutzen, P.J. (1970) The influence of nitrogen oxides on the atmospheric ozone content. Quarterly Journal of the Royal Meteorological Society 96:320-325.

Crutzen, P.J. (1971) Ozone production rates in an oxygen-hydrogen-nitrogen oxide atmosphere. Journal of Geophysical Research 76:7311-7327.

Dave, J.V., C.L. Mateer, and J.J. De Luisi (1981) An examination of the effect of haze on the short Umkehr method for deducing the vertical distribution of ozone. Pages 222-229, Proceedings of the Quadrennial

International Ozone Symposium, August 4-9, 1980. Boulder, Colo.: National Center for Atmospheric Research.

De Luisi, J.J., B.M. Herman, R.S. Browning, and R.K. Sato (1975) Theoretically determined multiple-scattering effects of dust on Umkehr observations. Quarterly Journal of the Royal Meteorological Society 101:325-331.

Duewer, W.H. and D.J. Wuebbles (1980) Effects of Speculative Reactions and Mechanisms on Predicted Ozone Perturbations. Paper presented at Fourteenth Informal Conference on Photochemistry, April 1. (Unpublished)

Hampson, J.F. (1964) Photochemical Behavior of the Ozone Layer, CARDE Technical Note 1627/64, Canadian Armament Research and Development Establishment, Valcartier, Quebec, 280 pp.

Heath, D.F., A.J. Krueger, and P.J. Crutzen (1977) Solar proton event: Influence on stratospheric ozone. Science 197:886-887.

Hudson, R.D., et al., eds. (1982) The Stratosphere 1981: Theory and Measurements. WMO Global Research and Monitoring Project Report No. 11. Geneva: World Meteorological Organization. (Available from National Aeronautics and Space Administration, Code 963, Greenbelt, Md. 20771.)

Hunt, B.G. (1966) The need for a modified photochemical theory of the ozonosphere. Journal of Atmospheric Sciences 23:88-95.

Johnston, H.S. (1971) Reduction of stratospheric ozone by nitrogen oxide catalysts from supersonic transport exhaust. Science 173:517-522.

Kaufman, F. (1980) Laboratory Measurements of Stratospheric Reactions: Recent Results and Their Interpretation. Paper presented at Fourteenth Informal Conference on Photochemistry, March 31. (Unpublished)

Keating, G.M., L.R. Lake, J.Y. Nicholson, and M. Natarajan (1981) Global ozone long-term trends from satellite measurements and the response to solar activity variations. Journal of Geophysical Research 86. (In press)

Keyser, L.F. (1980) Kinetics of the reaction $OH + H_2O_2 \rightarrow HO_2 + H_2O$ from 245°K to 423°K. Paper presented at Fourteenth Informal Conference on Photochemistry, April 1. (Unpublished)

Liu, S.C., D. Kley, M. McFarland, J.D. Mahlman, and H. Levy (1980) On the origin of tropospheric ozone. Journal of Geophysical Research 85:7546-7552.

McGhan, M., A. Shaw, L.R. Megill, W. Sedlacek, P.R. Guthals, and M.M. Fowler (1981) Measurements of nitric oxide after a nuclear burst. Journal of Geophysical Research 86:1167-1173.

McPeters, R.D., C.H. Jackman, and E.G. Stassinoupoulos (1981) Observations of ozone depletion associated with solar proton events. Journal of Geophysical Research 86:12071-12081.

Molina, L.T. and M.J. Molina (1981) Ultraviolet absorption cross sections of HO_2NO_2 vapor. Journal of Photochemistry 15:97-108.

Molina, M.J. and F.S. Rowland (1974) Stratospheric sink for chlorofluoromethanes: Chlorine-atom catalyzed destruction of ozone. Nature 249:810-812.

NASA/JPL Kinetics Panel (1981) Chemical Kinetic and Photochemical Data for Use in Stratospheric Modeling. Evaluation No. 4: NASA Panel for Data Evaluation. JPL Publication No. 81-3. Pasadena, Calif.: California Institute of Technology.

Reagan, J.B., R.E. Meyerott, R.W. Nightingale, R.C. Gunton, R.G. Johnson, J.E. Evans, and W.L. Imhof (1981) Effects of the August 1972 solar particle events on stratospheric ozone. Journal of Geophysical Research 86:1473-1494.

Reber, C.A. and F.T. Huang (1982) Total ozone-solar activity relationship. Journal of Geophysical Research 86. (To be published Feb. 1982).

Solomon, S. and P.J. Crutzen (1981) Analysis of the August 1972 solar proton event including chlorine chemistry. Journal of Geophysical Research 86:1140-1146.

Swider, W. and T.J. Keneshea (1973) Decrease of ozone and atomic oxygen in the lower mesosphere during a PCA event. Planetary and Space Science 21:1969.

Sze, N.D. and M.K.W. Ko (1981) The effects of the rate for OH + HNO_3 and HO_2NO_2 photolysis on stratospheric chemistry. Atmospheric Environment 15:1301-1307.

Tolson, R.H. (1981) Spatial and temporal variations of monthly mean total column ozone derived from 7 years of BUV data. Journal of Geophysical Research 86:7312-7330.

Weeks, L.H., R.S. Cuikey, and J.R. Corbin (1972) Ozone measurements in the mesosphere during the solar proton event of 2 November 1969. Journal of Atmospheric Sciences 21:1138.

Weiss, R.F. (1981) The temporal and spatial distribution of tropospheric nitrous oxide. Journal of Geophysical Research 86:7185-7196.

Wine, P.H., A.R. Ravishankara, N.M. Kreutter, R.C. Shah, J.M. Nicovich, R.L. Thompson, and D.J. Wuebbles (1981) Rate of reaction of OH with HNO_3. Journal of Geophysical Research 86:1105-1110.

Appendix B

STRATOSPHERIC PERTURBATIONS--THE ROLE OF DYNAMICS, TRANSPORT, AND CLIMATE CHANGE

Robert E. Dickinson
National Center for Atmospheric Research
Boulder, Colorado

INTRODUCTION

The purpose of this paper is to review the role of transport, dynamics, and climate change in the question of stratospheric perturbations, with emphasis on progress over the last two years. Atmospheric dynamics and thermal structure are major factors in quantitative evaluations of the possible changes in the concentrations of stratospheric ozone. The distribution of ozone itself below 25 km is controlled more by transport by atmospheric motions than by chemical sources and sinks. Furthermore, atmospheric transport between troposphere and stratosphere determines the concentrations of the various chemical families that determine the catalytic destruction of ozone.

In particular, the transport of organic chlorine species from the troposphere to levels above 25 km provides the radical chlorine species whose effect is of special concern here. The concentration of total odd chlorine species derived from photodissociation of chlorocarbons depends on the balance between production and downward transport to the troposphere. The longer-lived chlorocarbons such as F-11 and F-12 whose only loss is by photodissociation in the stratosphere have average lifetimes in the troposphere inversely proportional to their rate of transport into the stratosphere. Likewise, the concentrations of total stratospheric odd nitrogen as derived from N_2O generated in the troposphere are also controlled by atmospheric transport. Finally, the concentrations of water, which provides the OH radicals so crucial to ozone chemistry in the lower stratosphere, are determined by exchanges with the troposphere.

Atmospheric thermal structure is important for determining the rates of various photochemical processes. At lower temperatures, most chemical kinetic processes, including those responsible for the catalytic destruction of ozone, proceed at a slower rate. Consequently, lowering of temperatures in the upper stratosphere, for example, as a result of ozone loss or increase of carbon dioxide, tends to increase stratospheric ozone.

The atmospheric trace species discussed in this report are of concern not only because of possible changes in ultraviolet fluxes due to this impact on ozone change but also because of possible climate change. Climate change is possible either because of the ozone change or because of the direct radiative effects of the species. There have been no significant modifications in the last two years of our understanding of possible climate change due to the direct radiative effects of the CFMs. However, currently projected ozone change profiles imply a much larger change in the energy balance of the tropospheric energy balance than was inferred from ozone profile change estimates of two years ago.

PROGRESS IN QUANTITATIVE MODELS OF TRANSPORT

One-Dimensional Models

Current evaluations of possible ozone depletion are still primarily based on one-dimensional empirical diffusion transport models. Quantitative approaches for objectively obtaining optimum eddy diffusion coefficients K(z) for such models were discussed at length in NRC (1976) and NRC (1979a). The basic concept is to determine K(z) empirically to reproduce one or more of the long-lived stratospheric species, in particular, N_2O, CH_4, O_3 (below 25 km) or the CFMs. Stratospheric H_2O is poorly simulated by one-dimensional models; it is not expected that the global average profiles of all the above-mentioned tracer species would simultaneously be accurately modeled by any particular K(z). Eddy diffusion parameterizations are not inferred from known physical processes but rather are simply representations of the time scales for vertical transport as indicated by the profile of a given tracer. Insofar as all the tracers have somewhat different sources and sinks, they all are expected to have somewhat different vertical transfer rates.

Little progress has been made in the last two years in deriving improved K(z)'s, and it is believed that remaining uncertainties in transport inferred from one-dimensional models should be due more to the physical unreality of the approach than inaccuracies in the derivation of K(z). It was previously estimated (NRC 1979a) that projections of global average ozone depletion were uncertain by a factor of two due to inaccuracies in transport calculations. This estimate was somewhat subjective, but there is no current basis for improving it.

Current models provide reasonable agreement with the observed vertical distributions of both N_2O and CH_4, but they calculate concentrations of F-11 and F-12 above 20 km that are somewhat too large in comparison with that observed.

Two-Dimensional Models

It was reported previously (NRC 1979a) that a number of two-dimensional empirical transport models were on the verge of completion. About a dozen of these models are now operational, but at the time of the May 1981 NASA workshop only one such model had obtained a projection of steady state ozone depletion with currently recommended chemical rates. This projection did not depart significantly from those of one-dimensional models (Hudson et al. 1982). If such a model were to simulate latitudinally varying vertical profiles of O_3, H_2O, CH_4, and the CFMs, it could be regarded as providing a major improvement in the parameterization of transport over that given by one-dimensional models. If it also gave a reasonable simulation of stratospheric H_2O, it would be a remarkable success. Some current two-dimensional models appear to simulate the latitudinal-seasonal variation of total ozone quite well but not the latitudinal variations of stratospheric N_2O and CH_4 (Hudson et al. 1982).

Besides possibly improving estimates of global average ozone depletion, two-dimensional models can provide the latitudinal and seasonal patterns of ozone change. As reported in NRC (1979a), Pyle and Derwent (1980), and Hudson et al. (1982), the two-dimensional models indicate ozone depletions to be greatest at high latitudes in winter where there is the least hazard of excess UV. It is evident that multidimensional models are required for detailed studies of the impacts of ozone change even if the estimates they provide of global average ozone change are no better than those of one-dimensional models.

Three-Dimensional Models

Three-dimensional model studies of transport to the troposphere from the stratosphere have been carried out recently by Mahlman and his collaborators at the Geophysical Fluid Dynamics Laboratory in Princeton. No attempts have been made to include realistic chlorine chemistry. They have largely been concerned with the transport of various tracer species as inferred from winds generated from a past general circulation model simulation. In particular, they have analyzed in detail two model simulations of a tracer whose source is similar to ozone (Mahlman et al. 1980); they have used the second of these simulations to study the sampling errors for total ozone measurements in a global network of stations.

ADVANCES IN THEORETICAL UNDERSTANDING OF STRATOSPHERIC TRANSPORT

Considerable advances have been made in our theoretical understanding of stratospheric transport (e.g., Matsuno 1980, Pyle and Rogers 1980). Transport in the latitude-altitude plane depends on the phase relationships between poleward and vertical eddy velocities, and the relative magnitude of the photochemical source terms compared to advective transport by motions. The phase difference between poleward (v) and vertical (w) velocities depends on fluctuations in wave amplitude and dissipative processes perturbing the motions.

For a simple model of a stationary planetary wave, Pyle and Rogers show that the symmetric components of the diffusion coefficient tensor (i.e., K_{yy}, K_{zz}) for a particular species depend on the rate at which that species damps to photochemical equilibrium, and on the strength of its chemical coupling to other species. The latter term can so drastically change the inferred K's that only for quasi-conservative species or families of species does the assumption of a species-independent diffusion tensor seem approximately justified. Fortunately, it is the quasi-conservative constitutents whose distribution is determined by transport.

It is not currently known whether or not complexities of the motions not included in the simple planetary wave models are less important than the photochemical phase shifts considered by Matsuno and Pyle and Rogers.

CONNECTIONS BETWEEN STRATOSPHERIC OZONE, STRATOSPHERIC TEMPERATURE STRUCTURE, AND CLIMATE CHANGE

In discussing stratospheric ozone, it is important to recognize possible effects of changes in stratospheric temperature on ozone concentrations. Such changes will occur either due to changes in the ozone concentrations themselves, e.g., Penner and Luther (1981), or due to changes in the concentrations of the other species that are important for stratospheric radiative balance, i.e., CO_2 and H_2O. The concentration of H_2O in turn can be affected by changes in the temperature of the tropical tropopause. Our understanding of these feedbacks has changed since NRC (1979a) primarily because of the recent changes in the assumed chemical rate constants for the lower stratosphere and consequent ozone perturbations there.

In particular, small increases of O_3 in the lower stratosphere, as now inferred in steady state CFM scenarios, imply a warmer tropical tropopause (as does the direct radiative heating by the CFMs), hence likely increases in stratospheric H_2O concentrations. This water vapor-temperature feedback has not recently been examined quantitatively, but it should amplify, somewhat, the ozone depletion.

It has been argued in the past that atmospheric CO_2 would double in 50 years due to burning of fossil fuel. The stratospheric cooling due to such a doubling (10°K at 50 km according to Fels et al. (1980)) would increase O_3 by 2 to 4 percent (Hudson et al. 1982) compared to the ozone column without the cooling, given the odd-chlorine concentrations expected if current CFM releases were to continue indefinitely.

This CO_2 effect now appears to be much more important than the 2 percent effect suggested in NRC (1979a), because it is a much larger fraction of the anticipated ozone depletion (one-fourth to one-half of it). However, it should be noted that doubling of CO_2 in 50 years is no longer regarded as a credible scenario. Current scenarios for CO_2 growth (Rotty and Marland 1980) suggest only a 30 to 40 percent increase of CO_2 in 50 years.

There has been considerable progress in developing an understanding of possible changes in stratospheric temperature and winds consequent to changes in stratospheric radiative heating terms. In particular, Fels et al. (1980) studied the stratospheric response to either a 50 percent reduction of O_3 or a doubling of CO_2. They used both a three-dimensional general circulation model (GCM) and simpler radiative equilibrium models. They showed that a simple model that assumed pure radiative balance for the perturbation, an approximation also used by Ramanathan and Dickinson (1979), gave temperature changes closely resembling those predicted by the GCM. This conclusion is very important for the development of two-dimensional chemical models for it provides a simple means to include temperature feedback in photochemical sensitivity studies. The recommended procedure is to assume observed temperature structure plus whatever temperature changes are needed to balance changes in radiative heating due to changes in ozone.

The effects of various radiative perturbations on tropospheric climate continue to be a major concern in climate studies. Anticipated increases of CO_2 still give the largest effect. However, most other likely changes in atmospheric composition also lead to warming and therefore exacerbate the problem. In particular, an increase of CFM concentrations to 1 ppb F-11 and 2 ppb F-12 would heat the troposphere by about 20 percent, as much as would a doubling of CO_2 (NRC 1979b). It was inferred previously that the anticipated ozone decrease due to CFMs would provide a slight cooling due to a somewhat greater increase in thermal infrared cooling than the increase in solar heating. However, current projections of ozone change suggest ozone increases in the lower stratosphere, especially in the tropics where sensitivity to radiative changes is greatest (as shown by Ramanathan and Dickinson (1979) and Fels et al. (1980)). Hence the ozone change itself now also implies significant tropospheric warming; the change due to continuation of present emission would give about 5 to 10 percent as much warming as a doubling of CO_2 in the atmosphere.

REFERENCES

Fels, S.B., J.D. Mahlman, M.D. Schwarzkopf, and R.W. Sinclair (1980) Stratospheric sensitivity to perturbations in ozone and carbon dioxide: Radiative and dynamical response. Journal of Atmospheric Sciences 37:2265-2297.

Hudson, R.D., et al., eds. (1982) The Stratosphere 1981: Theory and Measurements. WMO Global Research and Monitoring Project Report No. 11. Geneva: World Meteorological Organization. (Available from National Aeronautics and Space Administration, Code 963, Greenbelt, Md. 20771.)

Mahlman, J.D., H. Levy II, and W.J. Moxim (1980) Three-dimensional tracer structure and behavior as simulated in two ozone precursor experiments. Journal of Atmospheric Sciences 37:655-685.

Matsuno, T. (1980) Lagrangian motion of air parcels in the stratosphere in the presence of planetary waves. Pure and Applied Geophysics 118:189-216.

National Research Council (1976) Halocarbons: Effects on Stratospheric Ozone. Panel on Atmospheric Chemistry, Committee on Impacts of Stratospheric Change, Assembly of Mathematical and Physical Sciences. Washington, D.C.: National Academy of Sciences.

National Research Council (1979a) Stratospheric Ozone Depletion by Halocarbons: Chemistry and Transport. Panel on Chemistry and Transport, Committee on Impacts of Stratospheric Change, Assembly of Mathematical and Physical Sciences. Washington, D.C.: National Academy of Sciences.

National Research Council (1979b) Protection Against Depletion of Stratospheric Ozone by Chlorofluorocarbons. Committee on Impacts of Stratospheric Change, Assembly of Mathematical and Physical Sciences and Committee on Alternatives for the Reduction of Chlorofluorocarbon Emissions, Commission on Sociotechnical Systems. Washington, D.C.: National Academy of Sciences.

Penner, J.E. and F.M. Luther (1981) Effect of temperature feedback and hydrostatic adjustment in a stratospheric model. Journal of Atmospheric Sciences 38:446-453.

Pyle, J.A. and R.G. Derwent (1980) Possible ozone reductions and UV changes at the earth's surface. Nature 286:373-375.

Pyle, J.A. and C.F. Rogers (1980) Stratospheric transport by stationary planetary waves--the importance of chemical processes. Quarterly Journal of the Royal Meteorological Society 106:421-446.

Ramanathan, V. and R.E. Dickinson (1979) The role of stratospheric ozone in the zonal and seasonal radiative energy balance of the earth-troposphere system. Journal of Atmospheric Sciences 36:1084-1104.

Rotty, R.M. and G. Marland (1980) Constraints on fossil fuel use. In Interactions of Energy and Climate, edited by W. Back, J. Pankrath, and J. Williams. Boston, Mass.: D. Reidel.

Appendix C

RECENT DEVELOPMENTS IN STRATOSPHERIC PHOTOCHEMISTRY

Steven C. Wofsy and Jennifer A. Logan
Division of Applied Sciences
Harvard University

INTRODUCTION

Studies of the stratospheric ozone layer are impeded by two characteristics common to many environmental questions. First, it is impossible to perform meaningful, controlled experiments to test the response of the system to changes in environmental parameters. Second, the chemistry of the system is very complex, involving labile species at low concentrations subject to transport processes that are not well understood. These difficulties force us to rely on simulations using theoretical models to assess possible perturbations to stratospheric ozone. The models are inevitably too simple to describe the complete physical system and yet are often so complicated that it may be quite difficult to understand the models and to draw model-independent conclusions from the results.

This paper examines recent models of stratospheric ozone and associated chemical species, with emphasis on developments subsequent to the earlier NRC study on the stratosphere (NRC 1979). The discussion relies primarily on calculations performed using our own one-dimensional model of the stratosphere (Logan et al. 1978, Wofsy 1978) and on results from two recent two-dimensional models (Miller et al. 1981; Steed et al. 1982; Ko, Sze, and co-workers reported in Hudson et al. 1982). This choice reflects our access to model results and our view that these models contain most of the essential features of other operational models.

EFFECTS OF NEW KINETIC DATA ON MODEL RESULTS

Species Concentrations

Stratospheric models in use during the previous NRC study (NRC 1979) appeared to underestimate by a factor of between 2 and 5 concentrations of NO and NO_2 below 25 km, and to overestimate the concentration of ClO by a factor exceeding 10 at the same altitudes. These discrepancies may be attributed to inaccurate values for kinetic data affecting calculation of the concentration of the OH radical. Below 25 km, NO and NO_2 are controlled by chemical exchange with HNO_3, the major odd-nitrogen species, with the main reactions being

$$NO_2 + OH + M \rightarrow HNO_3 + M \quad (1)$$
$$HNO_3 + h\nu \rightarrow NO_2 + OH \quad (2)$$
$$NO_2 + h\nu \rightarrow NO + O \quad (3)$$
$$NO + O_3 \rightarrow NO_2 + O_2. \quad (4)$$

Nitrogen dioxide and nitric oxide concentrations thus vary inversely as the concentration of OH,

$$[NO_2] \simeq \frac{J_2\,[HNO_3]}{k_1\,[M]\;[OH]} \quad (5a)$$

$$[NO] \simeq \frac{J_2\,[HNO_3]}{k_1\,[M]} \quad \frac{J_3}{k_4\,[O_3]} \quad \frac{1}{[OH]}\,, \quad (5b)$$

where [x] denotes the concentration of species x and $k_i(J_i)$ refers to the rate coefficient (photolysis rate) for the ith chemical reaction.

The concentration of ClO also is controlled by interchange with a more abundant species, HCl, but in this case ClO <u>increases</u> with OH. The principal reactions are

$$HCl + OH \rightarrow H_2O + Cl \quad (6)$$
$$Cl + O_3 \rightarrow ClO + O_2 \quad (7)$$
$$Cl + CH_4,\ H_2,\ H_2CO \rightarrow HCl + CH_3,\ H,\ HCO \quad (8\text{-}10)$$
$$ClO + NO \rightarrow Cl + NO_2, \quad (11)$$

which lead to the expression

$$ClO \approx \frac{k_6[HCl][OH]k_7[O_3]}{\{k_8[CH_4]+k_9[H_2]+k_{10}[H_2CO]\}k_{11}[NO]} \quad (12a)$$

$$\approx [OH]^2 \frac{k_6[HCl]k_4k_7[O_3]^2k_1[M]}{\{k_8[CH_4]+k_9[H_2]+k_{10}[H_2CO]\}k_{11}J_3J_2[HNO_3].} \quad (12b)$$

Hence [ClO] increases as $[OH]^2$. McConnell and Evans (1978) pointed out that model and observations could be brought into agreement if it was assumed that the model overestimated the concentrations of OH, and they noted that such an error could strongly affect estimates quoted in NRC (1979) for the response of ozone to enhanced levels of stratospheric chlorine or odd nitrogen.

New laboratory measurements lend support to the hypothesis advanced by McConnell and Evans (1978) and others (Turco et al. 1981). Wine et al. (1981) and Nelson et al. (1981) showed that the rate for the reaction

$$OH + HNO_3 \rightarrow H_2O + NO_3 \quad (13)$$

increases at low temperature. This reaction is the major sink for odd hydrogen below 25 km, as shown in Figure C.1. Rates for reactions involving peroxynitric acid ($HOONO_2$ or HNO_4) have also been revised recently as shown in Table C.1. Rates for formation of HNO_4 and for reaction between OH and HNO_4 appear to be faster than formerly believed,

$$HO_2 + NO_2 + M \rightarrow HNO_4 + M \quad (14)$$
$$HNO_4 + OH \rightarrow H_2O + \text{products} \quad (15)$$

(NASA 1981, Littlejohn and Johnston 1980, see also Hudson et al. 1982), whereas photolysis of HNO_4 may be slower than indicated by earlier studies,

$$HNO_4 + h\nu \rightarrow \text{products} \quad (16)$$

(Molina and Molina 1981). These results, if confirmed by further work, indicate that reaction (15) is a major pathway for loss of odd hydrogen (see Figure C.1).

Figure C.2 shows how calculated profiles for OH, HO_2, ClO, NO, and NO_2 (at noon) have changed in response to the new laboratory rate data. Model concentrations of OH have been lowered by about a factor of 3 at 20 km, NO and NO_2 have been increased by a

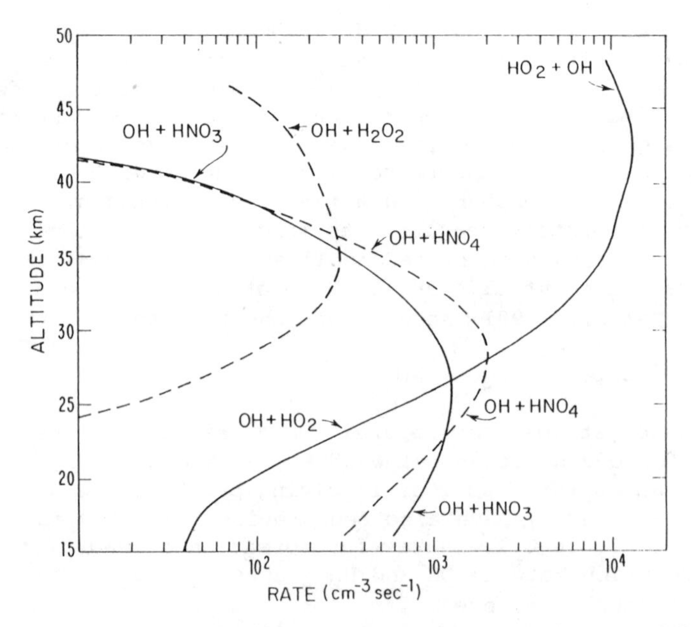

FIGURE C.1 Rates for loss of odd hydrogen, averaged over a 24-hour period. Profiles are shown for 30°N latitude at equinox. Results are from the Harvard one-dimensional model (Logan et al. 1978) using kinetic data from Hudson et al. (1982).

TABLE C.1 Reaction Rate Constants Used in Model Calculations

	Rate Constant (molecules cm^{-3} s^{-1})			Relative Change at 22 km Between 1979 and 1981
	Hudson and Reed (1979)	NASA (1981)	Hudson et al. (1982)	
$OH + HNO_3 \rightarrow$ products	8.5×10^{-14}	$1.5 \times 10^{-14} \exp(650/T)$	no change	$\times 3.4$
$OH + HO_2NO_2 \rightarrow$ products	5.0×10^{-13}	8.0×10^{-13}	4.0×10^{-12}	$\times 8$
$HO_2NO_2 + h\nu \rightarrow$ products	*a*	*b*	no change	$\times 0.7$
$OH + HO_2 \rightarrow H_2O + O_2$	4.0×10^{-11}	no change	8.0×10^{-11}	$\times 2$
$OH + H_2O_2 \rightarrow H_2O + HO_2$	$1.0 \times 10^{-11} \exp(750/T)$	$2.7 \times 10^{-12} \exp(145/T)$	no change	$\times 4.2$

[a] See Graham et al. (1978).
[b] See Molina and Molina (1981).

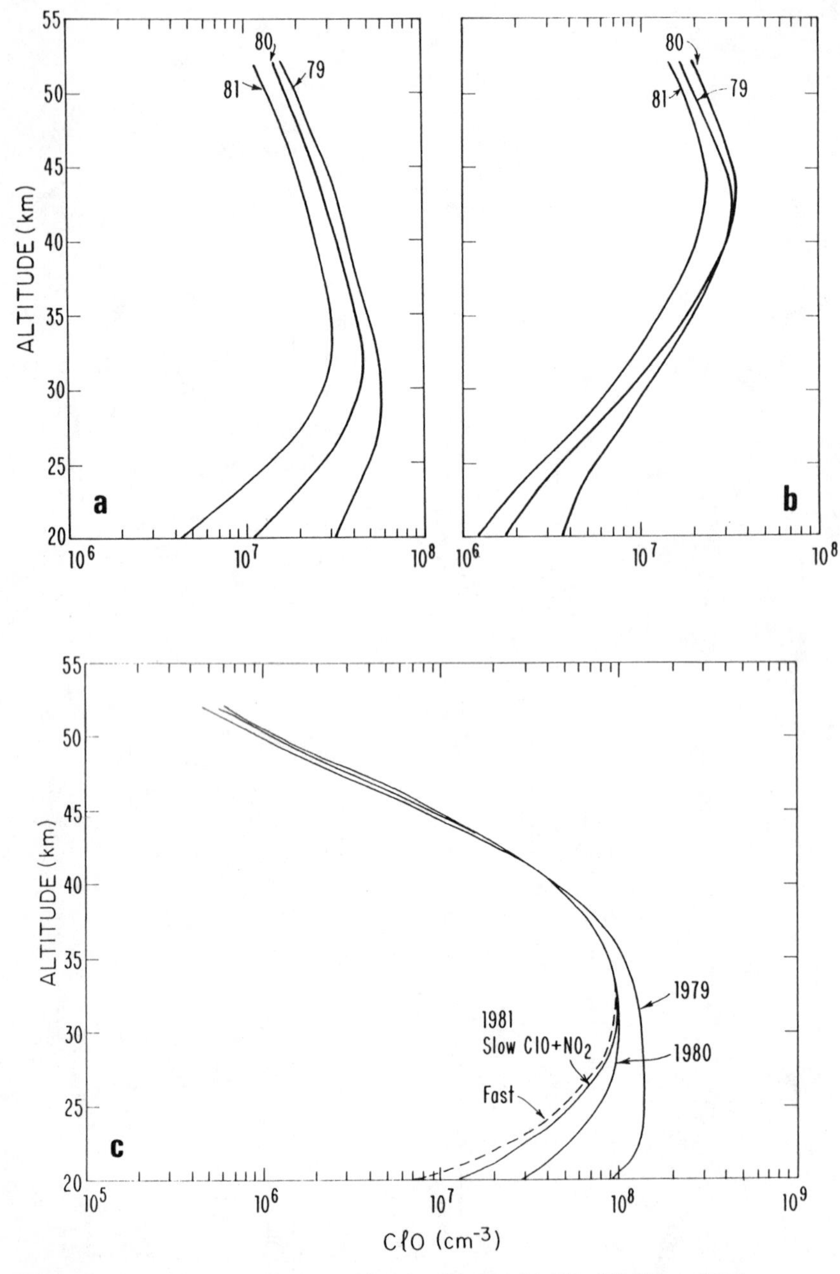

FIGURE C.2 Altitude profiles for (a) HO_2, (b) OH, (c) ClO, (d) NO and NO_2 at noon. The labels 1979, 1980, and 1981 indicate rate constant sets shown in Table C.1 (Hudson and Reed 1979, NASA 1981, and Hudson et al. 1982, respectively).

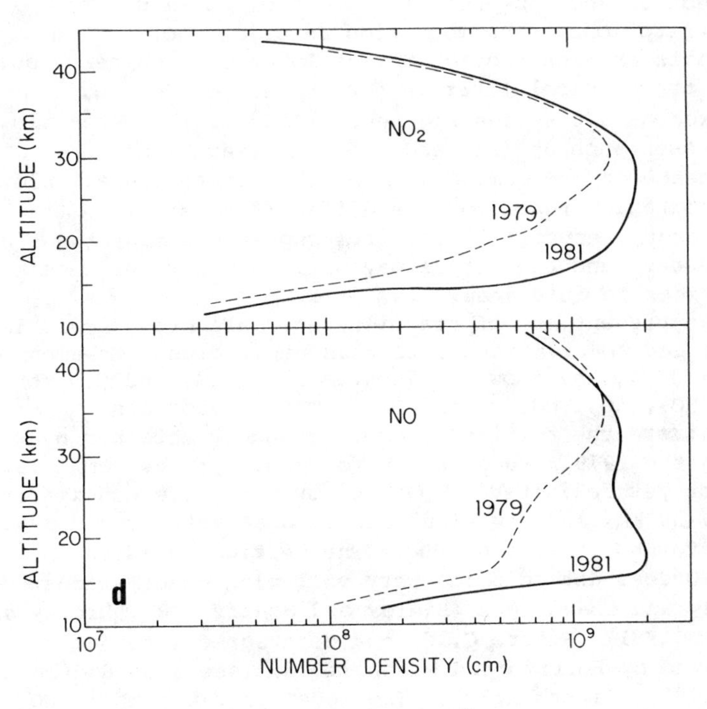

FIGURE C.2 (Continued)

similar factor, and calculated ClO concentrations have decreased by nearly a factor of 10.

It may seem surprising that relatively modest changes in the rates for (13) through (16) should have such dramatic effects on calculated profiles for OH. Chemical interchange among HO_x radicals is quite rapid in the lower stratosphere, with lifetimes for HO_2 and OH at noon about 50 and 10 s, respectively. The fast reactions establish the ratio of [OH] to [HO_2], but radical production and loss reactions control the absolute concentrations. Recombination reactions for HO_x radicals are inefficient in the lower stratosphere, such that the chemical lifetime for the sum of HO_2, H, and OH exceeds 500 s (see Figures C.1 and C.2). Hence slow processes such as (13) and (15) can exert a major influence on the composition of the stratosphere. Slow recombination reactions are difficult to study in the laboratory, especially for stratospheric temperatures and pressures, and the future may well hold further chemical surprises in this area.

The present set of reaction rate data brings calculations and observations into reasonably close agreement below 30 km, as shown in Figures C.3, C.4, and C.5 for OH, HNO_3, NO_2, NO, O, and ClO. The figures also illustrate the relatively poor agreement obtained by using the 1979 rate data. Unfortunately, the comparison is not yet definitive. Data on OH and O are nonexistent below 30 km, and few simultaneous observations are available for NO, NO_2, and HNO_3. The vertical gradient for NO does not coincide very well with observations by Ridley and co-workers (Ridley and Schiff 1981, Ridley and Hastie 1981) (Figure C.3d) but does agree with data obtained by Horvath and Mason (1978) (see also Hudson et al. 1982) (Figure C.3c). The model predicts more HNO_3 than is observed between 25 and 30 km. The apparent discrepancy observed for O (Figure C.3f) at low altitude may be attributed to differences for [O_3] and local albedo between the model and the particular observations. The model does predict accurate values of the ratio [O]/[O_3], as shown in Figure C.3g.

Observations of ClO require special consideration. Reported measurements are shown in Figure C.4 (Weinstock et al. 1981, Anderson et al. 1980). Summer data (solar declination of >0) fall in a rather narrow band, as predicted by the model, except for anomalous results obtained on June 15, 1979, and July 14, 1977. (The anomalous Bastille Day profile (July 14, 1977, Anderson

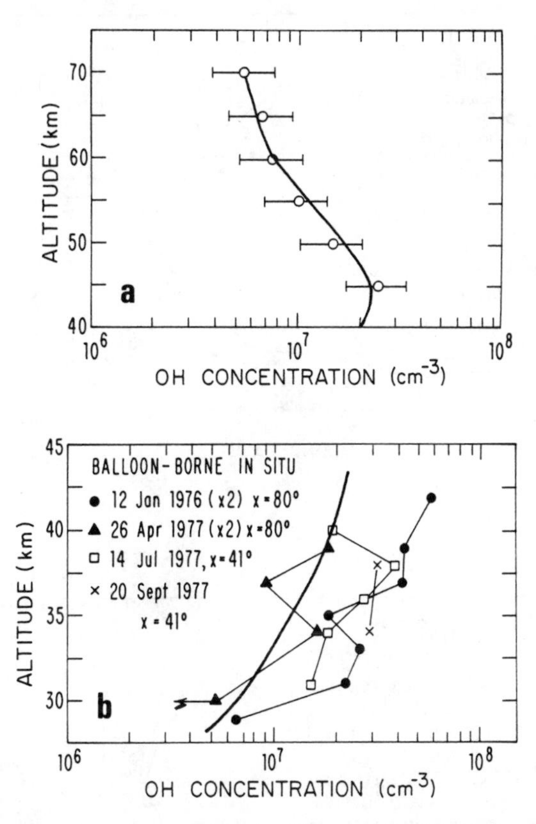

FIGURE C.3 Model results for OH in the (a) upper and (b) middle stratosphere; (c) HNO_3; (d) NO_2; NO in the (e) lower and (f) upper stratosphere; (g) $O(^3P)$; and (h) $[O]/[O_3]$ compared with measurements. The measurements are presented and discussed in Hudson et al. (1982). Calculations are appropriate for 30°N latitude at equinox and for solar zenith angles and local times as indicated.

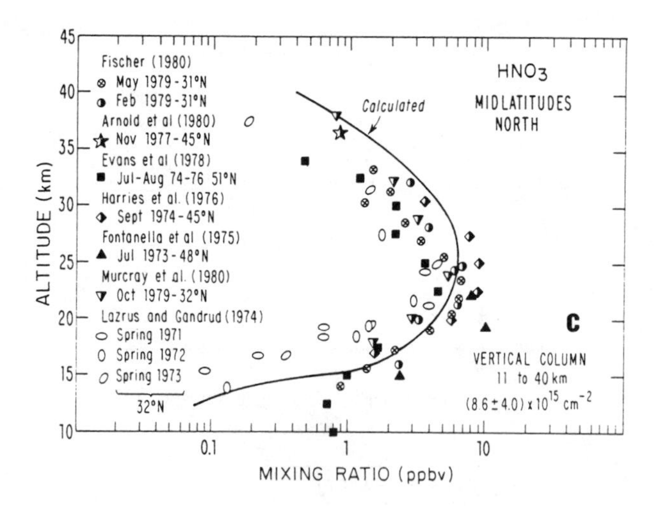

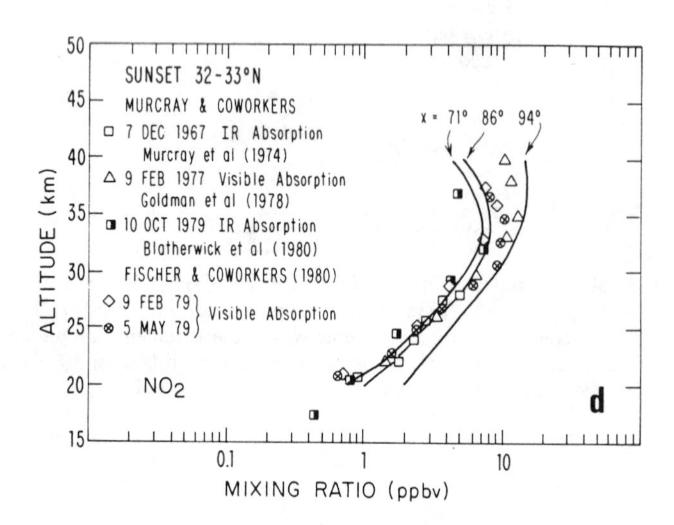

FIGURE C.3 (Continued)

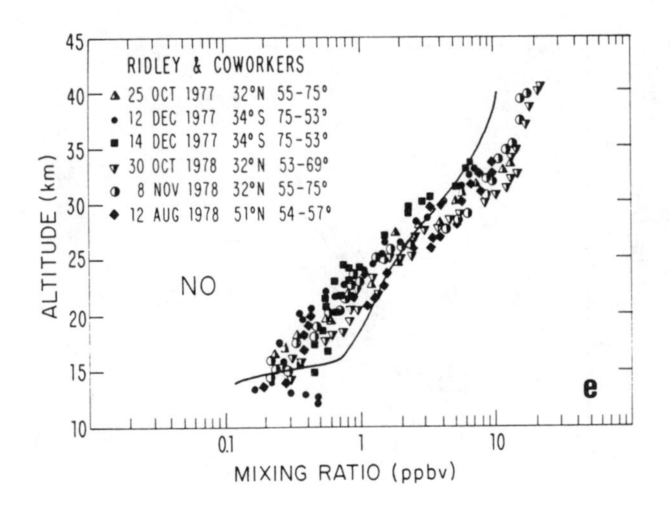

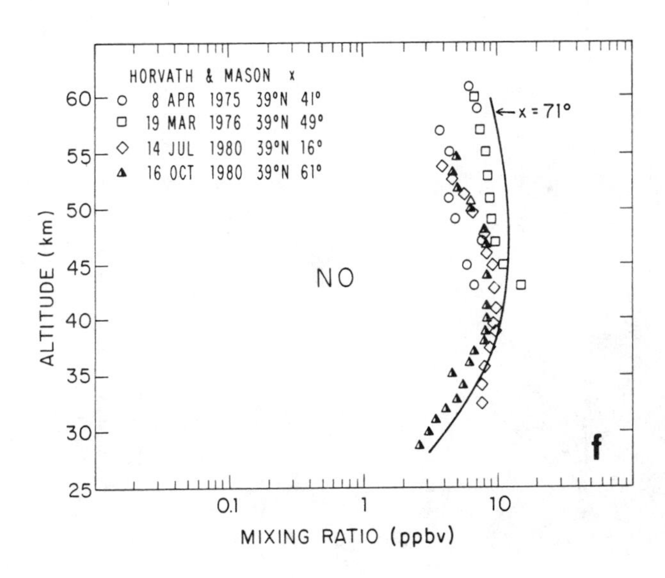

FIGURE C.3 (Continued)

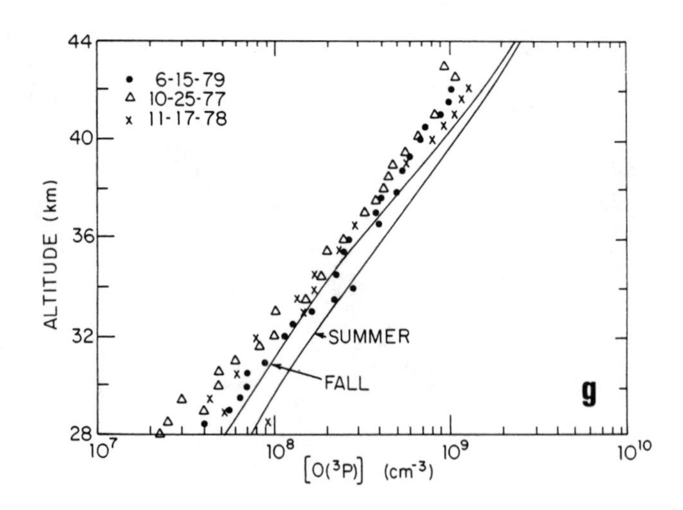

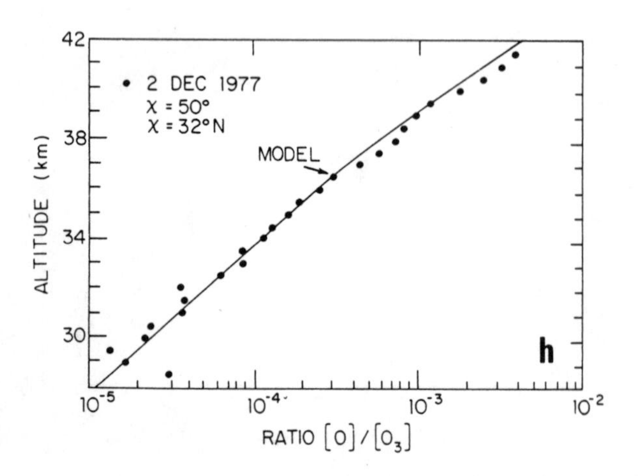

FIGURE C.3 (Continued)

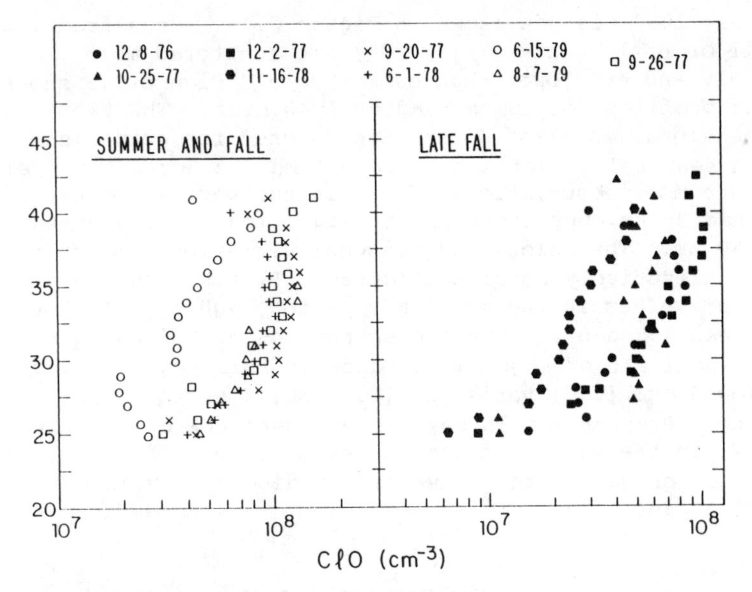

FIGURE C.4 Measurements of ClO concentration by in situ resonance fluorescence (Anderson et al. 1980).

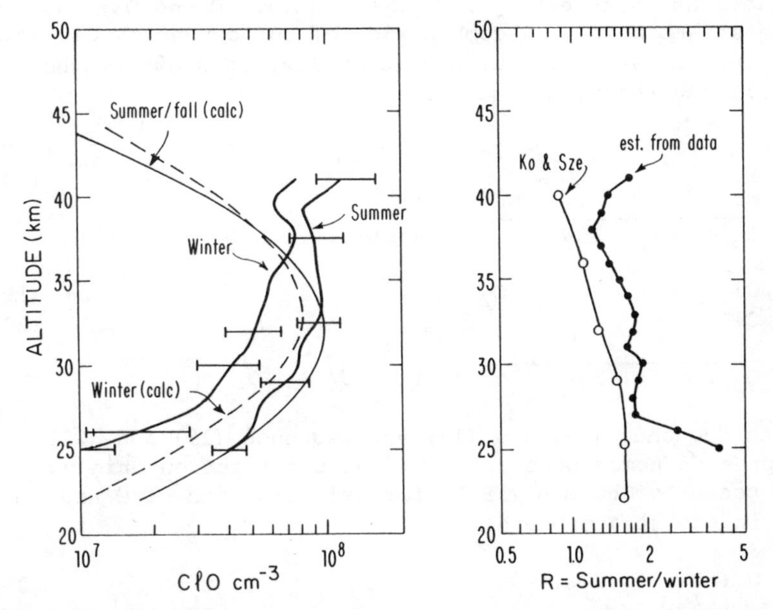

FIGURE C.5 Seasonal variations of ClO concentration. Mean profiles for summer and winter are derived from the measurements shown in Figure C.4, and calculated profiles are from the two-dimensional model of Ko and Sze (see Hudson et al. 1982).

et al. 1980) is not shown in Figure C.4). Data from late fall or early winter ($\chi < -10°$) are scattered more widely and are lower than summer data. Figure C.5 shows mean profiles for summer and early winter. The two-dimensional model of Ko and Sze is used for comparison with seasonal variations. Present models agree very well with summer observations of ClO in the key region between 25 and 35 km, but there may be significant disagreement at 40 km. The calculated seasonal variation appears to be qualitatively correct (Figure C.5), although the winter values in the model may be too high by a factor between 1.5 and 2. The concentrations of NO and O_3 in the model are of major importance in this regard, since below 30 km [ClO] varies as $[O_3]/[NO]$ (see equation (10)). Present models may overestimate the concentration of NO in the winter stratosphere, as discussed below, and this error may be the cause of the discrepancy in ClO during winter.

Rates for Catalytic Cycles

Figure C.6a shows calculated profiles for rates of reactions that destroy odd oxygen (i.e., O and O_3) in the stratosphere. Recombination of odd oxygen by ClO and NO_2 proceeds through catalytic cycles, as shown by the following reaction sequences:

$$NO + O_3 \rightarrow NO_2 + O_2 \qquad (17)$$
$$NO_2 + O \rightarrow NO + O_2 \qquad (18)$$
$$\overline{O_3 + O \rightarrow 2O_2} \qquad \text{(net)};$$

$$Cl + O_3 \rightarrow ClO + O_2 \qquad (19)$$
$$ClO + O \rightarrow Cl + O_2 \qquad (20)$$
$$\overline{O_3 + O \rightarrow 2O_2} \qquad \text{(net)}.$$

Both sequence (17) and (18) and sequence (19) and (20) represent homogenous catalysis of the reaction originally proposed by Chapman (1930) for recombination of O and O_3 in the stratosphere,

$$O + O_3 \rightarrow 2O_2 \qquad (21)$$

One of the striking features of Figure C.6a is the dominant role played by NO_2 (reaction (18)), a feature

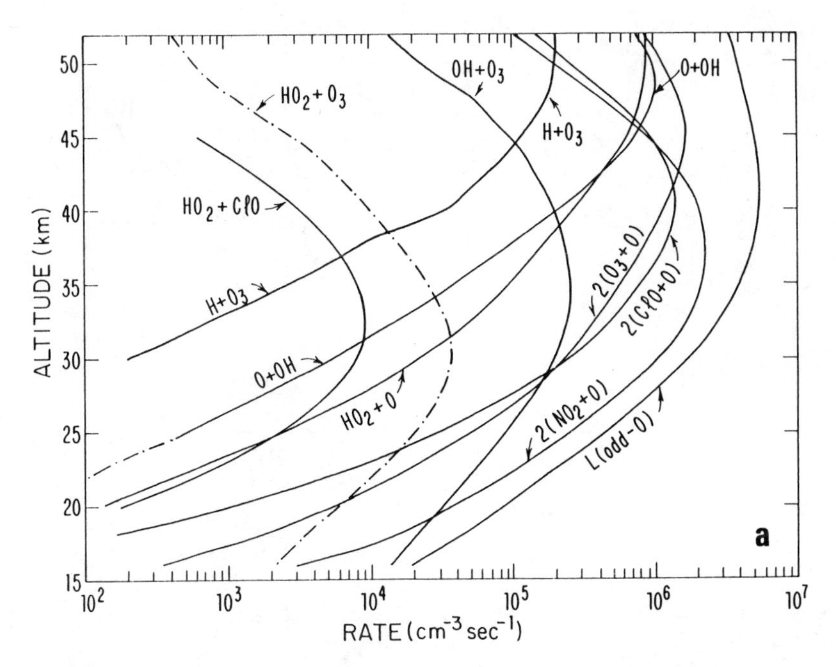

FIGURE C.6 (a) Rates for production and loss of odd oxygen, averaged over a 24-hour period. The model is the same used in Figure C.1. (b) and (c) Rates for production and loss of odd oxygen calculated using rate constant sets from Hudson and Reed (1979) and Hudson et al. (1982). Observed ozone profiles were used in these calculations.

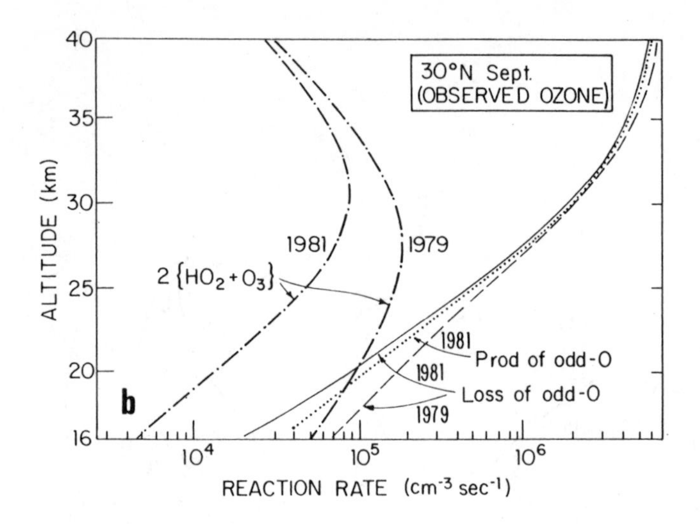

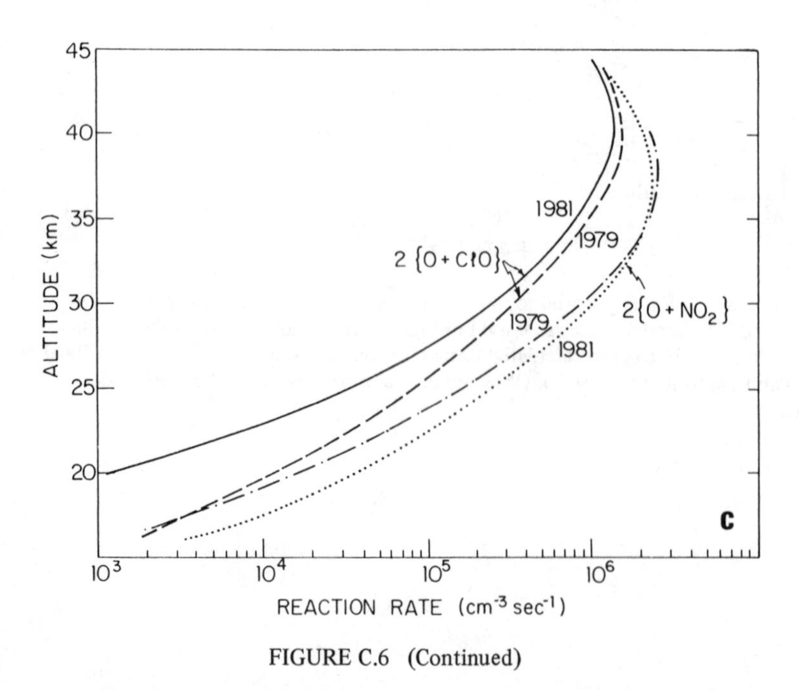

FIGURE C.6 (Continued)

that also characterized the earliest studies of stratospheric NO_x and Cl_x (Wofsy and McElroy 1974, Crutzen 1974, Stolarski and Cicerone 1974, Rowland and Molina 1975). Chlorine radicals influence ozone primarily at altitudes well above the ozone maximum, whereas odd nitrogen radicals are important throughout the stratosphere.

Loss profiles for odd oxygen obtained with the best rate data of 1979 (Hudson and Reed 1979) were significantly different from those shown in Figure C.6a, as may be seen in Figures C.6b and C.6c (see also Table C.1). The 1979 rates imply a major role for reactions of HO_2 between 16 and 25 km, and reduced contributions from reactions of NO_2. Ozone loss rates may exceed production rates below 25 km in this model. The principal cycles for HO_2 and OH at these altitudes are

$$OH + O_3 \rightarrow HO_2 + O_2 \quad (22)$$

$$HO_2 + NO \rightarrow OH + NO_2 \quad (23)$$

$$HO_2 + O_3 \rightarrow OH + 2O_2. \quad (24)$$

The sum of reactions (22) and (24) is

$$O_3 + O_3 \rightarrow 3O_2 \text{ (net)}. \quad (25)$$

The sequence (22) and (23) corresponds to little or no net destruction of odd oxygen, since most of the NO_2 formed in (23) will be rapidly cycled back to O by (3), and ozone will be regenerated via

$$O + O_2 + M \rightarrow O_2 + M. \quad (26)$$

The rate at which HO_2 catalyzes recombination of ozone (reaction (25)) is very sensitive to [OH]. This rate is given by the production rate for HO_2 via (22) multiplied by the fraction of HO_2 molecules that react with O_3, or

$$\text{net rate for cycle (25)} = 2k_{24}[HO_2][O_3]$$

$$= 2k_{22}[OH][O_3] \frac{1}{1 + k_{23}[NO]/k_{24}[O_3]} \quad (27)$$

$$\simeq \frac{2k_{22}k_{24}k_1k_4[OH]^2[O_3]^3[M]}{k_{23}J_2J_3[HNO_3]}. \quad (28)$$

Here we have incorporated equation (5b), introducing additional dependence on [OH], and we have exploited the fact that $k_{23}[NO] > k_{24}[O_3]$. In harmony with equation (28), the rate for (24) is reduced using new rate data by about a factor of 10 at 20 km, corresponding to the factor of ~3 reduction for [OH] shown in Figure C.2. The present model indicates approximate balance between ozone production and loss at 30° latitude, with production slightly exceeding loss below 25 km. The excess of ozone production over loss is somewhat greater at low latitudes and may reverse sign at high latitudes.

Catalytic recombination of two O_3 molecules, reaction (25), can potentially exert a major influence on the lower stratosphere, where low concentrations of O limit the rates for (18) and (20). The possible effect of (25) on stratospheric composition remains something of a puzzle, as it has since 1974, in part because transport and chemistry act on comparable time scales in the key altitude range of 20 to 25 km. As discussed below, a dominant role for (25) in stratospheric chemistry seems inconsistent with much of what we know about stratospheric ozone.

Global Distribution and Balance of Ozone

It has been known for some time that long-lived tracers in the lower stratosphere tend to be distributed along surfaces of preferred mixing that slant downward from equator to pole. The isopleths of long-lived radioisotopes (List and Telegadas 1969, Johnston et al. 1976), fine particles from volcanic eruptions (Lazrus and Gandrud 1974), and gases such as SF_6 (Krey et al. 1977), Kr^{85} (Telegadas and Ferber 1975), and N_2O (Goldan et al. 1980) exhibit such distributions. Figure C.7a shows the mean contours for these surfaces as deduced from data on Sr^{90} and $C^{14}O_2$ (see McElroy et al. 1976, Wofsy 1978, and Logan et al. 1978 for details). The isopleths move downward by about 5 km from the tropics to 30°N, fall an additional 3 to 4 km from 30° to 45°N, and move lower by about 1 km from 45°N to the subarctic. Suitable tracer data are available only below 30 km, with most observations below about 22 km.

Figure C.7b shows that observed ozone concentrations follow the preferred mixing surfaces at low altitudes and at high latitudes, but at high altitudes the data depart substantially from the isopleths of long-lived tracers.

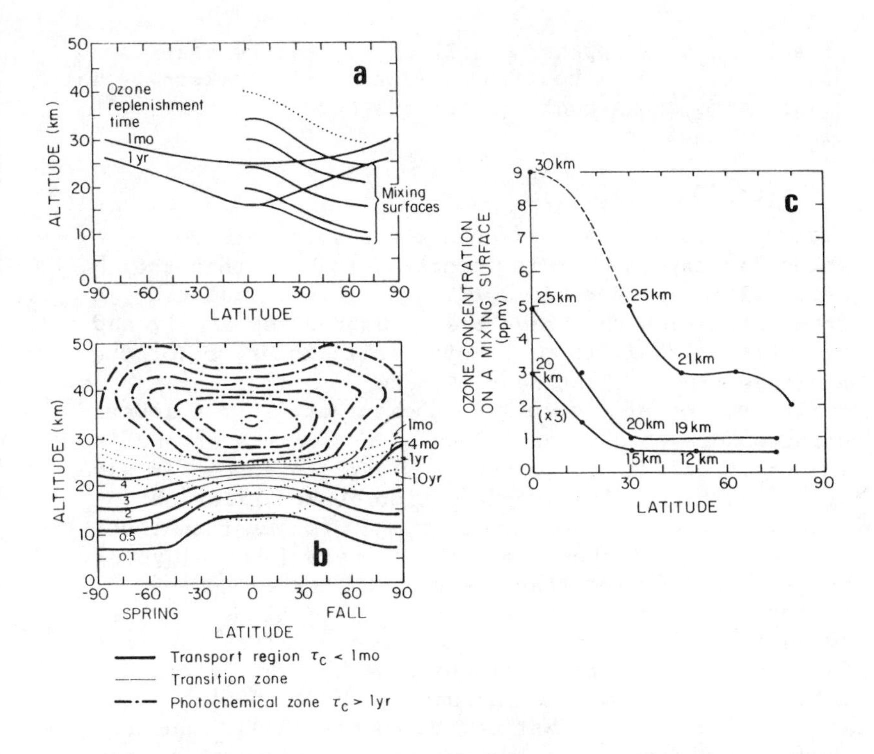

FIGURE C.7 (a) Latitude/altitude cross-section showing orientation of preferred mixing surfaces, deduced from data on the distribution of radioactive debris (List and Telegadas 1969, Wofsy 1978, Logan et al. 1978). Contours for ozone replenishment time are also shown. Note that a given mixing surface intersects a wide range of chemical lifetimes for ozone, with short lifetimes at low latitudes and long lifetimes at high latitudes. (b) Isopleths of ozone mixing ratio (ppmv) as functions of latitude. The data are taken from Johnston et al. (1976) for equinoctal conditions. Contours in the photochemical zone are represented as broken curves, in the transition zone as thin solid curves, and in the dynamically controlled zone as heavy solid curves. Contours of ozone replenishment time are shown as dotted curves for reference. Ozone follows the preferred mixing surfaces in the transport-controlled region, but departs from these surfaces in the photochemical region. (c) The ozone mixing ratio along several mixing surfaces as a function of latitude. Concentrations of ozone decrease along the mixing surfaces from the low-latitude source region to higher latitudes. Altitudes are indicated along each mixing surface at various latitudes.

Closed contours appear near 32 km, a feature that could not be produced without the influence of chemistry. The figure also shows contours for the ozone replenishment time, defined as

$$\tau_r = [O_3]/2J_{O_2}[O_2], \qquad (29)$$

where J_{O_2} is the 24-hour mean photolysis rate for molecular oxygen. It is important to note that the production rate for odd O ($2J_{O_2}[O_2]$) is calculated from quantities that have been measured repeatedly and that are unlikely to be substantially in error for the altitude range of interest (<25 km).

Inspection of Figure C.7b shows that, in the lower stratosphere, ozone behaves as a passive tracer and is uniformly distributed along the isopleths shown in Figure C.7a. At these altitudes the time constant for photochemical production of ozone (τ_r) is longer than the turnover time of the lower stratosphere (about 1 year). Where τ_r is shorter than the mixing time along the preferred surfaces, ozone is controlled almost completely by photochemical production and loss. Values for mixing times along preferred surfaces have been estimated to be 1 to 3 months from observations of the spreading rate for radioactive debris (List and Telegadas 1969). Hence above the contour τ_r = 1 month we may assume that, on average, the distribution of ozone is controlled by photochemistry and transport processes are unimportant. The transition zone between regions of photochemical dominance (τ_r < 1 month) and transport dominance (τ_r > 1 year) extends from 20 to 26 km at 30°N. About one third of the total ozone column lies in the transition zone. The zone is more extensive at low latitudes, whereas most of the ozone at high latitudes lies in the region of transport dominance.

The large concentrations of ozone observed at high latitudes are supplied by production at lower latitudes. If one follows the preferred mixing surfaces from high to low latitudes, they connect to the transition zone we have defined, where rates of transport and chemical production are similar. We anticipate therefore that ozone should exhibit a gradient along the preferred mixing surfaces in the transition zone, with highest concentrations at low latitudes. Figure C.7c confirms this result. We further expect that photochemical production of O_3 should exceed loss in the low-latitude source region (20 to 26 km), in order to supply the ozone that moves down this gradient. If we accept as reasonably accurate the O, NO_2, and ClO concentrations predicted by the present model, as shown in Figures C.3c, C.3d, and C.5, we cannot accept large rates for catalytic recombination of O_3 with O_3. We would otherwise calculate a net sink for O_3 where there should be a source. This result would appear to apply unless there exists an unknown process capable of dissociating the O_2 molecule. We conclude therefore that the set of kinetic rates used in 1979 predicted excessive rates for reaction (24), which were partially offset by an underestimate for the concentration of NO_2.

We may summarize the argument above as follows. The bulk of the world's ozone is stored at high latitudes, where it is essentially inert. This ozone appears to be supplied by transport along slant mixing surfaces from altitudes between 20 and 25 km at latitudes between 0° and 30°. If substantial recombination of O_3 with O_3 (reaction (25)) were occurring at these altitudes, chemical loss would significantly exceed production and it would be impossible to provide the source of high-latitude ozone. We argue that (25) cannot play a dominant role in the middle and low stratosphere. This conclusion will prove to be quite useful in our discussion of model simulations for perturbed conditions.

Ozone Response to Environmental Change

Figure C.8 shows calculated reductions of stratospheric ozone due to increased atmospheric burdens of chlorofluoromethanes (CFMs, Figure C.8a) and nitrous oxide (Figure C.8b). The calculations for CFMs compare model results for the present-day atmosphere (total chlorine 2.3 ppb) with a perturbed atmosphere containing 11.6 ppb of chlorine. The perturbed case corresponds to steady state conditions with constant release of CFMs at rates prevailing in 1977. Increased chlorine markedly reduces ozone above 26 km, with a small ozone increase predicted below that level. The distribution of ozone change reflects the height dependence of the rate for reaction of O with ClO (reaction (20), see Figure C.6a). Since most of the ozone change occurs where photochemistry is dominant, these results do not depend strongly on simulation of transport in the model.

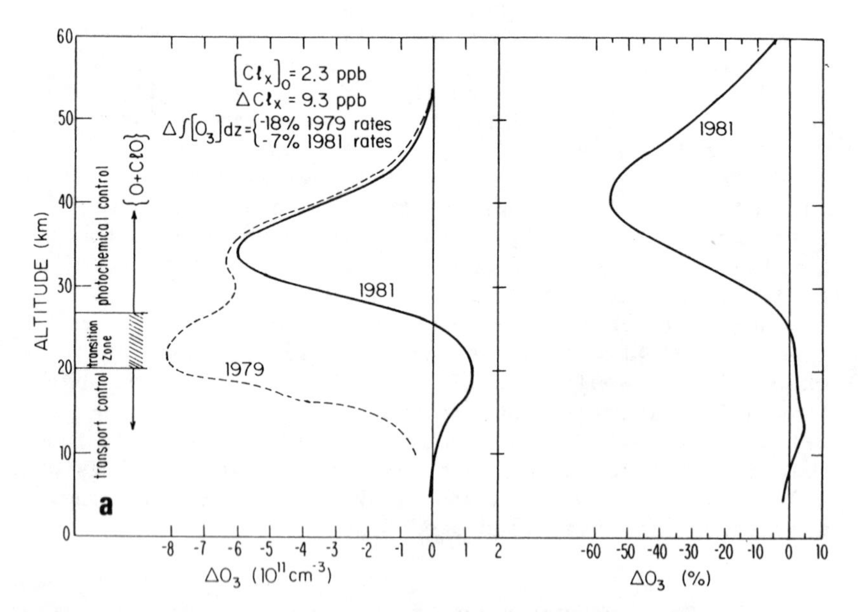

FIGURE C.8 (a) and (b) Perturbations to the concentration of ozone as a function of altitude for added Cl_X and N_2O, calculated using the Harvard one-dimensional model. Results are shown for (a) the change in Cl_X resulting from constant release of CF_2Cl_2 and $CFCl_3$ at 1977 release rates and (b) for a doubling of the N_2O mixing ratio (Hudson et al. 1982). The absolute change in O_3 is shown in the left-hand panel, and the percent change in O_3 in the right-hand panel. Rate constant sets, 1979 and 1981, are described in Table C.1. (c) History of model calculations since 1975 for the depletion of total ozone resulting from constant emission of CF_2Cl_2 and $CFCl_3$ (after Hudson et al. 1982).

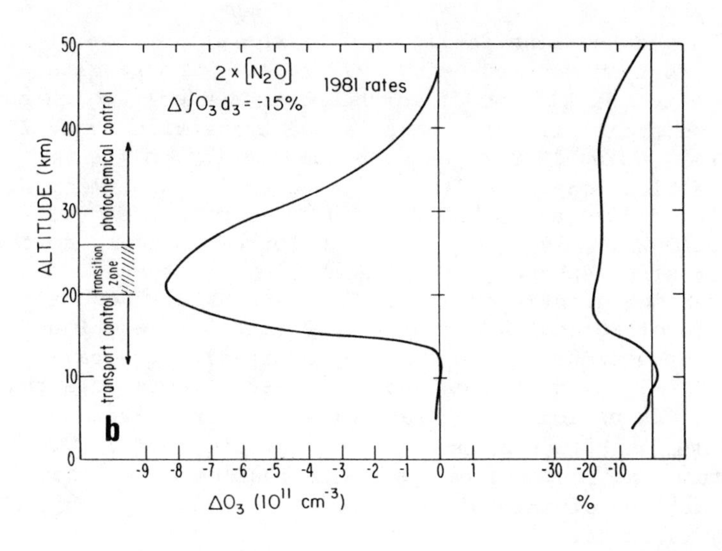

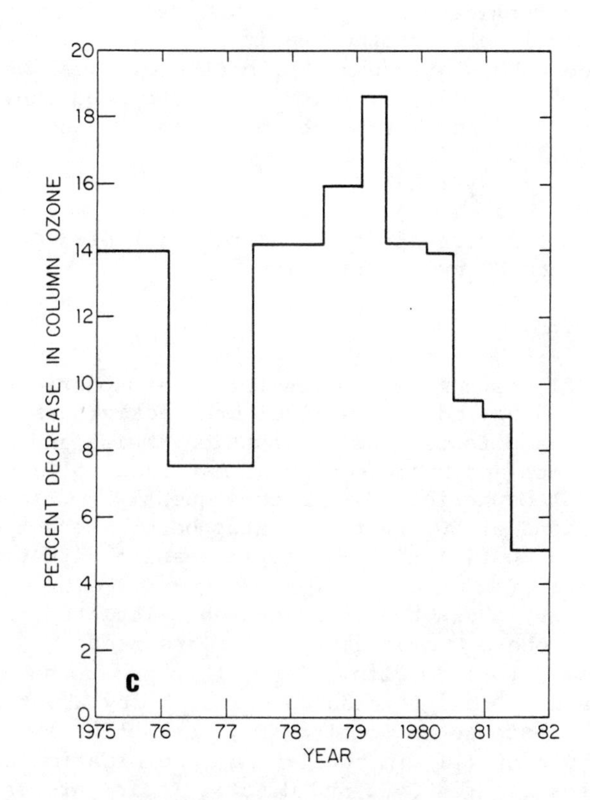

FIGURE C.8 (Continued)

The dashed line in Figure C.8a shows the same Cl_x-perturbation modeled using 1979 rates. The results are close to the 1981 model in the photochemical region but are substantially different in the transition zone, 20 to 26 km. Using 1979 rates, the calculation in the transition zone is quite sensitive to the treatment of transport processes and depends on a complex set of reactions involving $ClNO_3$. As chlorine is added to the lower stratosphere, formation of $ClNO_3$ removes increasing quantities of NO, NO_2, and HNO_3. Reduced levels of NO and HNO_3 amplify the rate for reaction (24) by increasing $[HO_2]$ and by decreasing the ratio $[NO]/[O_3]$, as discussed above. These effects more than offset reduction of the rate of (18), since (24) becomes the dominant loss process for O_3 in this model. The contrasting response of O_3 in the 1981 model reflects in part the diminished role for (24) obtained with present kinetic rates.

There is considerable interest in the response of O_3 to increased levels of atmospheric N_2O. Nitrous oxide is released to the atmosphere by microbiological processes (nitrification, denitrification) and by combustion, and it is removed in the stratosphere by photolysis,

$$N_2O + h\nu \rightarrow N_2 + O(^1D). \qquad (30)$$

Approximately 3 percent of the global flux of N_2O is converted into NO by the reaction

$$N_2O + O(^1D) \rightarrow NO + NO. \qquad (31)$$

Reaction (31) is a major source for stratospheric NO_x. It has been proposed that agricultural activities (McElroy 1976, 1980) and fossil fuel combustion (Weiss and Craig 1976, Pierotti and Rasmussen 1976) may lead to increased levels of atmospheric N_2O, and consequently to higher concentrations of NO_x in the stratosphere. Recent observations confirm that N_2O is increasing with time (Weiss 1981).

Figure C.8b shows the ozone change calculated for a doubling of the N_2O concentration, corresponding approximately to a doubling of NO_x throughout the stratosphere. Ozone is reduced _uniformly_ by about 15 percent in response to doubled NO_x, reflecting the dominant role of (18) in catalyzing recombination of O with O_3 (see Figures C.6a and C.6c). It is interesting to note that this calculation is also relatively insensitive to details of the model treatment of transport. For example, suppose we arbitrarily assume that transport is much faster than chemistry below 26 km, while chemistry is dominant above. The concentration below 26 km would be determined in this case primarily by the O_3 concentration at 26 km, which is reduced by 15 percent in the perturbation model. If we make the opposite assumption, that chemistry dominates from 26 to 20 km, the calculated ozone reduction is also about 15 percent. However, results for N_2O perturbations are likely to be quite different in two- or three-dimensional models, as compared to a one-dimensional model, since significant chemical changes occur in the transition zone that supplies ozone to high latitudes.

The calculations using 1981 rates indicate a much larger change for total ozone in response to increased NO_x, as compared to 1979 models. This result applies to NO_x introduced by enhanced N_2O, by high-flying aircraft, or by any other mechanism. The sensitivity to additional chlorine is reduced from 1979 models by about a factor of 2. In both cases the difference is due to the ozone response in the transition zone, and in both cases the results obtained with 1981 rates are less sensitive than the 1979 model to details of the transport parameterization. We argued above that the 1979 rate set produced spurious chemical losses for O_3 in the lower stratosphere; we now see that these loss processes distorted the calculated response of ozone to environmental change.

Figure C.8c reviews the history of calculated ozone depletion for continuous release of CFM at rates prevailing in 1977 (after Hudson et al. 1982). Each step on the curve corresponds to new kinetic information. The high values obtained between 1977 and 1978 reflect in part the influence of slow processes that combine O_3 with O_3. Lower values, near 6 percent, indicate the reduction in column ozone that results from chemical reactions occurring above 25 km. Model results for this region have changed little as new information has become available. Of course, ozone depletion above 25 km could have major indirect effects on the composition of the lower stratosphere, by inducing changes in the dynamics of the stratosphere. Considerable future work is required to define the effect of such coupling between chemistry and dynamics.

Present two-dimensional and one-dimensional models give very similar results for ozone reductions due to added chlorine. The global mean decrease in column ozone calculated by Miller et al. (1981) is very close to that predicted by the one-dimensional model (see also Steed et al. 1982). The reduction is nearly uniform over the globe, with variations smaller than ±20 percent about the mean. The calculated ozone reduction for a given change in Cl_x is insensitive to details of the transport, since most of the ozone change occurs high in the atmosphere. Thus the results of Miller et al. (1981) are consistent with the one-dimensional models discussed above.

The global mean ozone reduction due to increased N_2O is also very similar for one- and two-dimensional models (M. Ko et al., Atmospheric and Environmental Research Inc., personal communication, 1981). In this case, however, the meridional distribution of ozone depletion is not uniform, increasing from about 5 percent near the equator to about 25 percent in the subarctic. Poleward of 45° latitude ozone is strongly affected by increased NO_x between 15 and 30 km. As may be seen from Figure C.7b, the upper part of this region lies in the transition zone between the region of control by photochemistry and control by transport. The lower part of the affected region is supplied with ozone by transport from low latitudes. Thus ozone reductions predicted at high latitudes reflect slow consumption by reaction (18) in the lower stratosphere. Since models predict excessive NO_2 at high latitudes, especially during winter, ozone reduction by NO_x is probably overestimated above 30° latitude.

OUTSTANDING PROBLEMS OF PRESENT MODELS FOR THE STRATOSPHERE

We attempt in this section to identify important discrepancies between present models and observations of stratospheric composition. These problem areas merit attention in future theoretical and experimental work.

Long-lived Trace Gases

We examine here results from the one-dimensional model discussed above and from the two-dimensional models of Miller et al. (1981) and Sze and Ko (1981). Miller et al. (1981) use a diabatic circulation (Murgatroyd and

Singleton 1961, Dopplick 1972) to provide the field of mean motions, and they use Hunten's (1975) vertical diffusion coefficient. Horizontal diffusion coefficients are taken from Luther (1974). Sze and Ko (1981) use Luther's coefficients and a consistent set of winds derived from the dynamical study of Harwood and Pyle (1980). Transport in the model of Miller et al. (1981) is adjusted to reproduce the vertical distributions of N_2O and CH_4 at 30°-40°N latitude, as is the vertical diffusion coefficient used in the one-dimensional model (Logan et al. 1978). Comparison between one-dimensional and two-dimensional models is accomplished using the slant mixing surfaces to project the one-dimensional concentration profile to various latitudes (Wofsy 1978), with the one-dimensional profile taken to represent 30° latitude. This approach is equivalent to a two-dimensional model with infinite mixing rates (perfect mixing) along the surfaces of preferred mixing.

Figure C.9 shows removal lifetimes for CH_4, N_2O, CF_2Cl_2, CH_3Cl, and $CFCl_3$ as functions of altitude for 30°N at equinox. These gases originate in the troposphere and, with the possible exception of N_2O (Zipf and Prasad 1980), have no known sources in the stratosphere. Their chemical lifetimes decrease with altitude, reaching one year at 37, 33, 32, 29, and 26 km, respectively. The lifetime for $CFCl_3$ approaches meridional transport times near 26 km in the tropics.

Two-dimensional models produce meridional distributions of longer-lived gases very similar to those derived from the one-dimensional model, as illustrated for CF_2Cl_2 in Figure C.10. This result is hardly surprising, since Luther's diffusion coefficients are based in part on observed contours of potential temperature that are close to the isopleths observed for radioisotopes. The dispersion rates in the two-dimensional models are evidently large enough to ensure nearly perfect mixing along the preferred surfaces.

Observations are compared in Figure C.11 to model results for CH_4, N_2O, CF_2Cl_2, CH_3Cl, $CFCl_3$, and C_2H_6. Agreement is excellent for the longest-lived species, CH_4 and N_2O, reflecting in part adjustment of model parameters to fit these profiles. The models significantly underestimate the meridional gradient for CF_2Cl_2 and $CFCl_3$, but they simulate reasonably well the observed distribution of CH_3Cl. The results are particularly disappointing for $CFCl_3$. Both the vertical and the meridional gradients are

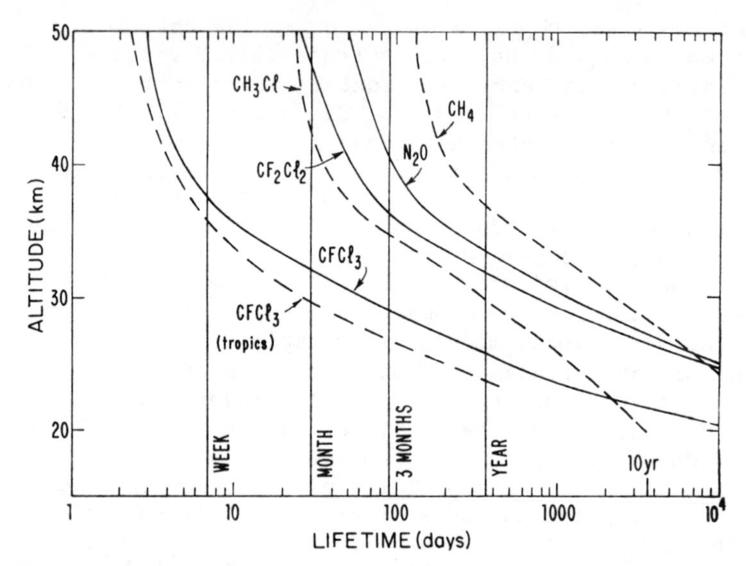

FIGURE C.9 Lifetimes of long-lived gases as a function of altitude. Results are appropriate for 30°N latitude at equinox and are from the Harvard one-dimensional model. Results are shown also for $CFCl_3$ in the tropics.

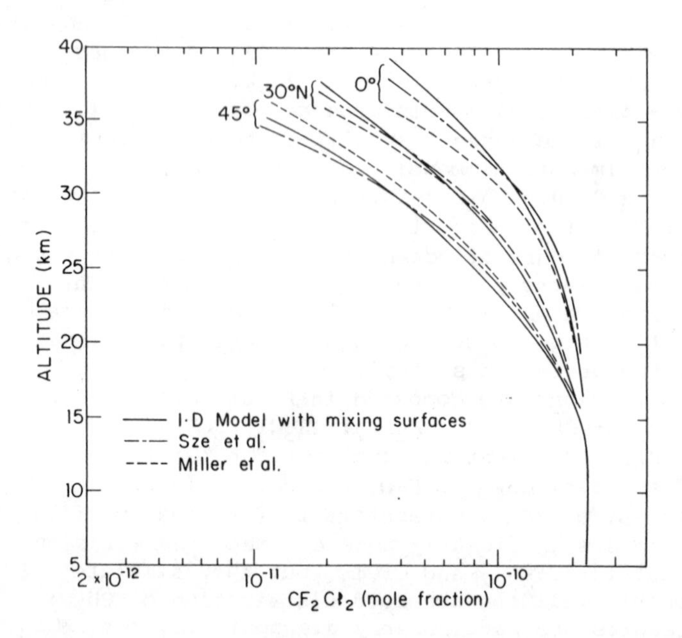

FIGURE C.10 Calculated profiles CF_2Cl_2 at 0°, 30°, and 45°N latitude.

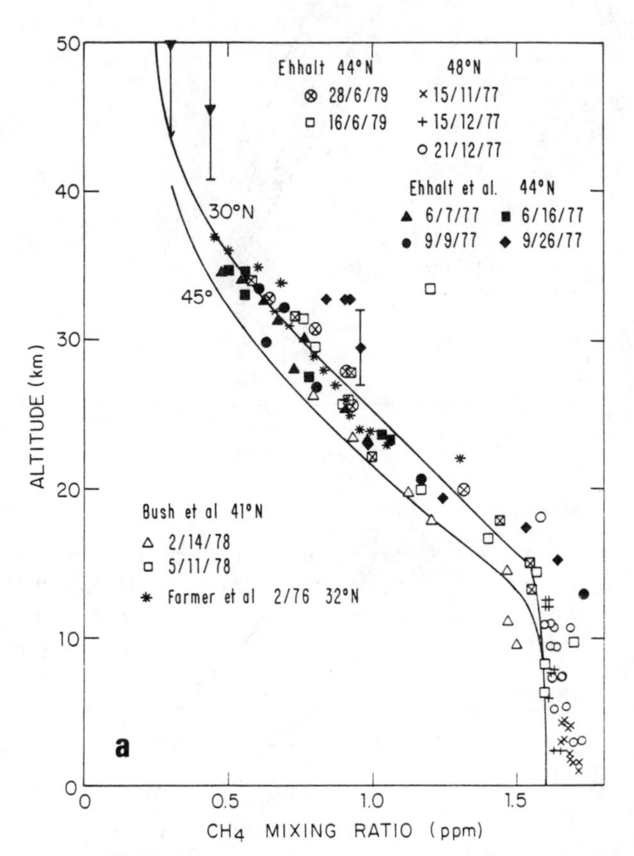

FIGURE C.11 Model results for (a) CH_4, (b) N_2O, (c) CF_2Cl_2, (d) $CFCl_3$, (e) CH_3Cl, and (f) C_2H_6 compared with observations. The measurements are discussed in Hudson et al. (1982), and model profiles are from the Harvard one-dimensional model with mixing surfaces, unless otherwise indicated.

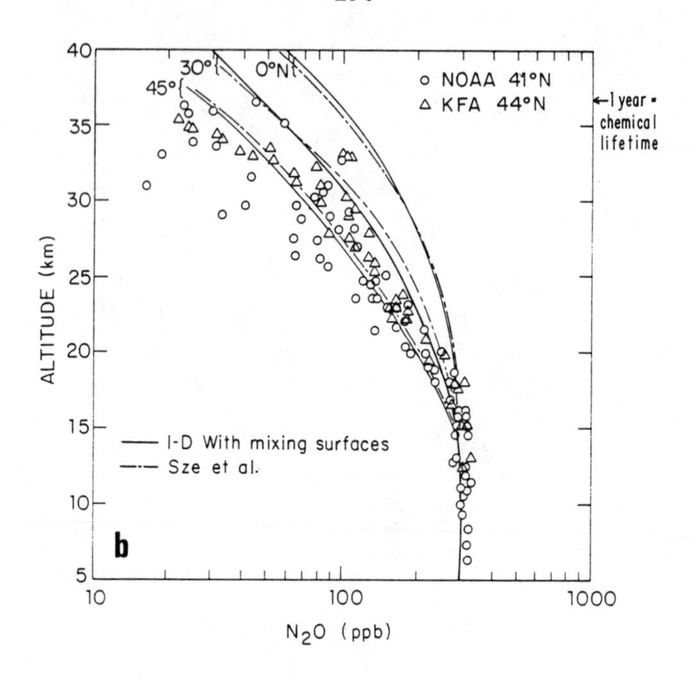

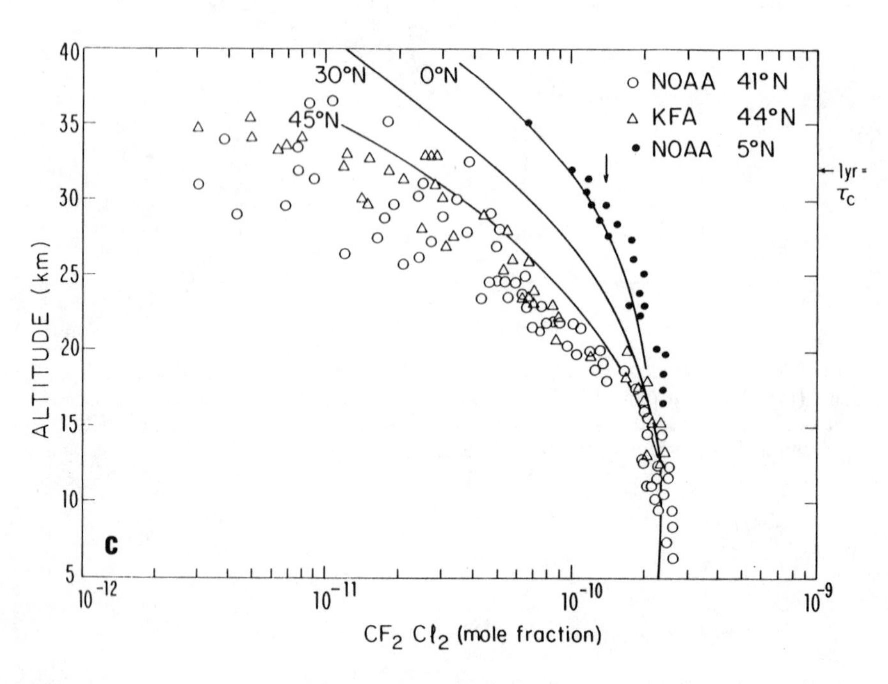

FIGURE C.11 (Continued)

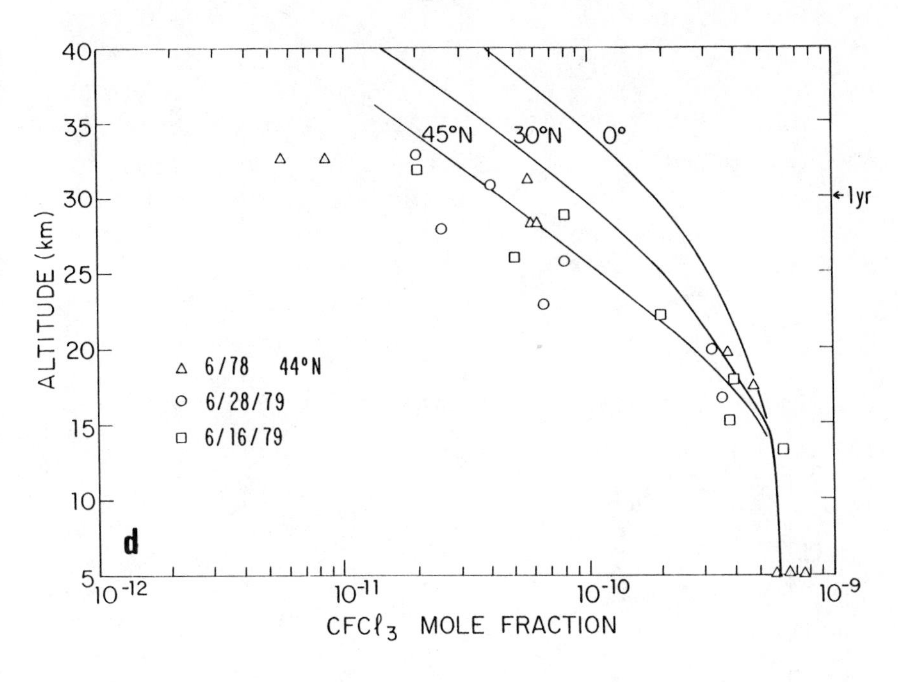

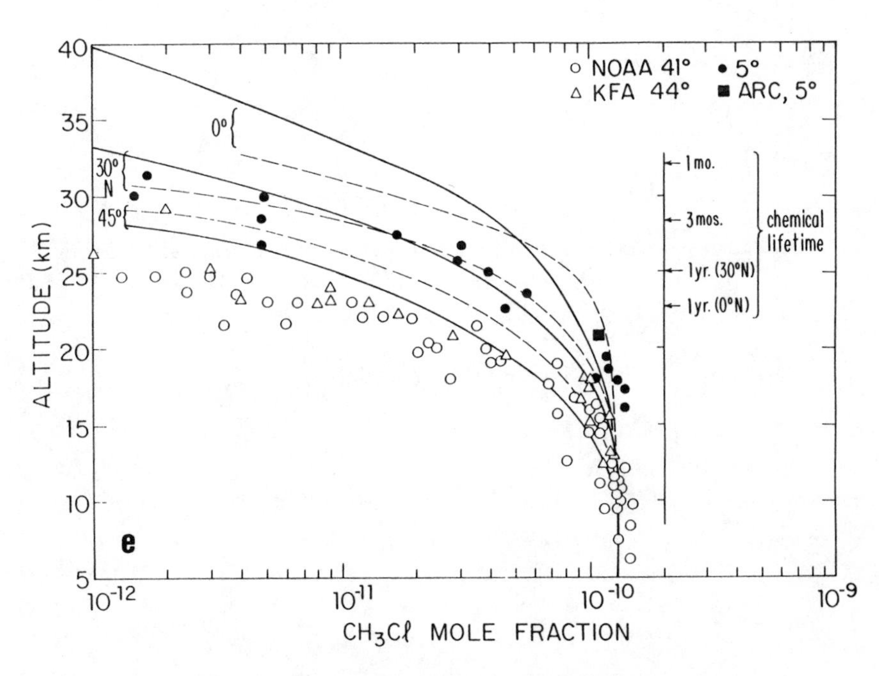

FIGURE C.11 (Continued)

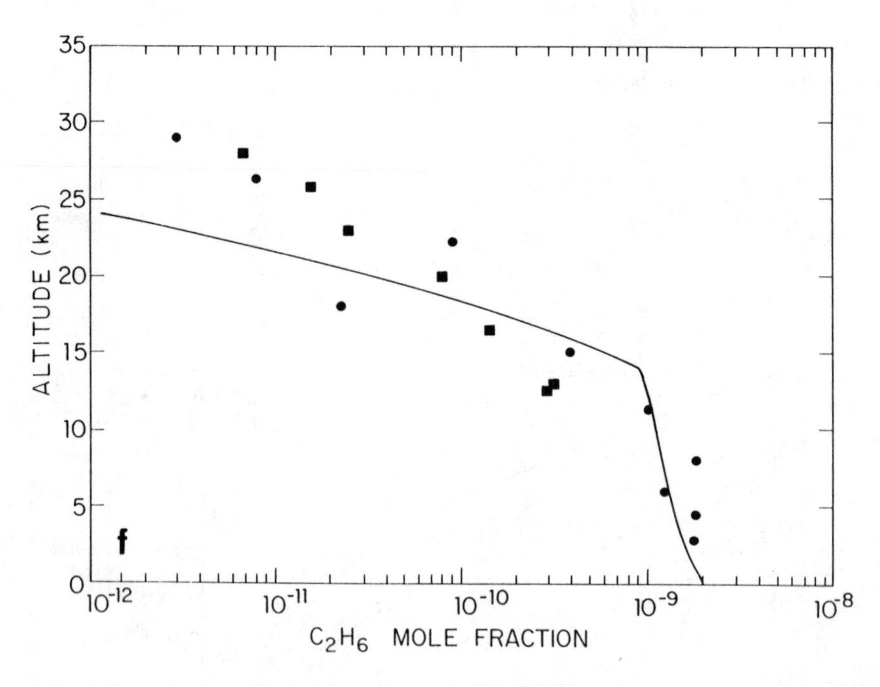

FIGURE C.11 (Continued)

incorrect, even in the lower stratosphere. The results in Figures C.10 and C.11 indicate that the two-dimensional and one-dimensional models agree better with one another than with the measurements of $CFCl_3$ and CF_2Cl_2. The models all appear to underestimate the rates for photolytic destruction of these gases in the lower stratosphere, leading to excessive concentrations at all latitudes and to an overestimate of global mean lifetimes. The models consequently predict excessive concentrations of chlorine, at steady state, in response to long-term industrial release. This matter is of considerable interest, and the difficulty cannot be blamed on the restrictive nature of one-dimensional models.

Chemistry at High Latitudes

There is a growing body of evidence that concentrations of NO and NO_2 are sharply reduced at high latitudes in winter (Noxon 1975, 1979, Coffey et al. 1981), although detailed concentration profiles are not available. Present chemical models (both one-dimensional and two-dimensional) predict column abundances of NO and NO_2 that are 3 to 5 times as large as those observed above 45° latitude. Hence in present models ozone is slowly consumed below 30 km during the period when it should be building up to the spring maximum. Simulations carried out by M. Ko and N.D. Sze (Atmospheric and Environmental Research, Inc., personal communication, 1981) confirm that this slow chemistry suppresses the spring maximum in ozone at least for their model. The influence is significant even at midlatitudes and extends into early summer. The discrepancy implies a major defect in our understanding of the chemistry of NO_x species. Detailed in situ measurements at high latitudes are necessary if we are to understand the chemistry of this important region. Data on ClO, NO, NO_2, and HNO_3 would be especially revealing. Laboratory measurements are needed to better define the chemistry of NO_x species at low temperature. One area of interest is the stability of weakly bound species like $ClNO_3$ and HNO_4, which may become major species at cold temperatures and low levels of light (cf. Prather et al. 1979, Fox et al. 1982).

Chemistry of Key Radicals

Atmospheric observations cannot now provide a definitive test for current models. Measurements of key species such as OH and O are lacking below 30 km, and few sets of simultaneous measurements exist for reactive species in the important families O_x, HO_x, NO_x, and Cl_x.

There are some hints that major discrepancies may emerge as better data are obtained. The difference between observed and calculated ClO at 40 km is particularly troubling, since the discrepancy lies near the peak for catalysis by reaction (20).

The distribution of ethane departs from the pattern exhibited by other atmospheric halocarbons and hydrocarbons, in that models predict significantly lower concentrations than observed in the middle stratosphere (see Figure C.11). Since reaction with Cl atoms is a major sink for C_2H_6 in the lower stratosphere, the observations suggest that models may overestimate the concentrations of Cl below 30 km. There are possible discrepancies for several other important species, including NO (see Figure C.3).

SUMMARY STATEMENT

Present models predict lower concentrations of stratospheric OH than models in use during the previous NRC study (NRC 1979). This change reflects new data on rates for reactions between OH, HNO_3, and HNO_4, which provide important pathways for recombination of odd hydrogen radicals in the lower stratosphere. Reduced estimates for OH concentrations imply sharply lower values for the concentration of ClO and higher values for NO and NO_2 below 35 km. Agreement between model results and observations is significantly improved by using new kinetic data, but several potentially important discrepancies remain.

Models predict that stratospheric ozone should decline by about 6 percent as the stratospheric chlorine concentration increases from current levels (3 ppb) to the asymptotic level (11 ppb) expected from industrial release of chlorofluorocarbons. Most of the ozone reduction is predicted to occur above 30 km, where transport is relatively unimportant. Hence, with current chemistry, results of chlorine perturbation studies are nearly model-independent.

Additions of odd nitrogen to the stratosphere produce relatively large reductions in stratospheric ozone, according to current models. This matter is of some concern since the abundance of atmospheric N_2O (the major precursor of NO_x) is increasing. Ozone reductions due to NO_x are distributed uniformly with altitude, affecting the ozone concentration as low as 20 km. Predictions for these ozone perturbations are quite sensitive to details of the transport mechanisms used in the model.

REFERENCES

Anderson, J.G., H.J. Grassl, R.E. Shetter, and J.J. Margitan (1980) Stratospheric free chlorine measured by balloon-borne in-situ resonance fluorescence. Journal of Geophysical Research 85:2869-2887.

Chapman, S. (1930) A theory of upper atmospheric ozone. Memoirs of the Royal Meteorological Society 3:103-125.

Coffey, M.J., W.G. Mankin, and A. Goldman (1981) Simultaneous spectroscopic determination of the latitudinal, seasonal, and diurnal variability of stratospheric N_2O, NO, NO_2, and HNO_3. Journal of Geophysical Research 86:7331-7342.

Crutzen, P. (1974) A review of upper atmospheric photochemistry. Canadian Journal of Chemistry 52:1569-1581.

Dopplick, T.G. (1972) Radiative heating of the global atmosphere. Journal of Atmospheric Sciences 29:1278-1294.

Fox, J.L., S.C. Wofsy, M.B. McElroy, and M.J. Prather (1982) A stratospheric chemical instability. (To be submitted to Journal of Geophysical Research.)

Goldan, P.D., W.C. Kuster, D.L. Albritton, and A.L. Schmeltekopf (1980) Stratospheric $CFCl_3$, CF_2Cl_2 and N_2O height profile measurements at several latitudes. Journal of Geophysical Research 85:413-423.

Harwood, R.S. and J.A. Pyle (1980) The dynamical behavior of a two dimensional model of the stratosphere. Quarterly Journal of the Meteorological Society 106:395-420.

Horvath, J.J. and C.J. Mason (1978) Nitric oxide mixing ratios near the stratopause measured by a rocket-borne chemiluminescent detector. Geophysical Research Letters 5:1023-1026.

Hudson, R.D. and E.I. Reed, eds. (1979) The Stratosphere: Present and Future. NASA Reference Publication 1049.

Greenbelt, Md.: National Aeronautics and Space Administration; N80-14641-14648. Springfield, Va.: National Technical Information Service.

Hudson, R.D., et al., eds. (1982) The Stratosphere 1981: Theory and Measurements. WMO Global Research and Monitoring Project Report No. 11. Geneva: World Meteorological Organization. (Available from National Aeronautics and Space Administration, Code 963, Greenbelt, Md. 20771.)

Hunten, D.M. (1975) Vertical transport in atmospheres. Pages 59-72, Atmospheres of Earth and the Planets, edited by B.M. McCormac. Boston, Mass.: D. Reidel.

Johnston, H.S., D. Katterhorn, and G. Whitten (1976) Use of excess carbon 14 data to calibrate models of stratospheric ozone depletion by supersonic transport. Journal of Geophysical Research 81:368-380.

Krey, P.W., R.J. Lagomarsino, and L.E. Toonkel (1977) Gaseous halogens in the atmosphere in 1975. Journal of Geophysical Research 82:1753-1766.

Lazrus, A.L. and B.W. Gandrud (1974) Stratospheric sulphate aerosols. Journal of Geophysical Research 79:3424-3430.

List, R.J. and K. Telegadas (1969) Using radioactive tracers to develop a model of the circulation of the stratosphere. Journal of Atmospheric Sciences 26:1128-1136.

Littlejohn, D. and H.S. Johnston (1980) Rate constants for the reaction of hydroxyl radicals and peroxynitric acid. EOS Transactions of the American Geophysical Union 61:966.

Logan, J.A., M.J. Prather, S.C. Wofsy, and M.B. McElroy (1978) Atmospheric chemistry: Response to human influence. Philosophical Transactions of the Royal Society 290:187-234.

Luther, F.M. (1974) Large Scale Eddy Transport. Lawrence Livermore 2nd Annual Report, DOT-CIAP Program, UCRL-51336-74. Livermore, Calif.: University of California Radiation Laboratory.

McConnell, J.C. and W.F.J. Evans (1978) Implications of low stratospheric hydroxyl concentrations for CFM and SST scenario calculations of ozone depletion. EOS Transactions of the American Geophysical Union 59:1078.

McElroy, M.B.(1976) Chemical processes in the solar system: A kinetic perspective. Pages 127-211, MTP International Review of Science, Series 2, Volume 9, edited by D.R. Herschbach. London: Butterworths.

McElroy, M.B. (1980) Sources and sinks for nitrous oxide. Pages 345-364, Proceedings of the NATO Advanced Study Institute on Atmospheric Ozone: Its Variation and Human Influences, edited by A.C. Aiken. October 1-13, 1979. U.S. Department of Transportation Report No. FAA-EE-80-20. Washington, D.C.: Federal Aviation Administration.

McElroy, M.B., J.W. Elkins, S.C. Wofsy, and Y.L. Yung (1976) Sources and sinks for atmospheric N_2O. Reviews of Geophysics and Space Physics 14:143-150.

Miller, C., D.L. Filkin, A.J. Owens, J.M. Steed, and J.P. Jesson (1981) A two dimensional model of stratospheric chemistry and transport. Journal of Geophysical Research 86:12,039-12,065.

Molina, L.T. and M.J. Molina (1981) UV absorption cross sections of HO_2NO_2 vapor. Journal of Photochemistry 15:97-108.

Murgatroyd, R.J. and F. Singleton (1961) Possible meridional circulations in the stratosphere and mesosphere. Quarterly Journal of the Royal Meterological Society 87:125.

National Aeronautics and Space Administration (1981) Chemical kinetic and photochemical data for use in stratospheric modelling. Evaluation Number 4, NASA Panel for Data Evaluation, JPL Publication 81-3. Pasadena, Calif.: Jet Propulsion Laboratory.

National Research Council (1979) Stratospheric Ozone Depletion by Halocarbons: Chemistry and Transport. Panel on Chemistry and Transport, Committee on Impacts of Stratospheric Change, Assembly of Mathematical and Physical Sciences. Washington, D.C.: National Academy of Sciences.

Nelson, H.H., W.J. Marinelli, and H.S. Johnston (1981) The kinetics and product yield of the reaction of OH with HNO_3. Chemical Physics Letters 78:495-499.

Noxon, J.F. (1975) NO_2 in the stratosphere and troposphere measured by ground based absorption spectroscopy. Science 189:547-549.

Noxon, J.F. (1979) Stratospheric NO_2. II. Global behavior. Journal of Geophysical Research 84:5067-5076.

Pierotti, D. and R.A. Rasmussen (1976) Combustion as a source of nitrous oxide in the atmosphere. Geophysical Research Letters 3:265-267.

Prather, M.J., M.B. McElroy, S.C. Wofsy, and J.A. Logan (1979) Stratospheric chemistry: Multiple solutions. Geophysical Research Letters 6:163-164.

Rasmussen, R.A. and M.A.K. Khalil (1981) Atmospheric methane: Trends and seasonal cycles. Journal of Geophysical Research 86:9826-9832.

Ridley, B.A. and D.R. Hastie (1981) Stratospheric odd-nitrogen: NO measurements at 51°N in summer. Journal of Geophysical Research 86:3162-3166.

Ridley, B.A. and H.I. Schiff (1981) Stratospheric odd nitrogen: Nitric oxide measurements at 32°N in autumn. Journal of Geophysical Research 86:3167-3172.

Rowland, F.S. and M.J. Molina (1975) Chlorofluoromethane in the environment. Reviews of Geophysics and Space Physics 13:1-35.

Steed, J.M., A.J. Owens, C. Miller, D.L. Filkin, and J.P. Jesson (1982) Two dimensional modelling of potential ozone perturbations by chlorofluorocarbons. Nature 295:308-311.

Stolarski, R.S. and R.J. Cicerone (1974) Stratospheric chlorine: Possible sink for ozone. Canadian Journal of Chemistry 52:1610-1615.

Sze, N.D. and M.K.W. Ko (1981) The effects of the rate of OH and HNO_3 and $HONO_2$ photolysis on the stratospheric chemistry. Atmospheric Environment 15:1301.

Telegadas, K. and G.J. Ferber (1975) Atmospheric concentrations and inventory of Krypton-85 in 1973. Science 190:882-883.

Turco, R.P., R.C. Whitten, O.B. Toon, E.C.Y. Inn, and P. Hamill (1981) Stratospheric hydroxyl radical concentrations: New limitations suggested by observations of gaseous and particulate sulfur. Journal of Geophysical Research 86:1129-1140.

Weinstock, E.M., M.J. Phillips, and J.G. Anderson (1981) In-situ observations of ClO in the stratosphere: A review of recent results. Journal of Geophysical Research 86:7273-7278.

Weiss, R.F. (1981) The temporal and spatial distribution of tropospheric nitrous oxide. Journal of Geophysical Research 86:7185-7196.

Weiss, R.F. and H. Craig (1976) Production of atmospheric nitrous oxide by combustion. Geophysical Research Letters 3:751-753.

Wine, P.H., A.R. Ravishankara, N.M. Kreutter, R.C. Shah, J.M. Nicovich, and R.L. Thompson (1981) Rate of reaction of OH with HNO_3. Journal of Geophysical Research 86:1105-1112.

Wofsy, S.C. (1978) Temporal and latitudinal variations of stratospheric trace gases: A critical comparison between theory and experiment. Journal of Geophysical Research 83:364-378.

Wofsy, S.C. and M.B. McElroy (1974) HO_x, NO_x and ClO_x: Their role in atmospheric photochemistry. Canadian Journal of Chemistry 52:1582-1591.

Zipf, E.C. and S.S. Prasad (1980) Production of nitrous oxide in the auroral D and E regions. Nature 287:525-526.

Appendix D

THE MEASUREMENT OF TRACE REACTIVE SPECIES IN THE STRATOSPHERE: A REVIEW OF RECENT RESULTS

J.G. Anderson
Harvard University

INTRODUCTION

The central objective of this report is to review critically the data base on trace species observations in the stratosphere for the specific purpose of testing predictions of global ozone depletion resulting from the release of compounds containing chlorine and nitrogen into the lower atmosphere. A corollary objective is to appraise prospects for significant advances in the next five years and to suggest a strategy for that research.

Achieving the first objective in a reasonably concise document must confront the often incompatible elements of data quality, quantity, and applicability to theory. For example, a large body of data may exist on a particular radical that is of demonstrably superior quality with respect to the analytical method, but that, if not taken at the proper time of day and referenced to the local tropopause height, may be uninterpretable in terms of a modeled distribution. We will deal with the sheer volume of information by referencing the recent WMO/NASA report document, "The Stratosphere 1981: Theory and Measurements," whenever possible while attempting to maintain reasonable continuity in this report (Hudson et al. 1982).

The species that are of interest to the stratospheric photochemistry of ozone are divided into groups and listed in Table D.1. The ordering of groups and of the species within each group in the table is rather arbitrary, but the choice seeks to represent the fact that the central objective of this report is an assessment of the effect of fluorocarbon release on stratospheric ozone. Thus, the photochemically active chlorine components are treated first.

TABLE D.1 Chemical Species of Interest in the Stratospheric Chemistry of Ozone

	Species							
Group	1	2	3	4	5	6	7	8
1	ClO	Cl	ClOO	OClO	HCl	HOCl	$ClONO_2$	
2	OH	HO_2	H	H_2	H_2O_2	H_2O		
3	$O(^3P)$	$O(^1D)$	$O_2(^1\Delta)$	$O_2(^1\Sigma)$	O^*_2 (other)	O_3		
4	NO	NO_2	N	NO_3	N_2O_5	$HONO_2$		
5	BrO	Br	BrO_2	OBrO	HBr	HOBr	$BrONO_2$	
6	FO	F	FO_2	OFO	HF			

A review of the data appears first. Then we examine how well the current data base constrains model predictions of ozone reduction. That analysis first summarizes uncertainties in the reaction rate constant data by defining a series of six "cases," tracing the impact of rate constant assumptions on the key free radicals and on the altitude dependence of odd oxygen destruction. The objective is first to correlate each case with the observed vertical distribution of the key free radicals to determine whether a consistent picture evolves, and, second, to identify the altitude regime in which the maximum impact on ozone occurs, resulting from changes in total chlorine or reactive nitrogen.

Finally we abstract from the analysis a series of questions that must be addressed by measurement of trace species in the stratosphere. The answers are essential for significant progress to be realized in the near future. Following each question is an appraisal of the prospects for progress in the next three years.

REVIEW OF DATA BASE ON TRACE SPECIES

Group 1: Reactive Trace Constituents Containing Chlorine

While the case linking fluorocarbons released at the earth's surface to the global distribution of ozone is made up of innumerable elements, the single most important observable in the stratosphere for a first-order appraisal of ozone destruction rates resulting from the decomposition of fluorocarbons is the concentration of the chlorine monoxide free radical, ClO. The reason for this is that ClO is the rate limiting (RL) chlorine constituent in the major catalytic cycles (see the recent discussion by Weubbles and Chang (1981)):

$$\begin{aligned} Cl + O_3 &\rightarrow ClO + O_2 \\ ClO + O &\rightarrow Cl + O_2 \ (RL) \\ \hline O + O_3 &\rightarrow 2O_2 \end{aligned}$$

$$\begin{aligned} Cl + O_3 &\rightarrow ClO + O_2 \\ ClO + HO_2 &\rightarrow HOCl + O_2 \ (RL) \\ OH + O_3 &\rightarrow HO_2 + O_2 \\ HOCl + h\nu &\rightarrow OH + Cl \\ \hline O_3 + O_3 &\rightarrow 3O_2 \end{aligned}$$

$$\begin{aligned} Cl + O_3 &\rightarrow ClO + O_2 \\ NO + O_3 &\rightarrow NO_2 + O_2 \\ ClO + NO_2 + M &\rightarrow ClONO_2 + M \\ \left.\begin{aligned} ClONO_2 + h\nu &\rightarrow NO_3 + Cl \\ NO_3 + h\nu &\rightarrow NO + O_2 \end{aligned}\right\} &(RL) \\ \hline O_3 + O_3 &\rightarrow 3O_2. \end{aligned}$$

ClO has thus been the focus of experimental attention since Molina and Rowland (1974) first linked fluorocarbon release to global ozone reduction. In addition, because ClO dominates the chlorine free radical system with respect to concentration, reaching nearly 1 part per billion (ppb) in the middle to upper stratosphere (its reactive partner, Cl, for example, reaches only 1 part per trillion [ppt] in the stratosphere), it is amenable to a broader class of observational techniques. Four other chlorine-containing constituents are of central importance: HCl, Cl, HOCl, and $ClONO_2$ (with possible isomeric forms).

Chlorine Monoxide (ClO)

Three methods have been successfully applied to the detection of stratospheric ClO (listed here in the chronological order of their application):

1. Balloon-borne in situ resonance fluorescence methods (Anderson et al. 1977, 1980; Weinstock et al. 1981).
2. Ground-based millimeter(mm)-wave emission spectroscopy of the ClO total column at 204 GHz (Parrish et al. 1981).

3. Balloon-borne, mm-wave emission spectroscopy of ClO at 204 GHz (Waters et al. 1981). Aircraft-borne observations by this group had previously established an upper limit on stratosphere ClO (Waters et al. 1979).

A fourth method, that of balloon-borne, laser heterodyne radiometry (see Menzies 1978, Menzies et al. 1981), has been applied to the problem, but ambiguities in spectral line position prevent an interpretation of the results.

While a clear consensus on several aspects of the stratospheric ClO distribution has not emerged, the last two years have witnessed several crucial steps toward a first-order understanding of [ClO] (where square brackets indicate concentration) at middle latitudes.

We consider first results from the two balloon-borne techniques that provide a direct determination of the altitude dependence of [ClO]. Figure D.1 summarizes 10 observations reported by Weinstock et al. (1981) obtained using method 1. All observations contained in Figure D.1 represent midday conditions at 32°N latitude; variations in solar zenith angle primarily reflect changes in solar declination.

The in situ observations fall into two classes; 8 of the 10 define an envelope with deviations limited to about ±50 percent about the observed mean; two of the observations, both obtained in July, fall clearly outside of the envelope and are not representative of the mean distribution of ClO at middle latitudes. Without independent substantiation, the two July observations cannot be included in the data base defining the mean distribution of ClO.

In Figure D.2, the envelope of in situ observations is superposed with the recent balloon-borne observations of Waters et al. (1981) using mm-wave emission techniques. Included in the in situ array is an observation (June 1, 1978) not included in the Weinstock et al. (1981) publication because it was obtained using an instrument with no previous flight history; the results are not at variance and are included for completeness.

The consistency in both absolute magnitude and gradient between the two techniques is one of the most important results to be achieved since the last NRC report (NRC 1979). It underscores the importance of using independent techniques to cross-calibrate observational methods for all of the key radicals involved directly in processes that control the rate of odd oxygen destruction.

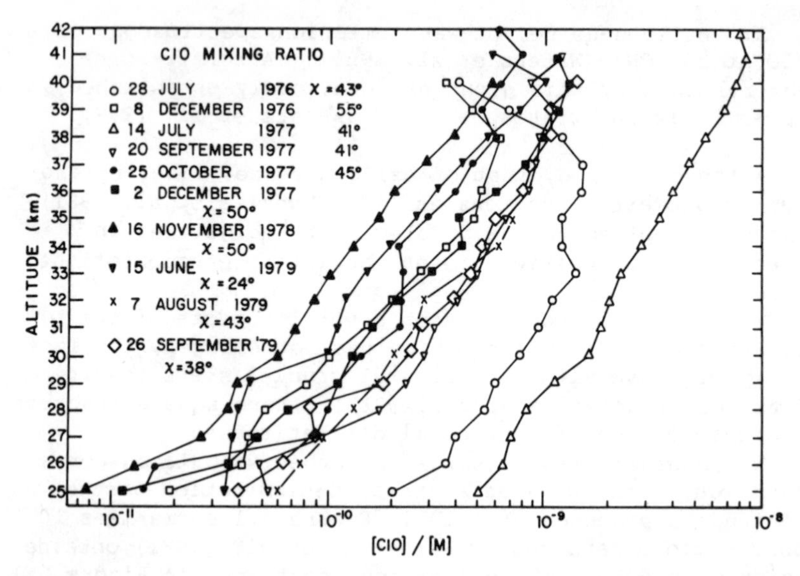

FIGURE D.1 Summary of the vertical distribution of ClO obtained between July 28, 1976, and September 26, 1979, using in situ resonance fluorescence methods (from Weinstock et al. 1981).

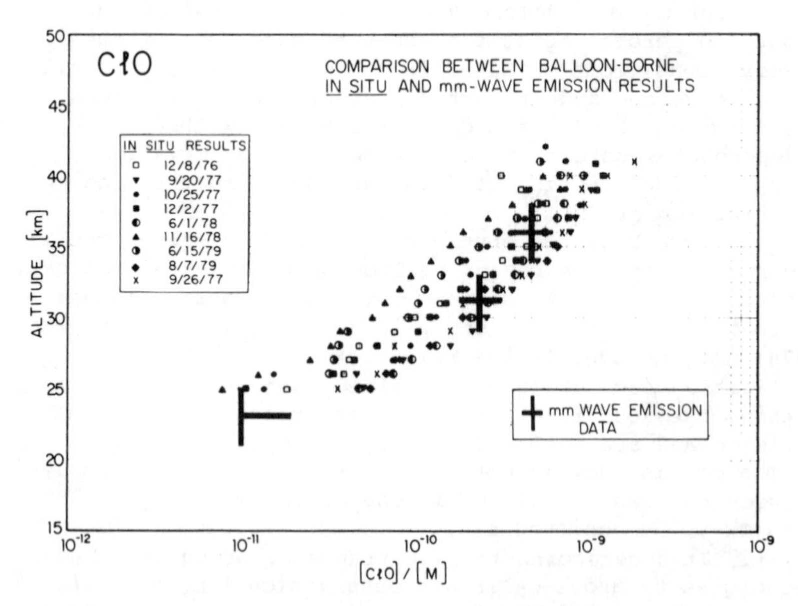

FIGURE D.2 Comparison between balloon-borne in situ and mm-wave emission observations of ClO (from Weinstock et al. 1981, Waters et al. 1981).

It also should be pointed out that while the envelope of ClO data appears to be rather well defined, the dispersion about the mean exceeds ±50 percent; the cited experimental uncertainty is ±30 percent. As we will see, when the results are applied to the problem of constraining model-predicted ozone reduction, this dispersion constitutes a serious impediment. In anticipation of that fact, we represent the nine in situ observations from Figure D.2 in a somewhat different way. Figure D.3 displays a composite of the data converted to absolute concentration to eliminate the steep gradient, and in each frame a single profile is highlighted against the background array. The variety in profile shape is significant, with clear evidence of vertical structure on the order of 2 km in some cases, but nearly absent in others. In addition, the top-side shape of [ClO] exhibits significant variation.

We summarize next the results recently reported by Parrish et al. (1981) using the ground-based, mm-wave emission technique noted earlier (method 2), which were obtained between 10 a.m. and 4 p.m. on 17 separate days (between January 10, 1980, and February 18, 1980) at 43°N latitude from the Five College Radio Astronomy Observatory, Amherst, Massachusetts. Such ground-based observations, which employ purely rotational transitions, are affected by collisional (pressure) broadening by approximately 4 MHz/mb at stratospheric pressures. This is both a blessing, in that low-resolution altitude information can be extracted from the emission line shape, and a curse, in that one must have a first-order estimate of the shape of the emitting layer in order to obtain the absolute column concentration for the observed brightness temperature as a function of frequency. In practice, however, the balloon-borne observations have provided the information on the layer shape, and thus absolute column measurements can be extracted. It should be noted, however, that even without knowledge of the shape of the emitting layer, some information on absolute concentration can be extracted.

Parrish et al. (1981) have taken the mean of seven in situ profiles, specifically those appearing in the envelope of Figure D.2, excluding the last profile obtained on September 26, 1979, and the June 1, 1978, data (which do not alter the conclusions to be drawn), scaled those results by 0.8, integrated the signal that would have resulted, and then overlayed that profile with the observed brightness as a function of frequency. The

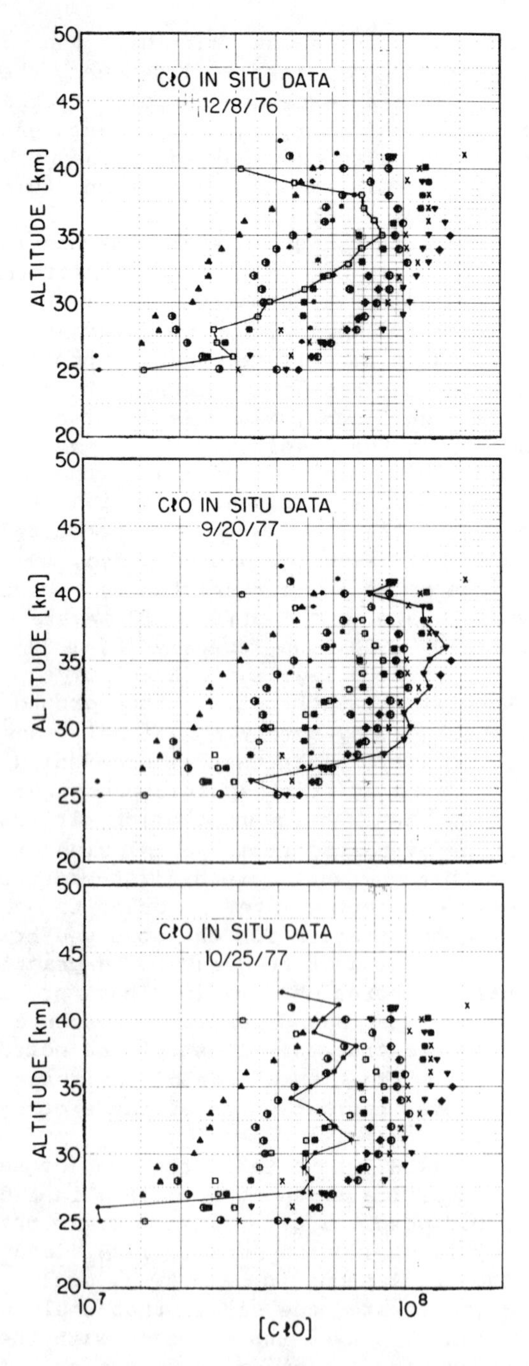

FIGURE D.3a Composite of the ClO profiles, 12/8/76, 9/20/77, and 10/25/77.

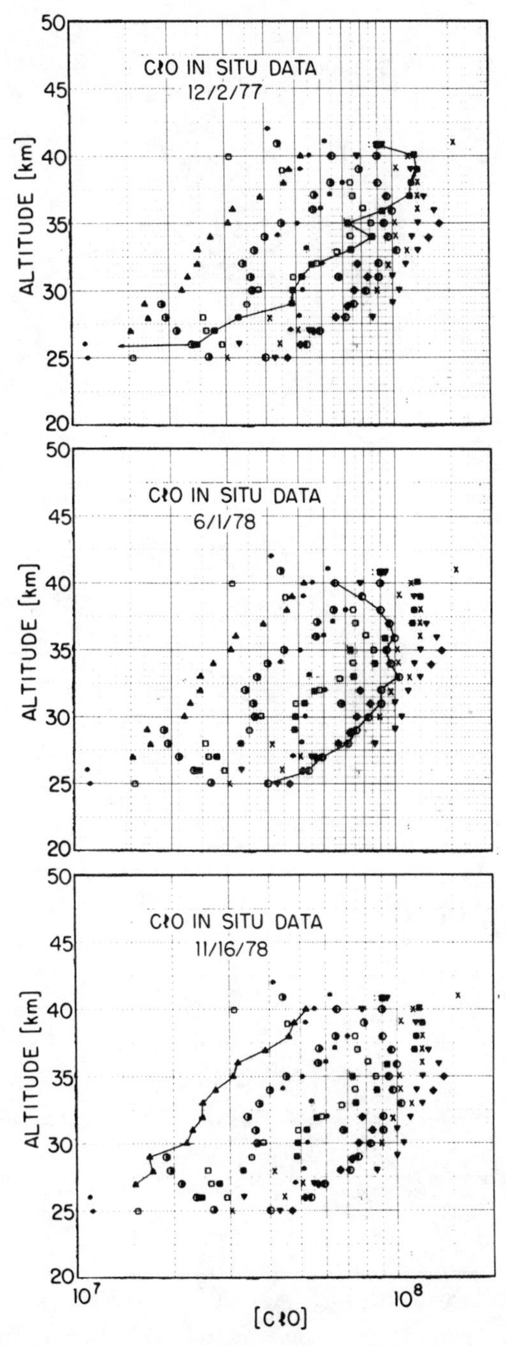

FIGURE D.3b Composite of the ClO profiles, 12/2/77, 6/1/78, and 11/16/78.

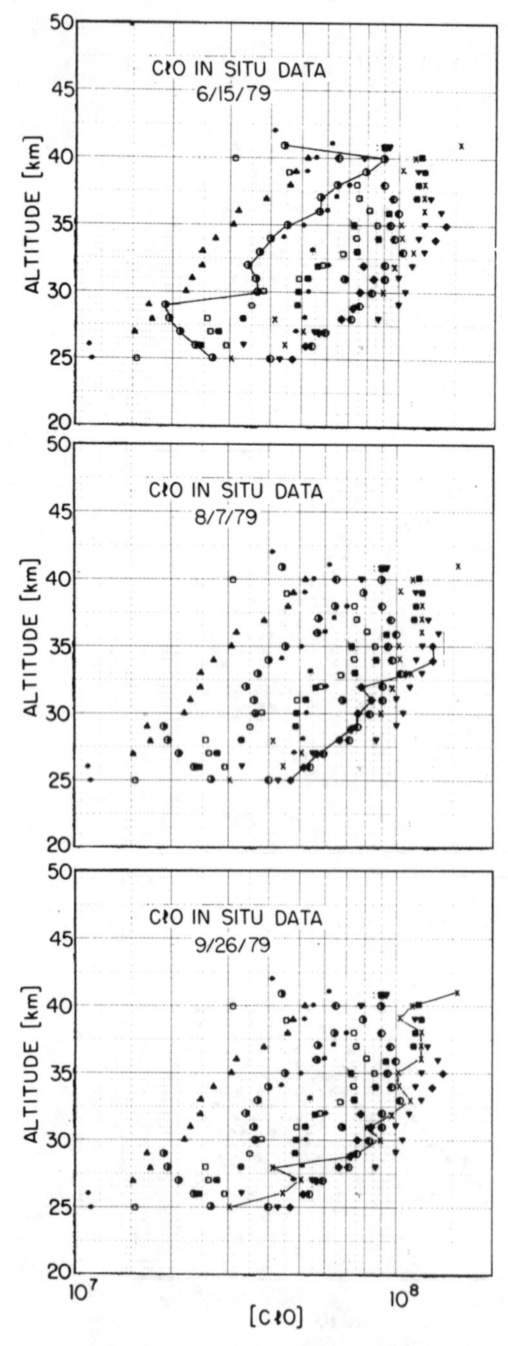

FIGURE D.3c Composite of the ClO profiles, 6/15/79, 8/7/79, and 9/26/79.

results are shown in Figure D.4. The first conclusion to be drawn is that substantial agreement exists with respect to absolute magnitude, since both techniques quote uncertainties of greater than or equal to 25 percent. However, it must be noted that the ground-based observations were done at a latitude 10° northward of the balloon measurements, and are confined to a relatively short period of time in midwinter. A broader data base and observations done in the same latitude band are clearly needed. Parrish et al. (1981) report that no single day of observation exceeded the average by more than a factor of 2.5, and tentative evidence for variations on the order of a factor of 2 in total ClO column density occurred on a time scale of a few days.

An inspection of Figure D.4 indicates a point of major importance: The mm-wave, emission line shape is consistent with the distribution determined by both balloon-borne techniques.

The ability of the ground-based observations to discriminate among the available model calculations is demonstrated in the three panels of Figure D.5. These figures compare the line shape that would be observed for three modeled cases: Case (a) with a mixing ratio of 2.7 ppb for total chlorine, a chemical reaction scheme comparable to that used for the previous NRC report, and an elevated stratospheric water vapor mixing ratio of 8 ppm (uniform from troposphere to stratosphere, as discussed in Logan et al. (1978)); Case (b) with 2.6 ppb for total chlorine and a "normal" mixing ratio for H_2O of 5 ppm (see Sze and Ko 1981); and, finally, Case (c) with 1.3 ppb for total chlorine and 5 ppm H_2O (see Crutzen et al. 1978). The point is not that those ground-based observations cast new light on the selection of a preferred combination of total chlorine and water; the determination of total chlorine (Berg et al. 1980) and H_2O (see Kley et al. 1980) had established that point. Rather, the line shape resulting from the calculated distribution of ClO using chemistry consistent with the previous NRC report (Case a) is distinctly broader than that observed by the mm-wave method. This reflects the larger concentration of ClO calculated by the model at lower altitudes in the stratosphere.

A reasonably thorough discussion of the experimental uncertainties associated with each of the methods discussed above appears in Chapter 1 of Hudson et al. (1982).

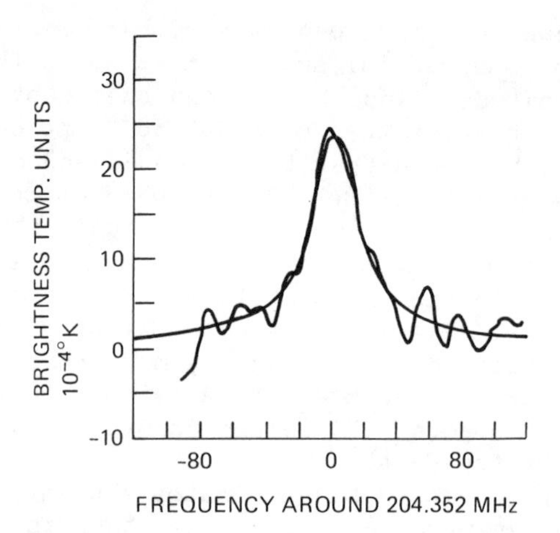

FIGURE D.4 An overlay of the ground-based mm-wave emission data of Parrish et al. (1981) and the signal that would result from an integral of the mean of the balloon-borne in situ observations multiplied by 0.8. The mean was taken excluding the July 28, 1976, and July 14, 1977, in situ ClO profiles.

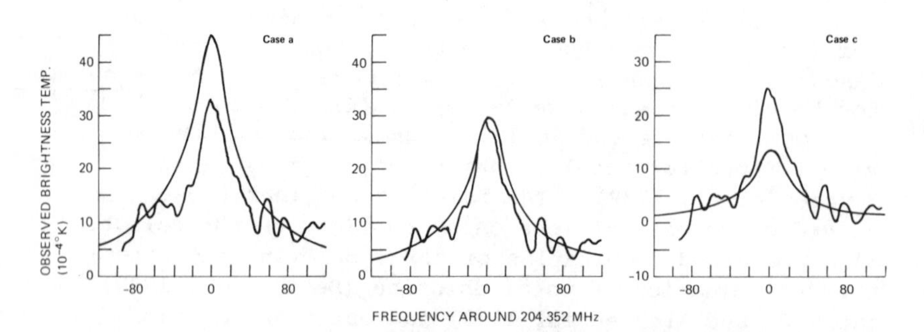

FIGURE D.5 A comparison between the ground-based mm-wave emission data of Parrish et al. (1981) and three modeled predictions: Case a from Logan et al. (1978) with 8 ppm H_2O throughout the stratosphere; Case b with 5 ppm H_2O and 2.3 ppb total chlorine from Sze and Ko (1981); and Case c for 5 ppm H_2O and 1.3 ppb total chlorine from Crutzen et al. (1978).

Chlorine (Cl)

There have been no further measurements of atomic chlorine since those reported by Anderson et al. (1977), which were noted in the last NRC report. However, the ratio of [Cl]/[ClO] was explicitly discussed in a recent paper (see Anderson et al. 1980); those results are summarized in Figure D.6.

A more complete discussion of atomic chlorine appears in Hudson et al. (1982), along with a detailed critique of experimental uncertainties. There are several reasons for the paucity of the data in this important area. The first is that attention has been focused on its rate limiting partner, ClO, and the second is that the exceedingly low concentrations of Cl make the observations exceedingly difficult. It is critical that progress be made in the study of atomic chlorine, particularly in conjunction with studies of ClO and HCl.

Chlorine Dioxide Radical (ClOO)

There have been no reported observations of the radical ClOO in the stratosphere, and there have been no concerted attempts to observe it. Because of its large cross section for photolysis, it is expected to exist at extremely small concentrations, well below currently available detection techniques.

Symmetrical Chlorine Dioxide (OClO)

The more stable form of chlorine dioxide has not been observed and at predicted concentrations of 10 to 100 cm^{-3} in the stratosphere will probably so remain in the foreseeable future.

Hydrochloric Acid (HCl)

Table D.2 summarizes the partitioning among the various chlorine compounds given our current understanding of the reactions that govern the chemical exchange of these constituents. The budget is clearly dominated by [HCl], and it has thus rightfully received a considerable amount of attention in experiments.

TABLE D.2 Partitioning Among Chlorine Compounds (Concentrations Correspond to Case 6 as Described in the Text)

$\Sigma = [HCl] + [ClO] + [ClONO_2] + [HOCl] + [Cl]$

Altitude	HCl/Σ	ClO/Σ	$ClONO_2/\Sigma$	$HOCl/\Sigma$	Cl/Σ	Σ
50	0.97	0.016	8.44×10^{-8}	0.002	0.007	0.4323×10^{8}
45	0.92	0.056	6.27×10^{-6}	0.01	5.3×10^{-3}	0.8258×10^{8}
40	0.78	0.15	0.0007	0.05	2.2×10^{-3}	0.16477×10^{9}
35	0.69	0.17	0.02	0.09	6.4×10^{4}	0.3332×10^{9}
30	0.72	0.10	0.10	0.06	1.4×10^{-4}	0.6692×10^{9}
25	0.83	0.03	0.11	0.01	3.0×10^{-5}	0.1164×10^{10}
20	0.95	0.008	0.034	0.001	7.5×10^{-6}	0.1098×10^{10}
15	0.98	0.004	0.008	0.001	2.9×10^{-6}	0.6454×10^{9}
10	0.98	0.004	0.001	0.009	2.07×10^{-6}	0.3717×10^{9}

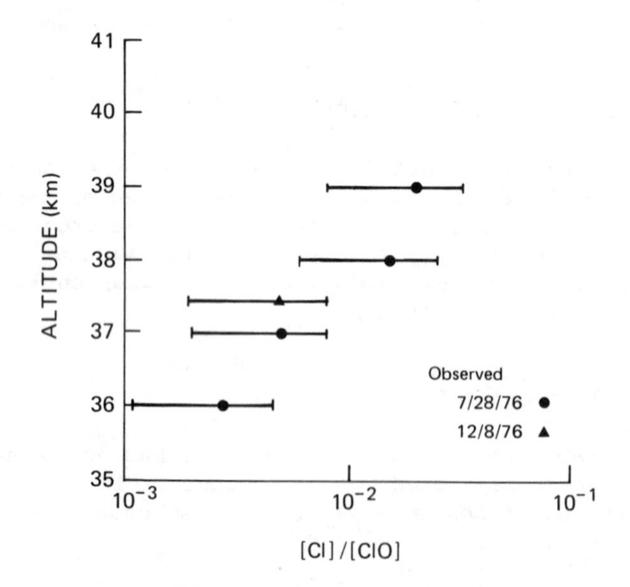

FIGURE D.6 Comparison between the observed and the calculated ratio of atomic chlorine to ClO (from Anderson et al. 1980).

Four remote sensing techniques and one in situ method have been employed from balloon platforms for the detection of HCl in the stratosphere. These include (1) high-resolution, middle infrared solar absorption (see, for example, Farmer et al. 1980, Zander 1980, Buijs et al. 1980); (2) mid-infrared emission (Bangham et al. 1980); (3) pressure-modulated infrared radiometry (Eyre and Roscoe 1977); and (4) far-infrared emission measurements with Fourier transform techniques (Chaloner et al. 1978). The only in situ method used thus far is the base impregnated filter collection method of Lazrus et al. (1977).

Results from those five data sets can best be summarized in two groups. First, the high-resolution middle infrared absorption data obtained by five independent research groups constitute a consistent data set that is reviewed in Figure D.7.

The uniformity of these middle IR results is not reflected in the survey of the other four methods, the results of which are reviewed in Figure D.8. In particular, the base impregnated filter data of Lazrus, which should provide an upper limit on [HCl], since any acidic chlorine compound should be collected, indicate significantly lower concentration in the critical 25- to 35-km altitude region. On the other hand, the pressure-modulated radiometer data lie considerably above the IR absorption data in the altitude region above 22 km. The far-infrared result of Traub lies below the band of middle IR absorption data, but well within the scatter of the results shown in Figure D.8.

Given the paramount importance of HCl in the total chlorine budget, and thus the need to understand in detail the distribution of HCl throughout the stratosphere, it is essential that those discrepancies be eliminated.

Hydrogen Oxychloride (HOCl)

Although HOCl has been searched for in the library of middle infrared sunset absorption data at 1238 cm^{-1} (D.G. Murcray, University of Denver, personal communication, 1981), no observable absorption has been found. This corresponds to an upper limit of 10 ppb at 25 km, which is approximately a factor of 1000 above current model predictions.

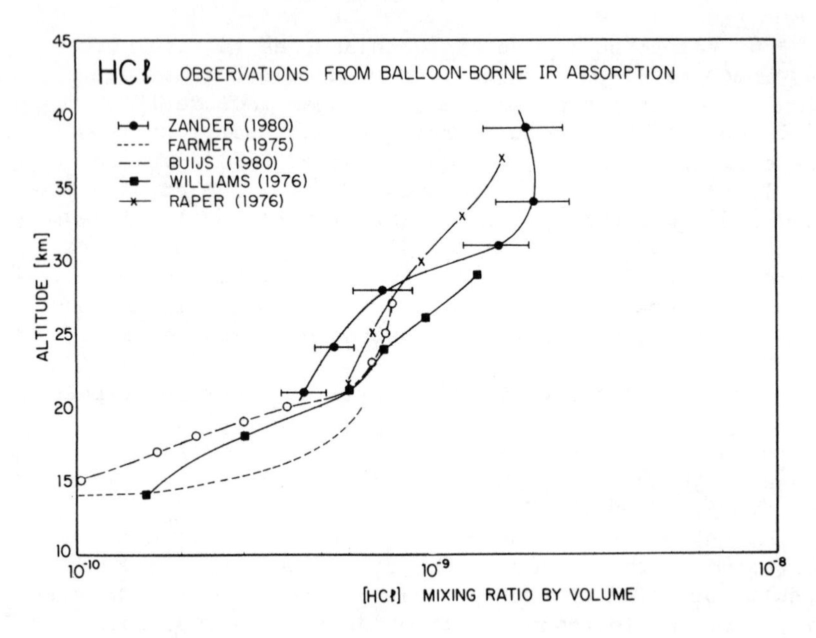

FIGURE D.7 HCl observations from balloon-borne IR absorption spectroscopy.

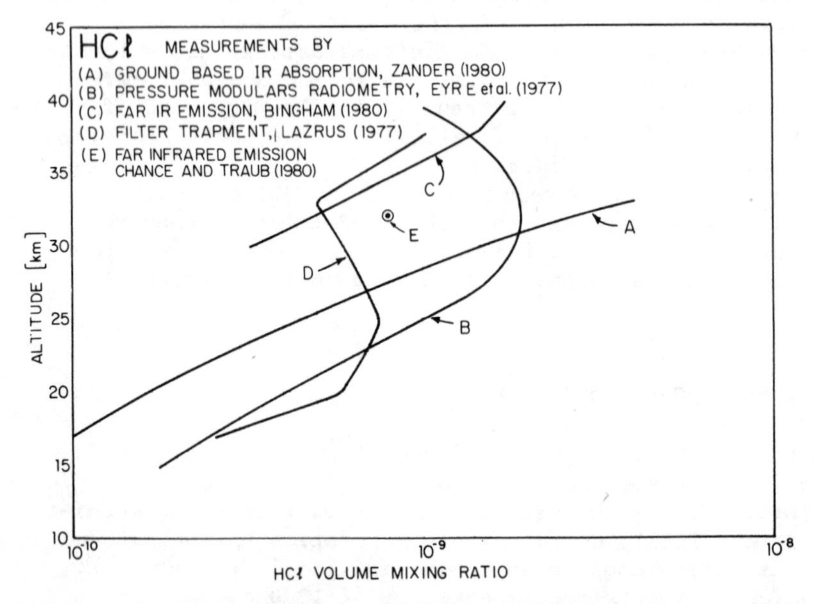

FIGURE D.8 HCl measurements by ground-based spectroscopy, pressure-modulated radiometry, far IR emission, and in situ filter collection.

The importance of HOCl to our understanding of stratospheric chlorine chemistry results from the fact that it is formed by the reaction,

$$HO_2 + ClO \rightarrow HOCl + O_2$$

and is thus a test of the coupling between the hydrogen and chlorine families. An unambiguous determination of its vertical concentration profile at low or middle latitudes would be of very significant value.

Chlorine Nitrate ($ClONO_2$)

A single observation of chlorine nitrate has been reported (see Murcray et al. 1979). As discussed in the WMO/NASA report (Hudson et al. 1982), the measurement is an exceedingly difficult one, given the broad nature of the $ClONO_2$ absorption feature and interferences from absorption bands of N_2O, CH_4, and H_2O that mask the chlorine nitrate feature. A review of these factors led the WMO/NASA panel to conclude that an upper limit of 1 ppb for $ClONO_2$ between 25 and 35 km was a defensible position at this time. Since the upper limit falls above current model predictions for chlorine nitrate in this altitude interval, the measurement cannot be used to establish whether isomers other than $ClONO_2$ are found in the recombination reaction,

$$ClO + NO_2 + M \rightarrow ClONO_2$$
$$\rightarrow \text{other isomeric forms.}$$

Group 2: Reactive Trace Species Containing Hydrogen

The Hydroxyl Radical (OH)

Hydroxyl has been observed in the stratosphere by four independent techniques noted in the chronological order of their application:

1. Solar flux induced resonance fluorescence observed by a rocket-borne spectrophotometer (Anderson 1971, 1975), which provides a local concentration measurement by determining the change in total column emission rates as a function of altitude.

2. Balloon-borne in situ molecular resonance fluorescence using a plasma discharge resonance lamp to induce fluorescence. The fluorescence chamber is lowered through the stratosphere on a parachute to control the altitude and velocity of the probe (Anderson 1975, 1980).

3. Ground-based high-resolution solar absorption by a PEPSIOS (Poly-Etalon Pressure Scanned Interferometer) instrument, which resolves a single rotational line in the (0-0) band of OH at 309 nm. The total column density of terrestrial OH between the instrument and the sun is observed and is dominated by the altitude interval of 25 to 65 km (Burnett 1976, 1977; Burnett and Burnett 1981).

4. Balloon-borne laser-induced detection and ranging (LIDAR) in which a pulsed laser system coupled to a telescope is used to observe the backscattered fluorescence from OH. The laser is tuned to the (0-1) band of the A-X transition at 282 nm and the fluorescence at 309 nm (the 0-0 band) is observed as a function of time following the laser pulse (Heaps et al. 1981).

Four methods have been employed for the detection of tropospheric OH:

1. Aircraft-borne laser-induced fluorescence wherein a contained atmosphere sample is passed through an enclosed detection chamber and is probed by a pulsed laser tuned to the (0-1) band of the A-X transition at 282 nm. Fluorescence is observed at 309 nm (Davis et al. 1976, 1979).

2. Aircraft-borne laser-induced fluorescence using an "open" optical arrangement in which a telescope is used to observe the backscattered fluorescence outside the boundary layer of the fuselage, but in the near vicinity of the aircraft (Wang et al. 1981).

3. Measurements of carbon 14 labeled CO oxidation rates by OH in which the sample is drawn into a Teflon-coated vessel of 10-liter volume. All reported observations were taken in the boundary layer (Campbell et al. 1979).

4. Long path (7.8 km) absorption of laser radiation at 308 nm (the Q(2) line of the $A^2\Sigma^+$, v = 0, $X^2\Pi$ v = 0 transition). The experiment employs a double pass (3.9 km per leg) optical arrangement in which the beam is returned by a spherical mirror to a double monochrometer located at the laser (Perner et al. 1976).

Although the subject of tropospheric OH is one of central importance to the photochemical structure of the atmosphere, it cannot be dealt with in adequate detail in this document. We extract from the above work, and from tropospheric lifetime studies of methyl chloroform (see Logan et al. 1981 and references therein) that the tropospheric contribution to total column OH does not exceed 5×10^{12} cm^{-2} and is thus a negligible contribution to the total OH column density measured from the ground.

Restricting the discussion to stratospheric OH, we first review the comparison between the in situ observations and the ground-based total column measurements; second, we summarize a considerable body of new information taken from the recent ground-based observations of Burnett and Burnett (1981).

For the purpose of summarizing the in situ results, Figures D.9, D.10, and D.11 present a three-panel display of (a) the upper stratosphere-mesosphere rocket data from Anderson (1975); (b) the stratosphere balloon data using in situ resonance fluorescence (Anderson 1980); and (c) a composite of the two data sets with an upper limit on the mean tropospheric OH concentration taken from the methyl chloroform lifetime studies and the tropospheric laser experiments noted above.

We note several features of the profile that will be referred to throughout this section. First, the total column concentration of OH determined from an integral of the in situ observations and an estimate of the upper mesospheric profile is 6.9×10^{13} cm^{-2}. The fractional contribution to this figure for each 15 km interval between 0 and 90 km is given in Table D.3. Second, the altitude interval over which the balloon and rocket data extend, 30 to 70 km, encompasses all but 13 percent of the total column concentration so that ground-based observations provide an excellent cross check on the absolute concentration determined in situ. Third, within the region between 30 and 70 km, the dominant source is $O(^1D) + H_2O \rightarrow 2OH$, and the dominant loss is $OH + HO_2 \rightarrow H_2O + O_2$. Thus the in situ observations and the comparison between the integrated in situ data and the total column observations relate primarily to the balance between these two reactions and do not involve in a sensitive way the question of OH reactions with nitric acid and pernitric acid.

Although the ground-based data will be discussed in detail in the remainder of this section, we extract from

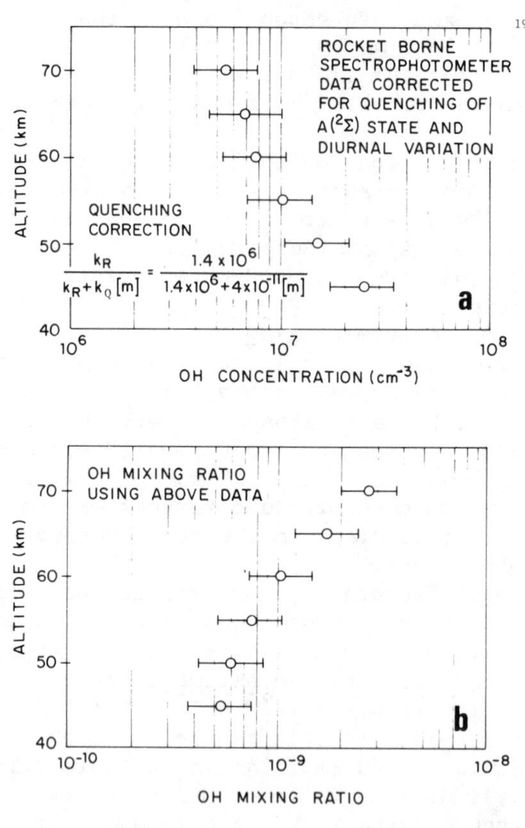

FIGURE D.9 (a) Concentration and (b) mixing ratio of OH in the upper stratosphere and mesosphere obtained by rocket-borne spectrophotometer (Anderson 1975). Data are corrected for collisional deactivation (see German 1975, 1976) and for diurnal behavior (see Logan et al. 1978). Zenith angle at time of rocket flight was 86°; data are corrected to midday steady state condition.

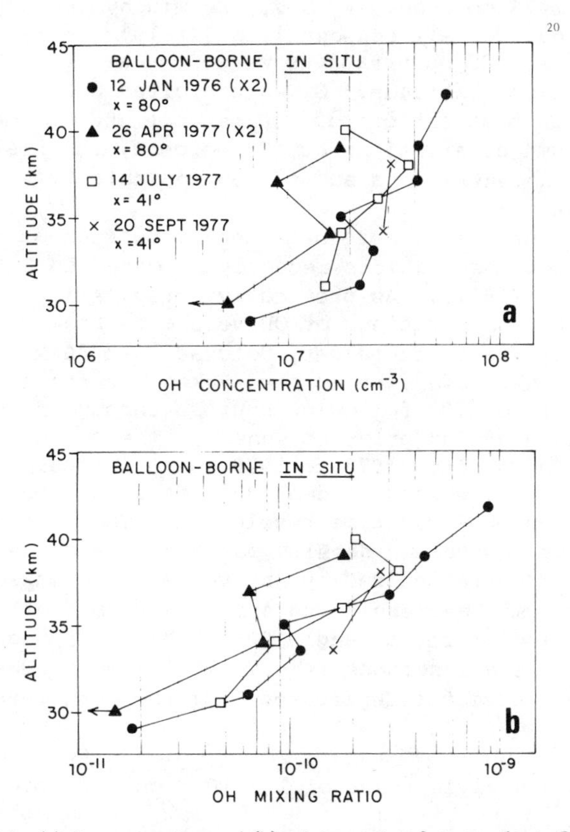

FIGURE D.10 (a) Concentration and (b) mixing ratio of stratospheric OH obtained in situ by molecular resonance fluorescence within a chamber lowered through the stratosphere at a controlled velocity on a parachute (Anderson 1980).

that discussion the key quantity for comparison with the in situ data, the total column density of OH observed by the PEPSIOS and reported in Burnett and Burnett (1981). Based on a total of 270 observing days extending from December 1976 to December 1979, the midday abundance of OH averaged over all seasons is 5.7×10^{13} cm^{-2}. All reported PEPSIOS observations were done at Fritz Peak, Colorado, 40°N latitude. Given the cited uncertainty of the in situ observations of ±30 percent and of the total column observations of ±25 percent, the observed absolute concentrations summarized in Table D.4, are consistent.

We turn next to a more detailed discussion of the ground-based observations recently reported in Burnett and Burnett (1981). As previously noted, all ground-based observations of OH were made from Fritz Peak Observatory, west of Boulder, Colorado, at 40°N latitude. All observations were taken between a solar zenith angle of 70° following sunrise through noon to a zenith angle of 70° prior to sunset. The period of observation was from 1976 to 1979, with a total of 270 observing days, which yielded 900 data sets with equal to or less than one-hour time resolution. The diurnal behavior of the column density was fit to a curve in sec χ, which is characterized by an overhead sun maximum of 7.1×10^{13} cm^{-2} decreasing to 4.9×10^{13} cm^{-2} at sec $\chi = 2$ (solar zenith angle 60°). Midday abundance averaged over all seasons is 5.7×10^{13} cm^{-2}. The following systematic departures from the mean were observed:

1. An annual increase of 1×10^{13} cm^{-2} in total column.
2. A gradual decrease of about 25 to 30 percent between spring and fall.
3. Diurnal oscillation observed with systematic changes of 30 to 40 percent that show a clear solar flux dependence on both a diurnal and an annual basis.

The observed and predicted diurnal behavior of total column OH with respect to shape and absolute magnitude is summarized in Figure D.12. A representative data set is also shown in Figure D.13, indicating the scatter about the mean.

Burnett and Burnett (1981) briefly discuss both an annual and a seasonal departure from the reported mean values. We consider first the observed year-to-year trends.

TABLE D.3 Contribution of Each Altitude Interval to the Integrated Column

Altitude Interval	Contribution to Total Integral	Fraction of Total Integral	Integrated Column Density
0-5	1.5×10^{12}	0.02	
15-30	3.4×10^{12}	0.05	
30-45	3.2×10^{13}	0.46	6.9×10^{13} cm^{-2}
45-60	1.9×10^{13}	0.28	
60-75	8.8×10^{12}	0.13	
75-90	4.3×10^{12}	0.06	

TABLE D.4 Summary of the Comparison Between the Integrated *In Situ* Results from Balloon and Rocket Data and the Ground-Based Total Column Observation

Composite of the *In Situ* OH Data	Ground-Based Total Column OH
6.9×10^{13} cm^{-2}	5.7×10^{13} cm^{-2}
Uncertainty: ±40 percent	Uncertainty: ±25 percent
Conditions: Midday, 32°N	Conditions: Midday, 40°N

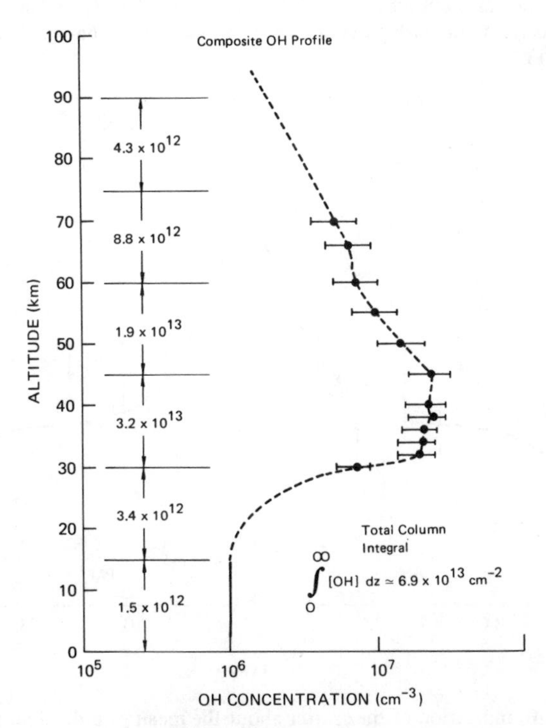

FIGURE D.11 Composite of OH profile.

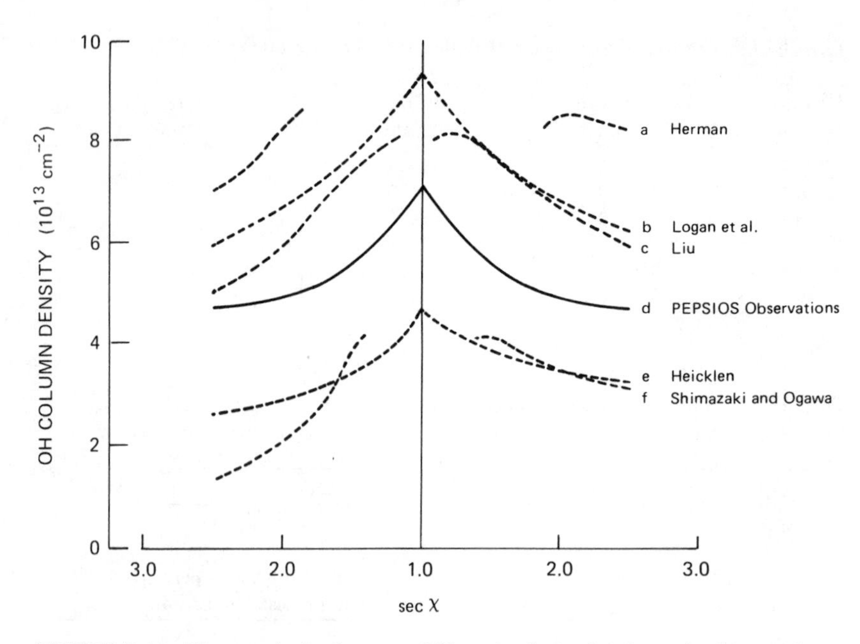

FIGURE D.12 The correlation between OH total column density and solar zenith angle expressed as sec X comparing the observed and modeled behavior (from Burnett and Burnett 1981).

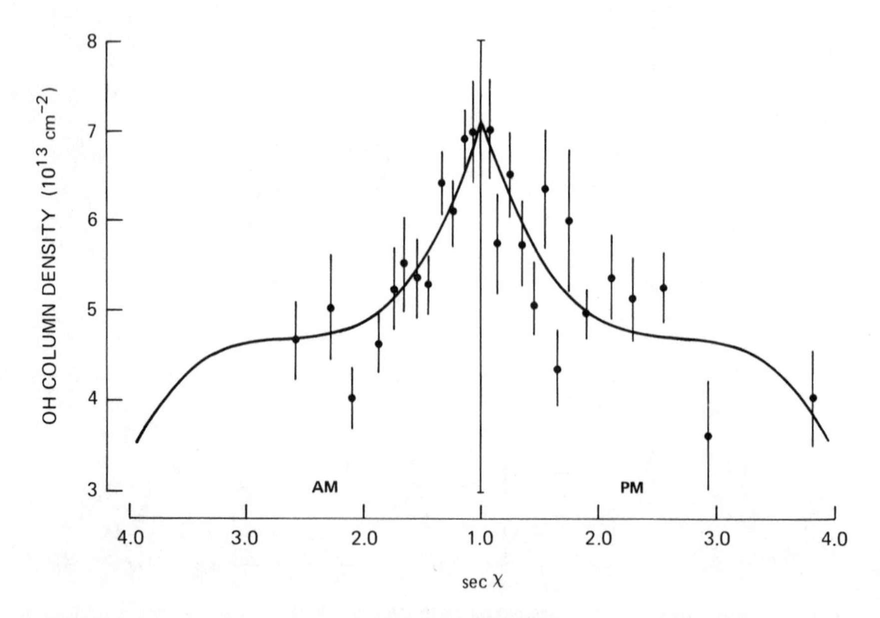

FIGURE D.13 An indication of the scatter about the mean of individual observations taken with the ground-based PEPSIOS (from Burnett and Burnett 1981).

A reanalysis of the December 1976 data (from Burnett 1977) using more advanced methods for baseline determination established $3.1 \pm 0.6 \times 10^{13}$ cm^{-2} for a midday mean. In Figure D.14, the evolution of the monthly mean from December 1976 to December 1979 is summarized. There is an apparent increase of approximately 1×10^{13} cm^{-2} per year during that three-year period. Correlating such an increase to the 11-year solar cycle was suggested by Burnett and Burnett (1981), but such a dramatic change seems difficult to rationalize and will require more extensive data coverage and a far more thorough analysis of solar cycle flux variations with an associated mechanistic hypothesis before it can be accepted.

The seasonal behavior in the total column of OH is a recurring and extremely interesting feature of the data in Figure D.14. There is the clear suggestion of a springtime maximum in OH and a fall minimum. One very important advantage to be gained from a more extensive geographic coverage with such ground-based observations would be an examination of this seasonal behavior as a function of latitude. The correlation of the dependence with other constituents such as O_3, H_2O, NO_2, and ClO would be of significant importance.

An unexpected and as yet unexplained aspect of the ground-based observations involves the appearance, particularly in the summer 1978 data set, of a zenith angle dependent pattern exemplified in Figure D.15. In particular, there is a distinct minimum in the observed column following local noon. The early afternoon decrease shows an abrupt drop of 4.5×10^{13} cm^{-2} with a subsequent increase of 3.5×10^{13} cm^{-2} followed by the conventional decrease into sunset. This feature has lead to a careful critique of the data by the authors, who believe that it is not an artifact of the data reduction or of instrumental performance. The oscillatory behavior shown in Figure D.15 persisted into the late summer of 1978, but was not apparent in the 1979 data set, which was taken with the same instrument at the same site using the same data reduction method. There was also a systematic progression of the position of the maximum and minimum as the 1978 season progressed.

In summary, an analysis of (a) balloon and rocket data on OH in the stratosphere and (b) ground-based total column observations provide the following conclusions:

1. There is substantial agreement among the three techniques; the in situ data provide a consistent picture

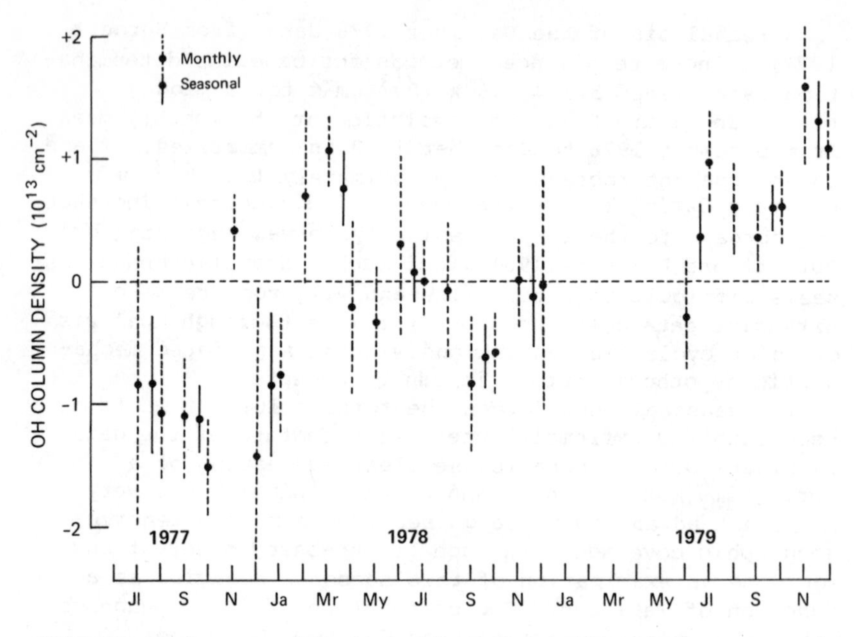

FIGURE D.14 Summary of the observed fluctuations in OH total column measured from the ground.

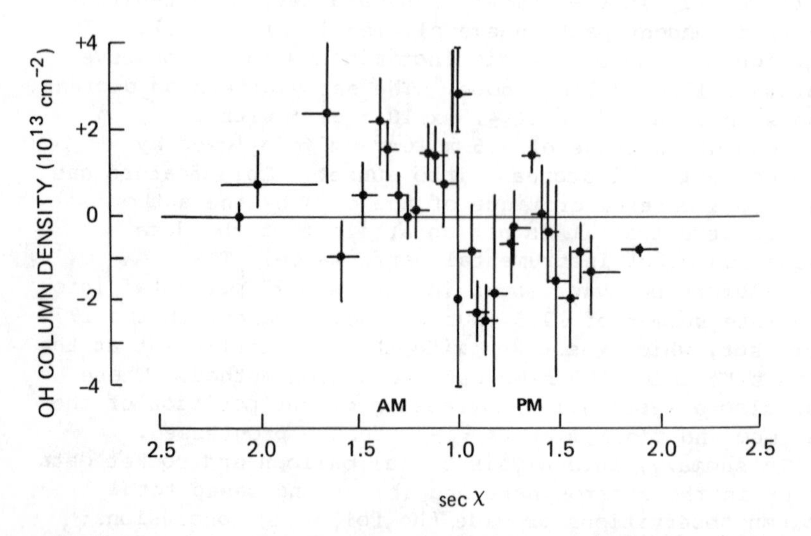

FIGURE D.15 Typical example of the oscillatory diurnal behavior of the OH total column, which was a characteristic signature of the summer 1978 data and does not appear to be an instrumental artifact (from Burnett and Burnett 1981).

of the altitude dependence of [OH] between 30 and 70 km, implying a peak concentration at 40 km of 2.4 x 10^7 cm^{-3} and a total column density at midday of 6.9 x 10^{13} cm^{-2}. The midday total column abundance determined from the ground is 5.7 x 10^{13} cm^{-2}, as summarized in Table D.4.

2. There is a systematic increase of approximately 1 x 10^{13} cm^{-2} per year between December 1976 and December 1979 and a suggestion of a yearly spring maximum and fall minimum. The spring-to-fall decrease is approximately 30 percent.

3. It is clear that knowledge of the OH distribution between 15 and 30 km, which is absent, is of the highest priority. This need results not only from the fact that HO_x becomes an increasingly important component of the odd oxygen destruction rate below 30 km, but also because the photochemical partitioning of chlorine and nitrogen depends currently on the OH concentration.

The Hydroperoxyl Radical (HO_2)

Two techniques thus far have been used for the detection of HO_2 in the stratosphere:

1. Balloon-borne cryogenically captured matrix isolation followed by laboratory detection of HO_2 by EPR methods. The experiment is carried out by drawing a stratospheric sample into an evacuated flask, collecting the sample on a "cold finger" at a given balloon float altitude, closing the flask, and returning it to the laboratory for analysis (Mihelcic et al. 1978).

2. Balloon-borne chemical conversion-molecular resonance fluorescence detection wherein HO_2 is converted to OH by the rapid bimolecular reaction HO_2 + NO → OH + NO_2. The product OH is then detected by molecular resonance fluorescence using a microwave-sustained plasma discharge lamp to induce fluorescence in the (0-0) band of the $A^2\Sigma - X^2\Pi$ transition at 309 nm. Chemical conversion and detection are done within a chamber lowered through the stratosphere at a controlled velocity on a parachute (Anderson 1980, Anderson et al. 1980).

A total of four HO_2 observations have appeared in the literature, one by the matrix isolation technique, and three by the resonance fluorescence method. Those observations are summarized in Table D.5, in chronological order.

As noted in Table D.5, the Mihelcic sample collection was initiated immediately following sunrise at a solar zenith angle of 85°. The conversion to midday for comparison with models and other observations was carried out using the diurnal calculation of Logan et al. (1978). That correction factor is significant--a factor of 2--and attempts to account for the period over which the sample was collected.

The data summarized in Table D.5 are presented graphically in Figure D.16. There is significant scatter evident in those observations that should not be attributed to atmospheric variability until (a) the signal-to-noise ratio of the observations is improved, and (2) simultaneous observations of photochemically related species such as OH or H_2O demonstrate a correlation in concentration fluctuations.

Atomic Hydrogen (H)

There are no reported observations of atomic hydrogen in the stratosphere. Although atomic resonance scattering can detect concentrations of H in the range of 10^5 cm^{-3}, current models predict a distribution shown in Figure D.17 so it is detectable only above 45 km.

Molecular Hydrogen (H_2)

Molecular hydrogen has received careful attention for more than 10 years. The most comprehensive work is that by Ehhalt and coworkers (see Ehhalt et al. 1977, Schmidt 1978, Schmidt et al. 1980). The tropospheric distribution is uniform with a slight interhemispheric asymmetry at the tropopause. The average mixing ratio in the northern hemisphere is 0.576 ppm by volume, and in the southern hemisphere 0.552 ppm.

Hydrogen Peroxide (H_2O_2)

The only reported observation of H_2O_2 is a tentative detection by Waters et al. (1981), which is noted here primarily to indicate that an analytical technique is under development that should provide empirical evidence

TABLE D.5 Summary of Experimental Parameters for the HO_2 Observations

	Altitude					Experimental Uncertainty
	37	35	33	31	29	
Mihelcic et al. (1980), Launch Date 8/8/76, Latitude 53°N, Solar Zenith Angle 85°						factor of 3
Observed NO_2 mixing ratio corrected to midday				1.0×10^{-10} [a]		
Anderson et al. (1980), Launch Date 9/20/77, Latitude 32°N, Solar Zenith Angle 41°						±45%
Resonance fluorescence S_{HO_2}	17	9	13	8	2	
HO_2 mixing ratio	7.1×10^{-10}	1.84×10^{-10}	8.5×10^{-11}	$\leqslant 8.0 \times 10^{-11}$	$< 7.0 \times 10^{-11}$	
Detection threshold	1.1×10^{-10}	8.2×10^{-11}	7.0×10^{-11}	7.6×10^{-11}	6.4×10^{-11}	
Anderson et al. (1980), Launch Date 10/25/77, Latitude 32°N, Solar Zenith Angle 45°						±45%
Resonance fluorescence S_{HO_2}	3	3	9	5	–	
HO_2 mixing ratio	3.6×10^{-10}	8.7×10^{-10}	6.9×10^{-10}	2.1×10^{-10}	1.0×10^{-10}	
Detection threshold	1.6×10^{-10}	1.5×10^{-10}	1.5×10^{-10}	1.0×10^{-10}	1.0×10^{-10}	
Anderson et al. (1980), Launch Date 12/2/77, Latitude 32°N, Solar Zenith Angle 50°						±45%
Resonance fluorescence S_{HO_2}	33	25	20	22	16	
HO_2 mixing ratio	3.4×10^{-10}	2.3×10^{-10}	4.2×10^{-10}	3.4×10^{-10}	1.7×10^{-10}	
Detection threshold	8.1×10^{-11}	6.0×10^{-11}	4.6×10^{-11}	8.8×10^{-11}	8.7×10^{-11}	

[a] Datum taken at an altitude of 31.8 km.

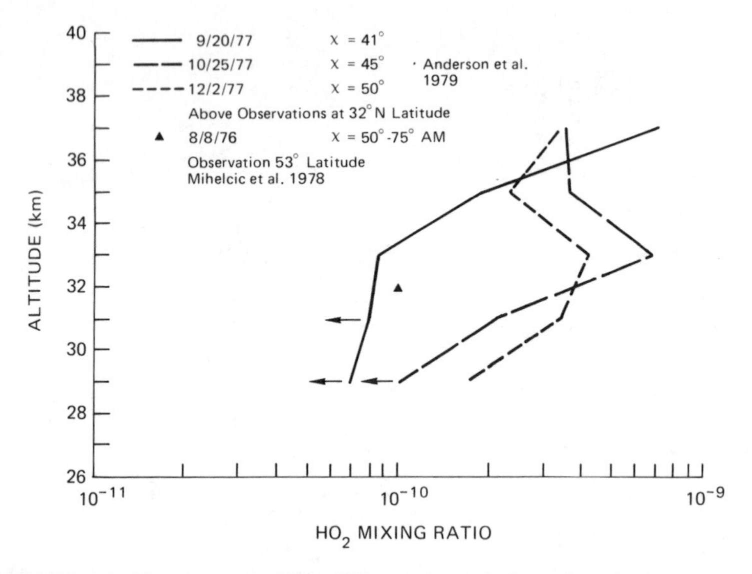

FIGURE D.16 The observed midday HO_2 mixing ratio from three in situ resonance fluorescence balloon flights (chemical conversion to OH followed by resonance fluorescence) and one in situ sample collection experiment (cryogenic sampling with EPR analysis).

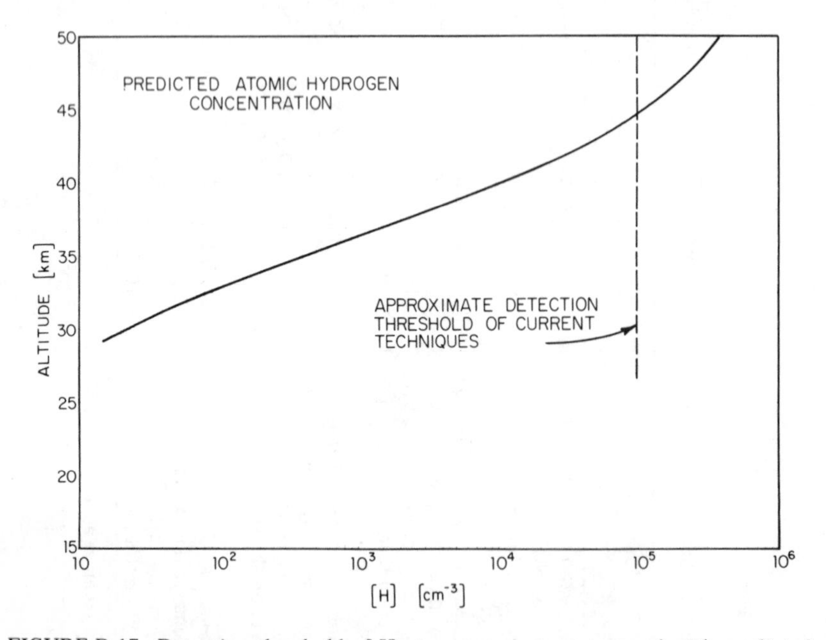

FIGURE D.17 Detection threshold of H atom experiment compared with predicted concentrations. H atom density is only approximate but parallels Cases 5 and 6 described in text.

in the near future (Figure D.18). Given that hydrogen peroxide is principally formed in the reaction

$$HO_2 + HO_2 \rightarrow H_2O_2 + O_2$$

and destroyed by photolysis, it constitutes an important component of HO_x chemistry and observational evidence is clearly needed.

Water Vapor (H_2O)

The subject of stratospheric H_2O is of sufficient size to preclude its treatment in this summary. The topic is, however, treated in considerable detail in Hudson et al. (1982).

Group 3: Oxygen

Atomic Oxygen in the Ground State ($O(^3P)$)

Atomic oxygen, $O(^3P)$, is of particular interest to the photochemistry of stratospheric ozone. First, it is believed to be in strict photochemical steady state with ozone through the rapid exchange reactions,

$$O_3 + h\nu \xrightarrow{J_{O_3}} O(^3P) + O(^3_2\Sigma)$$

$$\rightarrow O(^1D) + O_2(^1\Sigma)$$

$$O(^1D) + M \rightarrow O(^3P) + M$$

$$O(^3P) + O_2 + M \xrightarrow{k_1} O_3 + M$$

such that the ratio

$$[O(^3P)]/[O_3] = J_{O_3}/k_1[M][O_2]$$

should be obeyed throughout the stratosphere. Second, atomic oxygen is the reactive partner with ozone in the direct recombination step,

$$O(^3P) + O_3 \rightarrow O_2 + O_2,$$

which establishes the rate of odd oxygen destruction apart from any catalyzed recombination. Third, atomic oxygen is the odd oxygen reactant in virtually all catalytic rate limiting steps in the middle and upper stratosphere.

There are six reported observations of $O(^3P)$ in the stratosphere, all obtained using balloon-borne, parachute descent, in situ atomic resonance (Anderson 1980). These results are given in Figure D.19.

Several points are readily apparent from Figure D.19. First, there is both local structure within and absolute displacement among observed distributions that exceed, respectively, the precision and accuracy of the measurements. It should also be noted that the local structure does not consistently appear. For example, the profiles observed on October 25, 1977, and December 2, 1977, display a small degree of local structure, typically less than ±20 percent variation over an interval of ±1 km above approximately 34 km. Below that altitude, significantly greater local structure is apparent, though seldom more than ±50 percent. On the other hand, the remaining four observations exhibit at least one example of major (factor of 2) variation over a ±2-km interval with an increasing structural development below the 33- to 35-km interval. Although this local structure makes a detailed profile-by-profile comparison with modeled distributions difficult, a comparison with the mean of the observed $O(^3P)$ distribution can be made. Thus, in Figure D.20, we display the observed mean.

Atomic Oxygen in an Excited State $(O(^1D))$

There are no reported observations of $O(^1D)$ in the stratosphere.

Singlet Delta Molecular Oxygen $(O_2(^1\Delta))$

Although there are currently no known reaction mechanisms that involve $O_2(^1\Delta)$, its predicted concentration, substantiated by a limited number of rocket observations, is such that its number mixing ratio reaches 1 ppm at 50 km (dropping by 2 orders of magnitude in the interval between 50 and 35 km), and it is thus a potentially impor-

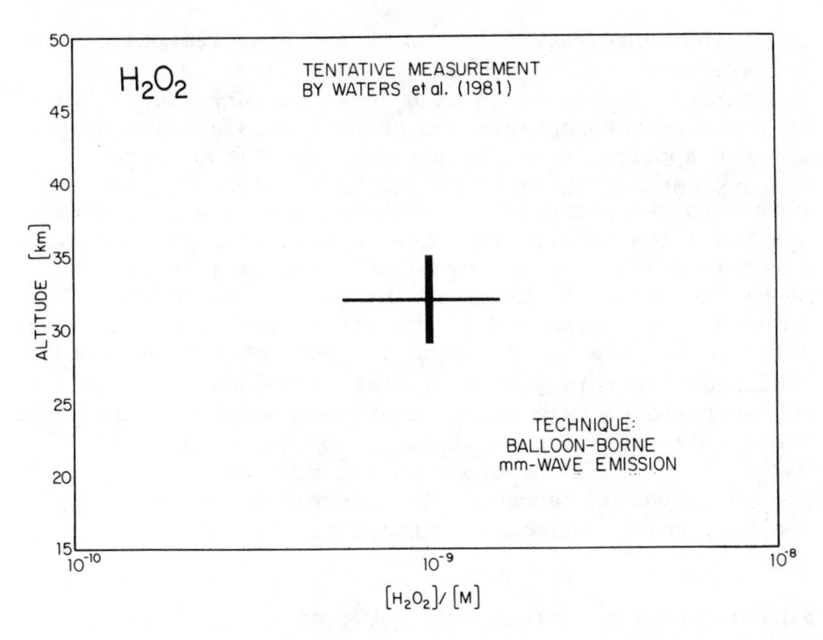

FIGURE D.18 Tentative measurement of H_2O_2 by Waters et al. (1981) using balloon-borne mm-wave emission techniques.

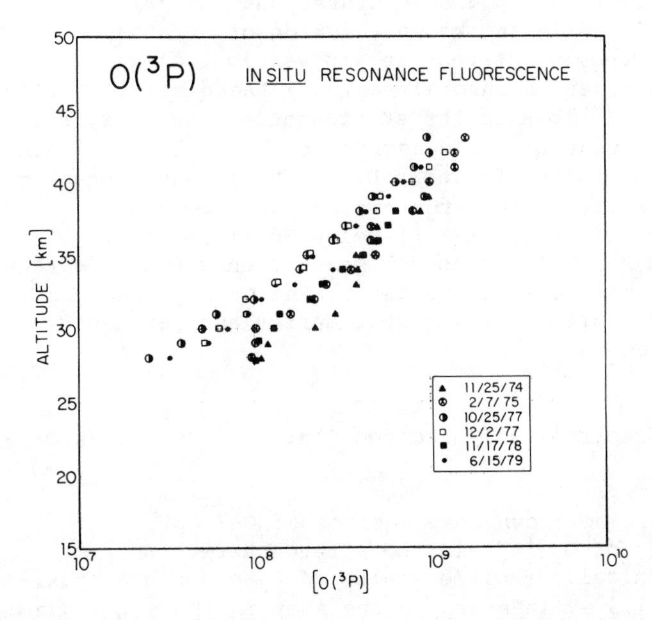

FIGURE D.19 Summary of the in situ $O(^3P)$ data obtained by atomic resonance fluorescence methods (Anderson 1980).

tant minor species. Numerous rocket measurements of the infrared atmospheric system of $O_2(^1\Delta)$ have been made and interpreted in terms of $O_2(^1\Delta)$ concentrations. Most of these are mesospheric and auroral studies, and only a few are applicable to the stratosphere. Two rocket measurements of the day airglow in the 1.27-μm band with a rocket photometer are shown in Figure D.21 (Evans and Llewellyn 1970). They are in essential agreement. Aircraft measurements (Noxon 1968) and balloon measurements (Evans et al. 1969) of the integrated dayglow intensity are in agreement with these rocket measurements. Below 30 km, new balloon ascent measurements would be required to obtain good estimates of $O_2(^1\Delta)$ concentrations. New measurement techniques such as photoionization mass spectrometry could be applied. $O_2(^1\Delta)$ is produced by ozone photolysis and resonance phosphorescence, is quenched by molecular species, and is reasonably simple to model.

Singlet Sigma Molecular Oxygen ($O_2(b^1\Sigma)$)

The $O_2(b^1\Sigma)$ state contains nearly 2 eV of excess energy over the $^3\Sigma$ ground state, but $O_2(b^1\Sigma)$ is not involved in any known reaction of stratospheric significance. Interest in its vertical distribution has been confined to auroral regions; there are few measurements applicable to the stratosphere. Results of the only relevant rocket measurements by Wallace and Hunten (1968) are given in Table D.6. The observations have a large radiative transfer correction; hence the concentrations of $O_2(b^1\Sigma)$ below 50 km are quite uncertain. Balloon ascent measurements would be required to obtain more accurate data. The main production processes are $O(^1D)$ energy transfer and resonance fluorescence.

Other Electronically Excited States of Molecular Oxygen (O_2*)

There are no known observations of $O_2(^3\Delta_u)$, $O_2(^3\Sigma_u^+)$, or $O_2(^1\Sigma_u^-)$ in the stratosphere. These electronically excited states of O_2 contain approximately 4.5 eV in energy above that in the $^3\Sigma_g^-$ ground state, and are thus of potential importance to all of the free radical reactions in the stratosphere. The quantum

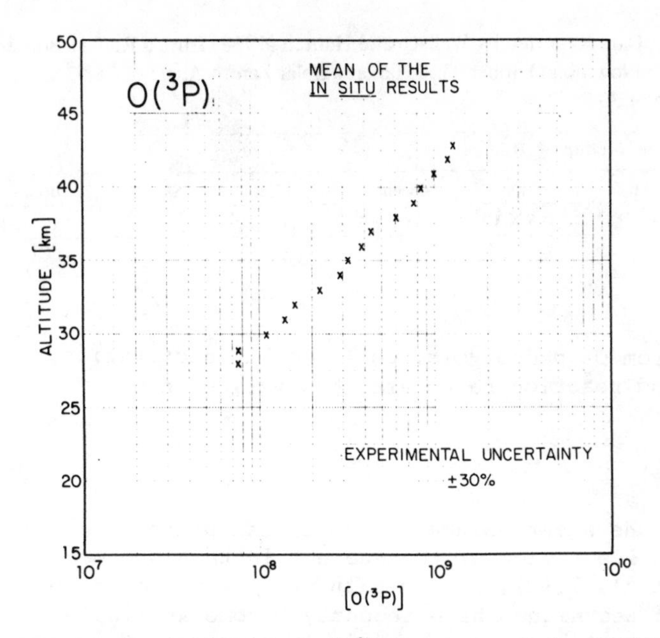

FIGURE D.20 Mean of the six in situ $O(^3P)$ observations displayed in Figure D.19.

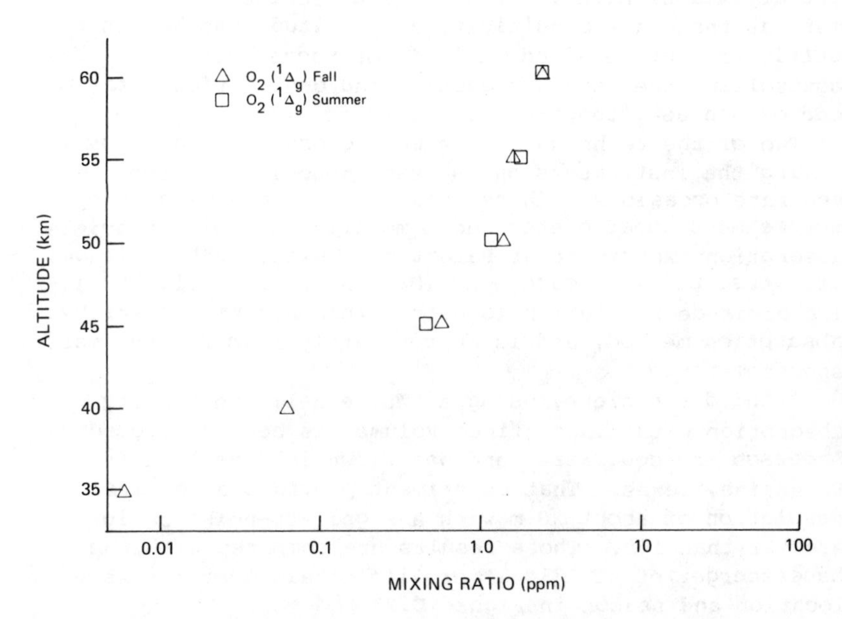

FIGURE D.21 Fall and summer profiles of $O_2(^1\Delta g)$. The uncertainty estimate is ±20 percent (Evans and Llewellyn 1970).

TABLE D.6 Data Reported by Wallace and Hunten (1968) from a Rocket-Borne Spectrometer Flown on October 11, 1966, at a Solar Zenith Angle of 75.5° and at a Latitude of 33°N

Molecule (State)	Concentration at Altitude (cm^{-3})					
	35 km	40 km	45 km	50 km	55 km	60 km
$O_2(^1\Sigma)$	0.4×10^5	0.5×10^5	1.8×10^5	1.8×10^5	1.6×10^5	1.2×10^5

yields from O_3 photolysis, collisional deactivation rates, and reaction rates are, however, unknown.

Ozone (O_3)

Although the sheer volume of ozone data prevents a comprehensive review of the subject in this document (see Hudson et al. 1982), new in situ results using three different techniques have recently become available. Those observations are of critical importance to the subject of trace species observations, for they signify the arrival of highly accurate (and precise) in situ methods that have sensitivity and altitude resolution sufficient for detailed analysis of those factors controlling the local production and destruction rates of odd oxygen as a function of altitude.

Two of the techniques have been cross-calibrated by flying the instruments on the same gondola on three separate occasions. Those results, obtained with an open source mass spectrometer and a modified Dasibi ultraviolet absorption experiment at Palestine, Texas, 32°N latitude, are presented in Figure D.22 (Mauersberger et al. 1981). The altitude resolution is better than 0.5 km for the UV absorption method, and is approximately 1 km for the mass spectrometer.

A third technique, using a "White cell" to amplify absorption within a confined volume has been developed by Anderson and coworkers, and was flown in June 1981 from Palestine, Texas. That experiment provides a vertical resolution of about 30 m with a signal-to-noise ratio greater than 100. Those results are compared with the Mauersberger et al. (1981) profile obtained at the same location and season in Figure D.23.

These data provide the opportunity to examine the ratio of $O(^3P)$ to O_3 between 28 and 40 km. Those

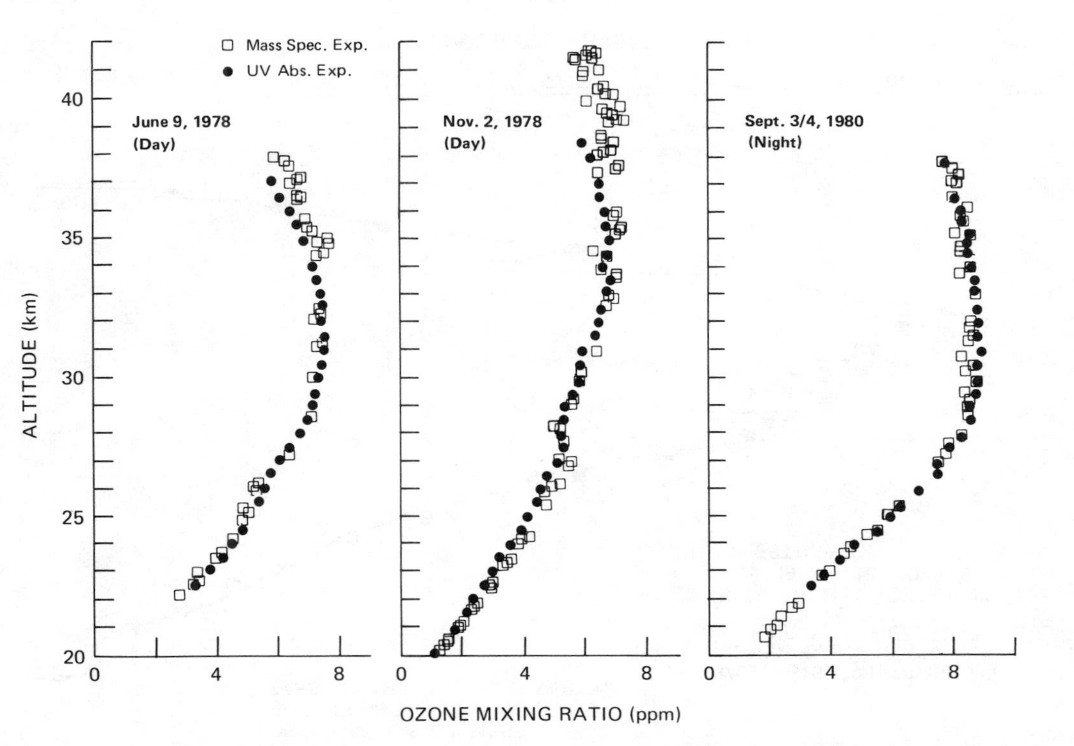

FIGURE D.22 A comparison of ozone volume mixing ratios measured simultaneously during balloon descent by the mass spectrometer beam experiment and UV absorption instrument (Mauersberger et al. 1981).

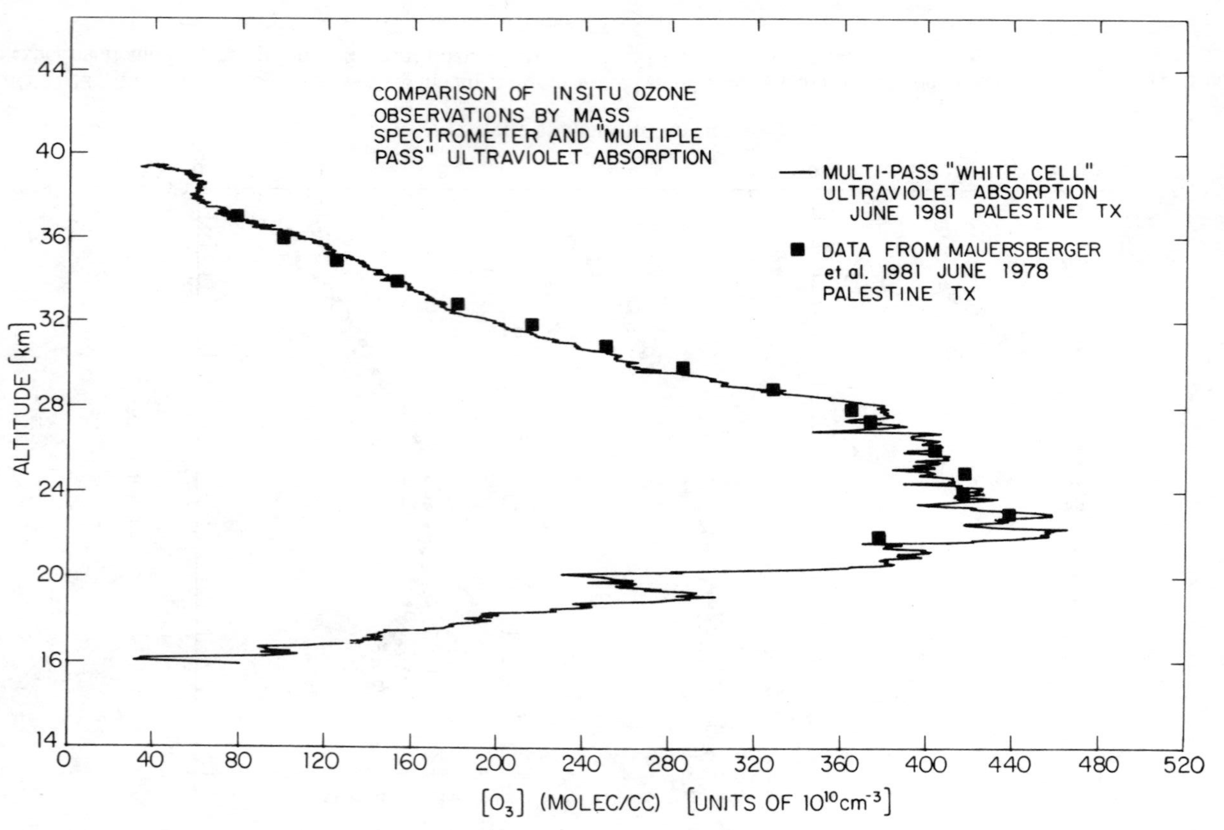

FIGURE D.23 Comparison between the in situ multipass "White cell" data and the in situ mass spectrometer data.

results will be discussed at the end of the section on the relationships between observational data and predictions of models.

Group 4: Reactive Trace Constituents Containing Nitrogen

Nitric Oxide (NO)

Concerns raised nearly 10 years ago about stratospheric ozone depletion resulting from supersonic transport flights above the tropopause placed early emphasis on measurements of NO and NO_2 in the stratosphere. NO has emerged as the most extensively studied radical in the stratosphere in terms of the variety of techniques applied, the number of observations reported, and the latitude coverage available. A critique of the nitric oxide data base is thus not a question of interpreting a limited number of observations, but rather a problem of selecting those observations that (a) are of demonstrated analytical quality, and (b) are useful for testing hypotheses in the chemical reaction schemes used in the stratospheric models.

We adopt here an approach that closely parallels that taken by the WMO/NASA review committee on trace species: Only those data on NO will be accepted that have been obtained by research groups who have repeatedly applied the method and have done the laboratory calibration tests required for a defensible absolute calibration.

Measurement techniques for NO can be separated into the usual categories of in situ and remote. Remote absorption techniques, however, are of little value for a detailed comparison with model calculations because remote techniques are confined to the period at and immediately following sunset/sunrise because of the very small optical depth of NO. NO is rapidly converted to NO_2 at sunset by the reaction

$$NO + O_3 \rightarrow NO_2 + O$$

and reformed rapidly from NO_2 by the direct photolysis

$$NO_2 + h\nu \rightarrow NO + O$$

at sunrise. Emission techniques in the infrared need not suffer from this problem, but there is an insufficient data base for these methods to warrant selection at this time. For an extensive treatment of this subject, the reader is referred to the WMO/NASA report (Hudson et al. 1982).

We are left, therefore, principally with in situ measurements to test the model-calculated NO height profile under midday conditions. Several techniques have contributed significantly to the in situ data base: balloon-borne chemiluminescence (Ridley and Howlett 1974, Drummond et al. 1977, and an extensive series of reports by Ridley and co-workers), aircraft chemiluminescence (see Loewenstein et al. 1978a,b and references therein), rocket-borne chemiluminescence (Horvath and Mason 1978), photoionization mass spectroscopy (Aiken and Maier 1978), spin flip laser absorption (Patel et al. 1974), and balloon-borne pressure-modulated radiometry (Chaloner et al. 1978).

Since the techniques listed above have not been cross-calibrated to isolate unknown systematic instrumental discrepancies, on the one hand, we must explore the overall consistency of the data sets, but on the other, any detailed analysis of vertical profiles is most effectively approached by considering a given set of internally consistent results. We simply note, therefore, that within the quoted experimental uncertainties of the observations there are no large discrepancies, but only one data set is extensive enough to warrant detailed consideration with respect to the vertical profile of NO in the stratosphere, the work of Ridley and co-workers (Roy et al. 1980, Ridley and Schiff 1981, Ridley and Hastie 1981).

The observations of particular interest are a series of six flights made with a chemiluminescence instrument that incorporated direct in-flight calibration procedures to eliminate the possibility of heterogeneous removal in the inlet/chamber section of the instrument. The excellent internal consistency of the data set is evident in Figure D.24.

Note, in particular, the nearly coincident half-filled symbols, which represent data gathered from three different flights at 32°N in the fall, but in two different years. Also, the data taken in two flights at nearly identical latitude in the southern hemisphere and equivalent season are only slightly lower than the corresponding northern hemisphere results. As noted in Hudson et al. (1982), since differences between the results of the flights are very nearly equal to variations within any one of the flights, the data present no evidence of systematic patterns over the ranges of season, latitude, and hemisphere. However, it is essential to caution against extracting statistically significant conclusions from but six observations. What can be concluded, however, is that very significant progress has been made in the analytical area: high-quality in situ observations of NO are technically feasible.

It will become apparent when attempting to use these data, that differences between profiles of a factor of 2 are of considerable importance when applying the NO results to modeled distributions. Thus, in anticipation of this, we replot the data points from Figure D.24 with higher resolution in the abscissa; those results are given in Figure D.25 in terms of absolute calibration.

This is the data set that will be used later to determine whether the NO_x data set provides a quantitative constraint on predictions of ozone depletion resulting from changes in N_2O.

We can quickly summarize the conclusions extracted from a consideration of the other data on NO in Figure D.26, which presents the range of observed NO as a banded region extending from the tropopause to the lower-middle stratosphere.

Diurnal Variation of NO. The qualitative features that one would expect for the diurnal behavior of NO based upon the current mechanistic links partitioning the reactive nitrogen family, as summarized in the flow diagram of Figure D.27, have been confirmed:

- NO decays following sunset at a rate comparable to that expected from the conversion to NO_2 via

$$NO + O_3 \rightarrow NO_2 + O_2.$$

- NO increases rapidly at sunrise as one would expect from the direct photolysis of NO_2
- NO increases slowly throughout the day consistent with the formation of NO + NO_2 from the back conversion of N_2O_5, which serves as a temporary reservoir.

The most extensive set of diurnal observations is from the in situ studies of Ridley and co-workers using balloon-borne chemiluminescence. The sunset measurements (Ridley and Schiff 1981) are shown in Figure D.28, and an analogous data set for sunrise is shown in Figure D.29.

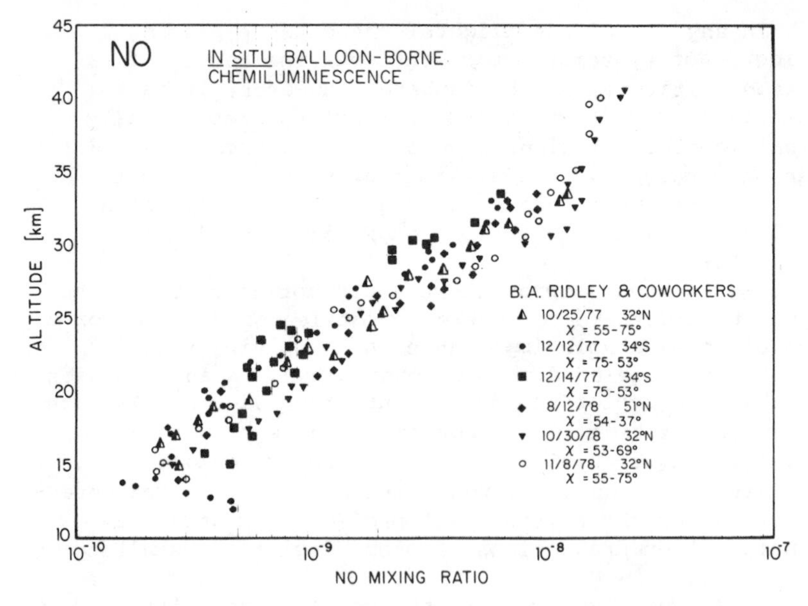

FIGURE D.24 Summary of six in situ NO observations by Ridley and co-workers obtained with a chemiluminescent probe (Roy et al. 1981, Ridley and Schiff 1981, Ridley and Hastie 1981).

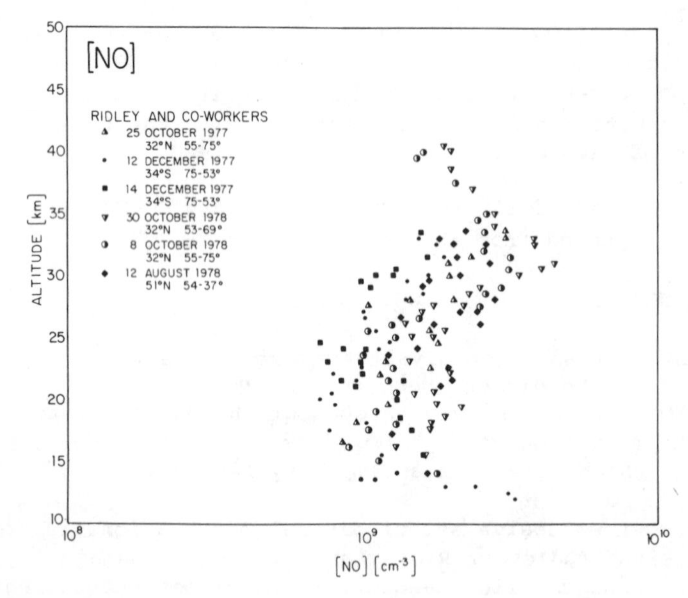

FIGURE D.25 Nitric oxide data of Figure D.24 converted to absolute concentration.

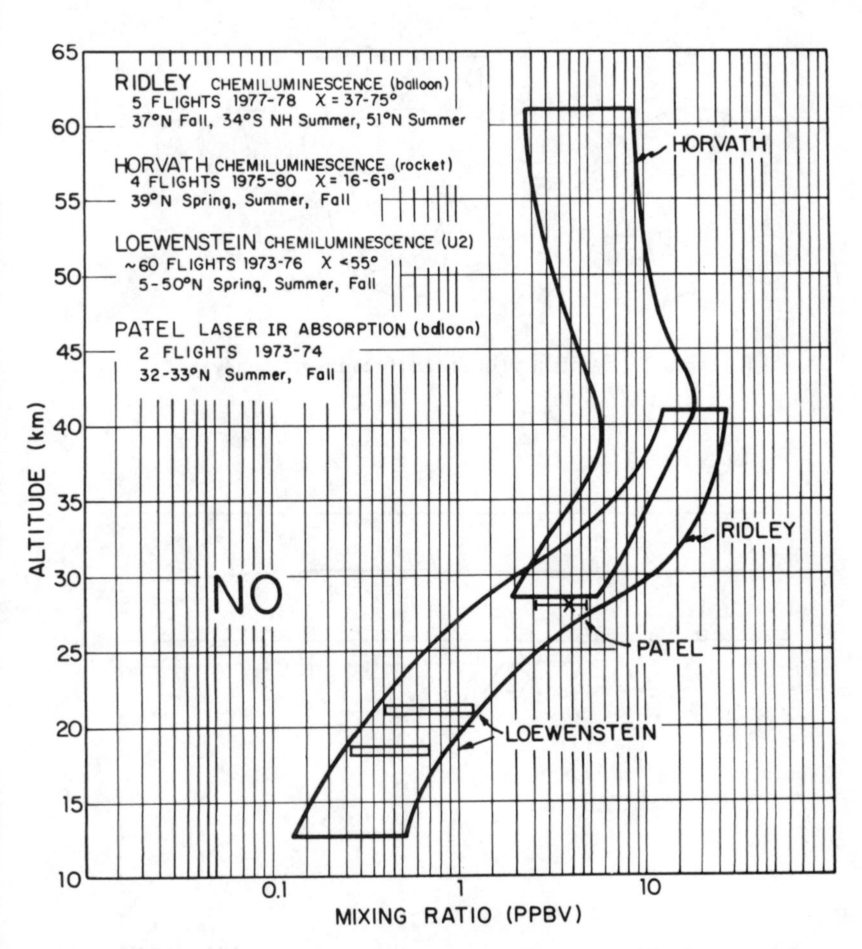

FIGURE D.26 In situ NO mixing ratio measurements reported by four research groups, which encompass the stratosphere and mesosphere.

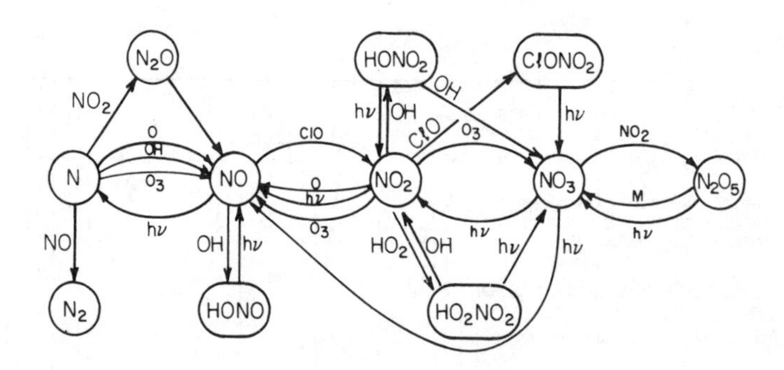

FIGURE D.27 Partitioning of the reactive nitrogen family.

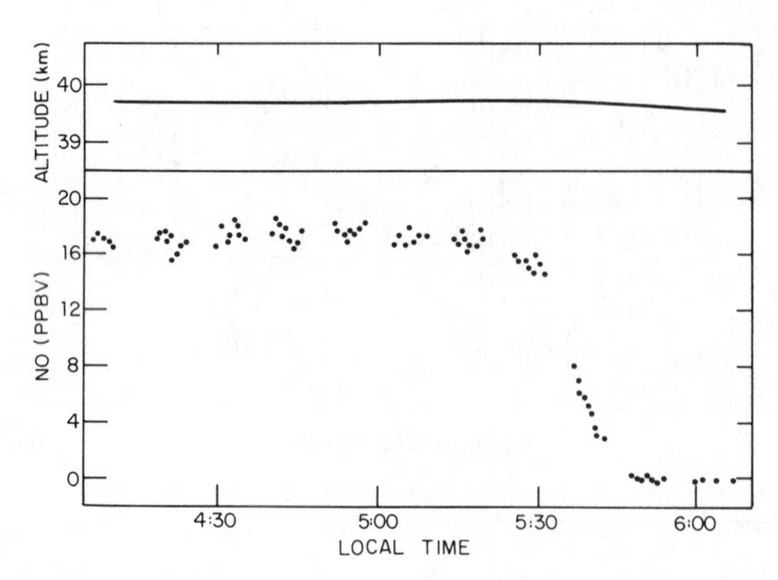

FIGURE D.28 Sunset observation of NO determined in situ by Ridley and Schiff (1981).

The sunrise data were obtained in 1975, before the improved inlet and calibration procedures were added, but the asymptote in the NO mixing ratio is consistent with that obtained by the improved analytical techniques, so the temporal behavior is almost certainly representative. The detail revealed by those two data sets is superb and demonstrates in principle how such data can be used to test rates of production and destruction within any given chemical mechanism.

Patel et al. (1974) have employed the laser Raman spin flip technique to the NO diurnal dependence. The results, not inconsistent with those presented above, are discussed in Hudson et al. (1982).

What these diurnal data do make clear is that the rapid fall off of NO at sunset obviates the possibility of using long-path absorption techniques for detailed measurements of nitric oxide.

Nitric Oxide Seasonal Variability. There is currently an insufficient data base on NO to establish any statistically meaningful seasonal dependence. However, the aircraft measurements of Loewenstein et al. (1978b) at 18 and 21 km extending over a period of four years have revealed two significant effects. The first is a rather sharp winter minimum that lasts just 2 to 3 months at 40°N latitude in the vicinity of 21 km. The second is a rather broad summer maximum that exhibits a duration of 7 to 8 months at 40°N, 21-km altitude. The observations are summarized in Figure D.30. A similar trend has been found at 18 km, although fewer data are available at that altitude.

Given that the observed winter minimum is a factor of 3 to 4 lower than the summer maximum, clearly exceeding the cited uncertainties and demonstrated reproducibility of the balloon-borne chemiluminescent technique, it is unfortunate that vertical profiles of NO at 40°N during December and January are not available. A great deal could be learned from such a data set. Vertical column (integrated) measurements of NO from aircraft (Coffey et al. 1981) include higher ratios of NO in summer than in mid-February by a factor of 1.4. However, as Figure D.30 reveals, the deep minimum occurs before mid-February. In addition, the almost certain altitude dependence of the effect may well erode the vertically integrated effect.

It is crucial to point out that such a deep, short-lived minimum presents an excellent opportunity to carry out diagnostic experiments to elucidate response of the

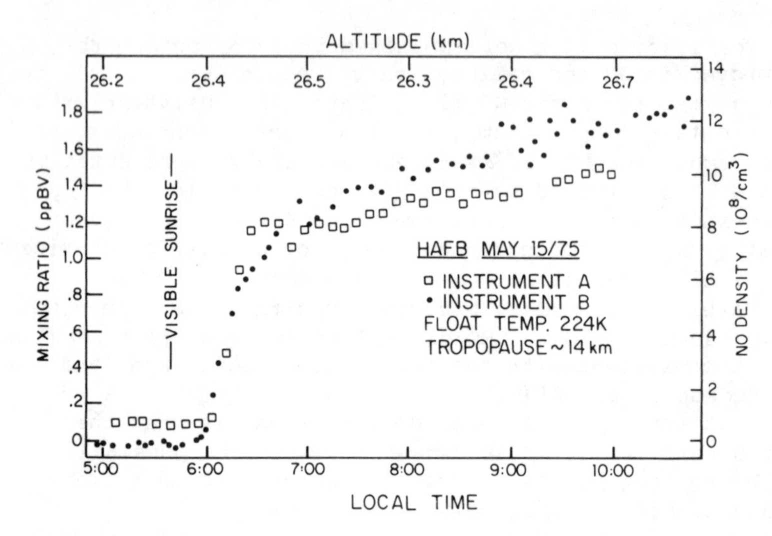

FIGURE D.29 Sunrise in situ measurements of NO by Ridley using two chemiluminescent instruments on a balloon platform at 33°N latitude.

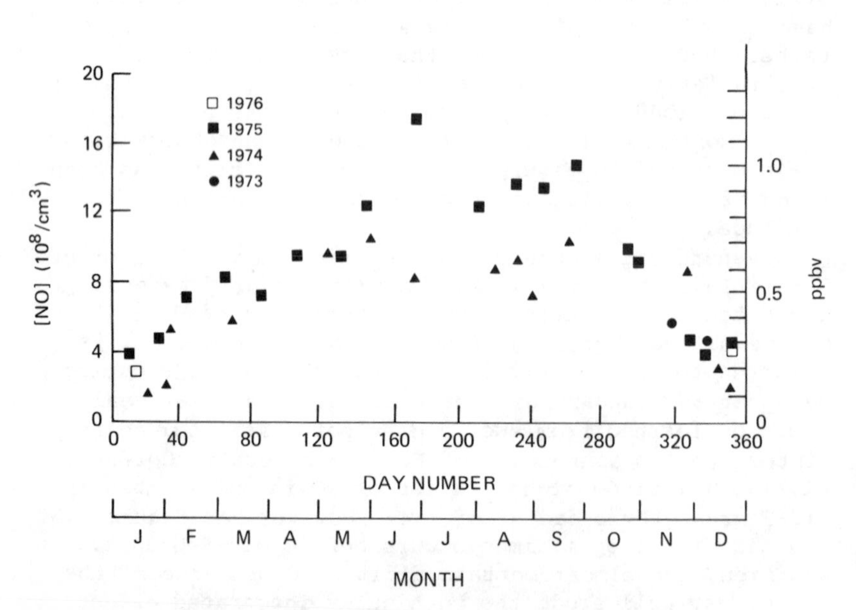

FIGURE D.30 Nitric oxide seasonal data (122°W, 40°N) summary at 21.3 km. The in situ NO measurements of Loewenstein et al. (1978) were obtained with a chemiluminescent instrument flown near 40°N, 122°W, and 21.3 km.

NO_x systems to perturbation, and the subsequent response of O_3 to these changes.

There is another marked seasonal/latitude dependent feature discovered by Loewenstein et al. (1978b). Specifically, above approximately 65°N latitude in the 18-km altitude aircraft flights, [NO] transits from a winter maximum to a fall minimum with a concentration excursion of more than an order of magnitude. Between 5° and 50°N latitude in contrast, the variation of NO is not large in the lower stratosphere. The vertical column data of Mankin and co-workers (see Coffey et al. 1981) exhibit very little latitude variation between 5° and 45°N latitude.

Nitrogen Dioxide (NO_2)

Given our current picture of the reactions that control the rate of odd oxygen production and destruction in the stratosphere, summarized in Figure C.6a of Appendix C, the concentration of NO_2 is the single most important radical for determining whether we have a quantitative understanding of the odd oxygen production/destruction budget. An inspection of Figure C.6a in Appendix C reveals that the NO_x catalytic cycle

$$NO + O_3 \rightarrow NO_2 + O_2$$
$$NO_2 + O \rightarrow NO + O_2$$
$$\overline{O + O_3 \rightarrow O_2 + O_2}$$

constitutes 60 to 70 percent of the total loss rate for odd oxygen between 20 and 35 km. We are thus particularly interested in the NO_2 data base as a test of ozone destruction rates.

The first fact that emerges from an investigation of available NO_2 data is that there are no in situ observations of the molecule, so that conclusions we draw will be based exclusively on remote sensing techniques. The remote sensing techniques, predominantly visible and infrared absorption, are strictly limited to the sunrise and sunset periods. This does not prevent conclusions from being drawn that are accurate to within a factor of 2 in the altitude range from 25 to 35 km, but at lower and higher altitudes the conversion of NO to NO_2 at sunset dramatically shifts the ratio such that the modeled diurnal behavior becomes critically involved.

This is summarized quantitatively in Figure D.31, which indicates that the sunset/noon ratio of [NO_2] approximates 1.5 between 20 and 35 km. Thus sunset observations of NO_2 are very difficult to interpret accurately in terms of midday steady state NO_2 concentration. It is also essential to realize that $[NO_2]_{90°}/[NO_2]_{30°}$ is greater than unity primarily because of the slow build-up of NO_2 during the day resulting from N_2O_5 decomposition. Thus we must correct the sunset observation back to midday if we are to compare them with modeled distributions close to local noon as is done for all other radicals.

We select for particular attention the data set collected at 34°N latitude by infrared solar absorption (Murcray et al. 1974, Goldman et al. 1978, Blatherwick 1980) and by visible absorption (Fischer et al. 1982) because of the extensive analysis afforded the IR results and because of the existence of two independent measurement techniques. It is also true, as we will discuss later in this section, that seasonal and latitude data have the greatest probability of being unbiased by variability on any given day.

We display in Figure D.32, therefore, the results from five different flights obtained between 1967 and 1980 using two independent methods, IR and visible absorption, all taken at sunset. Also displayed are the infrared pressure-modulated radiometer (PMR) data of Roscoe et al. (1981), which represent a time average from approximately 2 hours before local noon to sunset. We would expect these results to be about 20 to 25 percent below the sunset data, all other things being equal, because of the diurnal behavior of NO_2. Figure D.32 presents the mixing ratio data, a figure identical to that which appears in the WMO/NASA report with the pressure-modulated radiometer data added, and Figure D.33 presents the same set of data expressed in terms of absolute concentrations in an expanded abscissa.

Given the diurnal correction of about 25 percent to the pressure-modulated radiometer data, the extremely limited number of results in Figure D.33 yield surprisingly consistent results. The range in mixing ratio appears to be less than a factor of 2 using three independent experimental techniques. When one begins to probe within this factor of 2 envelope, it is advantageous to examine the absolute concentration data, which removes the gradient in NO_2 with altitude and offers a more discriminating examination of the data spread, as shown in Figure D.33.

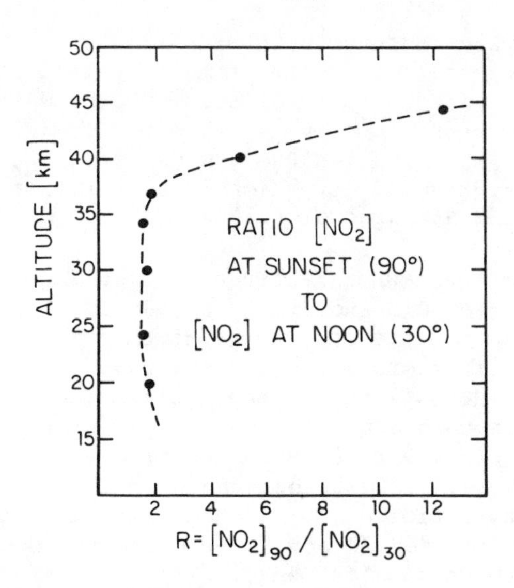

FIGURE D.31 Calculated ratio of $[NO_2]$ at sunset to that at noon.

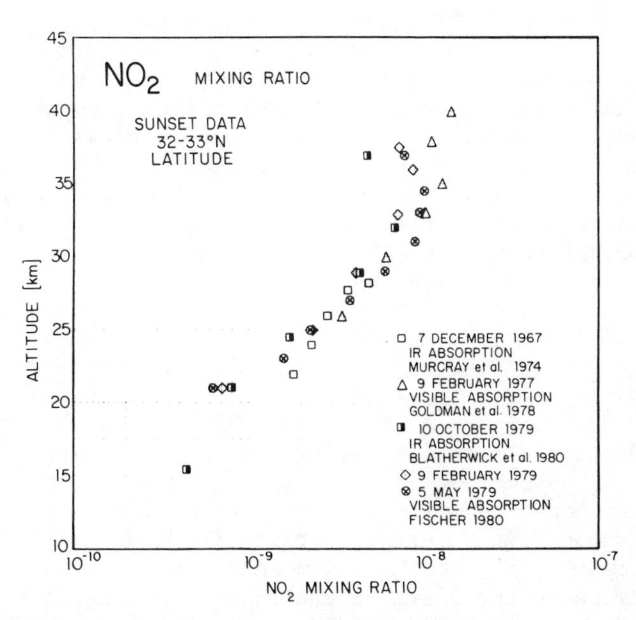

FIGURE D.32 Summary of sunset mid-latitude NO_2 data between 20 and 40 km. The midday to sunset mean reported by Roscoe et al. (1981) is included for completeness.

When we compare these results with model calculations, it becomes clear that it is the level of detail shown in Figure D.33 that is needed to discriminate between predicted ozone reduction levels between 2 and 15 percent. In addition, given the dominance of NO_x catalyzed destruction of O_x, which is rate limited by NO_2, factors of 2 are crucial to the question of odd oxygen balance.

We summarize the higher latitude data in Figure D.34 obtained at 45° to 50°N and at 51° to 58°N.

The same comments are applicable to the higher latitude data: There is a serious shortage of coverage in seasonal and diurnal dependence, but what data there are show a remarkable consistency.

The most obvious feature extracted from a comparison of Figures D.30 and D.32 is that there is an indication of larger mixing ratios of NO_2 at high latitude in the altitude region between 20 and 30 km where the concentration of nitrogen dioxide peaks. The difference between 32°N and 51° to 58°N corresponds to a 50 percent increase in mixing ratio over this latitude range. It is essential to verify this difference, and the seasonal dependence of it, preferably with the same array of cross-calibrated techniques on the same observation platform.

We turn next to an exceedingly important component of our experimental picture of global NO_2--the ground-based data set obtained by Noxon (see, for example, Noxon 1978, 1979, 1980; Noxon et al. 1979) using the visible absorption technique he pioneered. Noxon reports NO_2 vertical column densities that are a factor of 2 larger at night than during the day, which confirms in a semiquantitative way the conversion of NO_2 to N_2O_5 via NO_3 as outlined in our previous discussion of NO:

$$NO \underset{O,\ h\nu}{\overset{ClO,\ O_3}{\rightleftharpoons}} NO_2 \underset{h\nu}{\overset{O_3}{\rightleftharpoons}} NO_3 \underset{M,\ h\nu}{\overset{NO_2}{\rightleftharpoons}} N_2O_5$$

In addition, Noxon has defined the seasonal and latitudinal morphology of stratospheric NO_2 in a four-year series of data collection at four northern latitude sites. Those results are summarized in Figure D.35. The regularity of the winter minimum and summer maximum is dramatic, as is the distinct factor of 5 change in NO_2 total column at latitudes of 50°N over the period of a year.

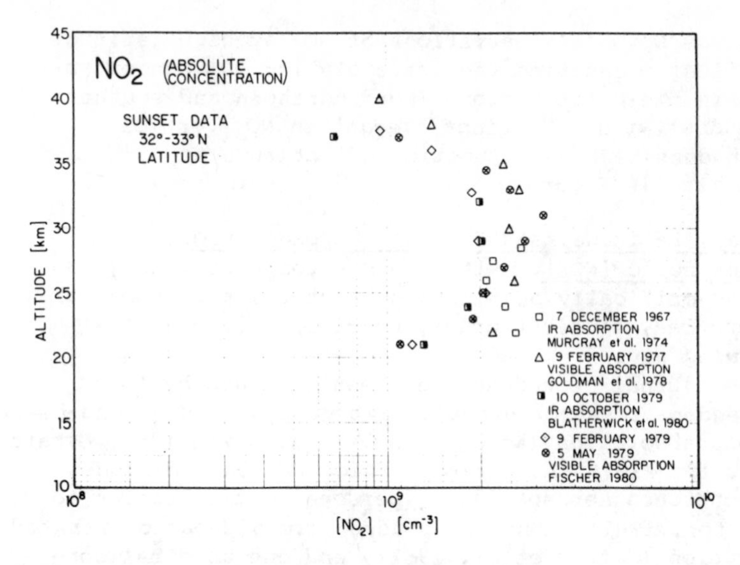

FIGURE D.33 Data from Figure D.32 converted to absolute concentration and presented with an expanded abscissa. The PMR data are not included.

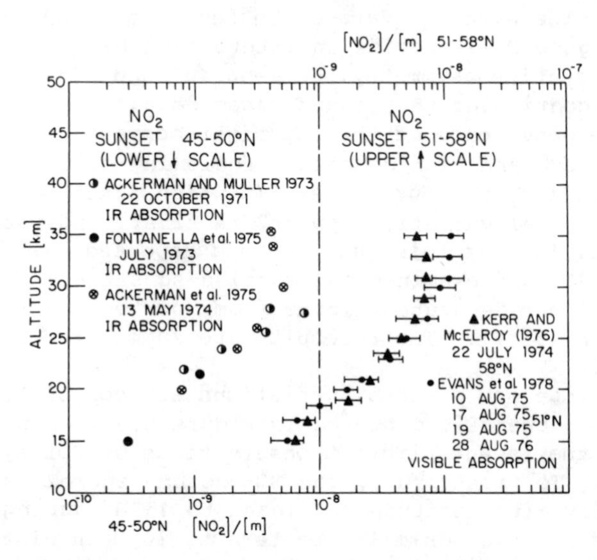

FIGURE D.34 NO_2 sunset 45°-50°N (lower scale) and NO_2 sunset 51°-58°N (upper scale).

It was Noxon's observations of NO_2 at high latitude that first identified the extremely low NO_2 concentrations in the polar regions (both northern and southern latitudes) with a distinct "ledge" in NO_2 vertical column densities as a function of latitude at 45°N, as summarized in Figure D.36.

Diurnal, Latitude, and Seasonal Dependence of Stratospheric [NO_2]. Although stratospheric [NO_2] depends critically on all these factors, the evolving picture can, to first order, be deconvoluted in the following way.

The diurnal dependence has been examined by four independent research groups. Two groups used ground-based visible absorption (Noxon et al. 1979, Noxon 1980, Girard et al. 1978/1979, A. Girard, Office National d'Etudes et de Recherches Aerospatiales, personal communication to D. Albritton, 1981). One group used aircraft-based infrared absorption (Coffey et al. 1981), and one used balloon-borne visible absorption (Evans et al. 1978). Both ground-based data sets report a factor of 2 larger NO_2 column amounts at night than during the day at middle latitudes. The aircraft data of Coffey et al. (1981) shown in Figure D.37 provide an interesting picture of the sunset-sunrise asymmetry, between 40° and 50°N latitudes, confirming the ground-based data.

The time dependence of the day/night conversion is highly altitude and time dependent following sunset. The obvious question, of course, is: In which altitude region does this diurnal variation appear? A first, and very informative, look at this question was reported by Evans et al. (1978), who examined the sunrise-sunset NO_2 profiles in four separate flights from Yorkton, Saskatchewan (51°N). Their results are summarized in Figure D.38.

The latitude and seasonal variations are convoluted, but the basic trend is captured in Figure D.39, which represents the recent aircraft observations of Coffey et al. (1981). To first order, the NO_2 column increases monotonically with latitude (at least to 45°N) during summer, and follows a similar pattern to 30°N in winter. At higher latitudes during the winter, however, there is a strong divergence to much lower total column concentration. This time-dependent transition between these two cases is, at least in part, exemplified by the exceedingly interesting "cliff" features discovered by Noxon and summarized previously in Figure D.36.

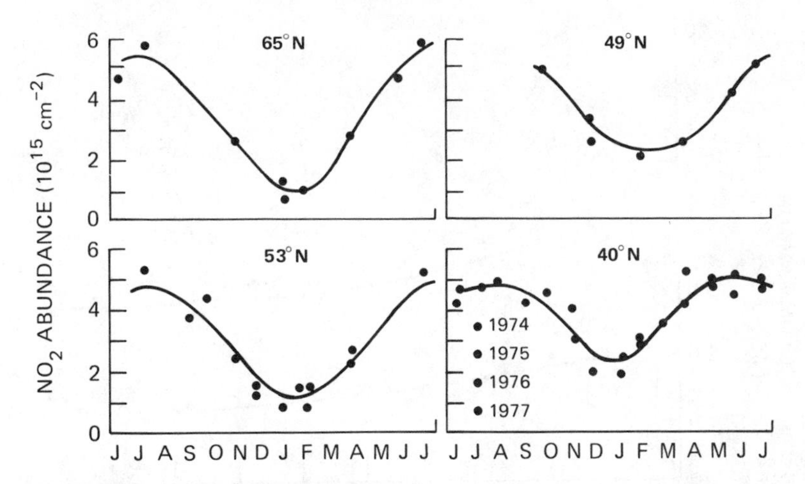

FIGURE D.35 Seasonal variation of late afternoon NO_2 at four latitudes, as given by the ground-based visible absorption spectroscopic measurements of Noxon (1979). The abundance should be multiplied by 1.25 (Noxon 1980).

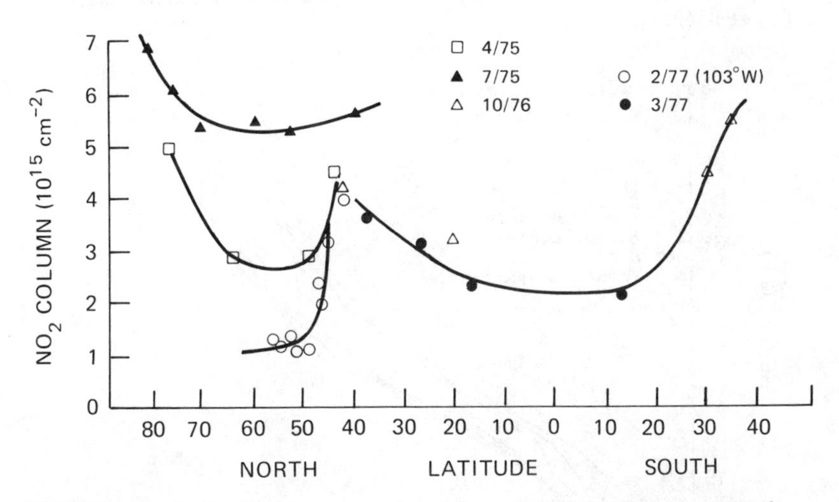

FIGURE D.36 Latitudinal and seasonal variations of the late afternoon vertical column of NO_2, as measured by Noxon (1979) using ground-based visible absorption techniques. The values represented by the open and solid circles should be multiplied by 1.6, and all others by 1.25 (Noxon 1980).

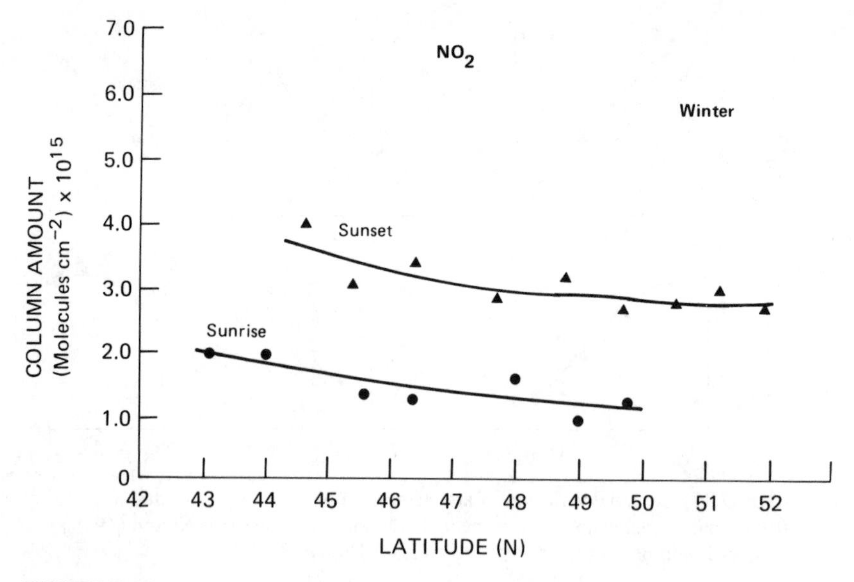

FIGURE D.37 Sunrise and sunset vertical-column measurements of NO_2 by Mankin and co-workers, who used an infrared absorption apparatus on an aircraft platform (Coffey et al. 1981).

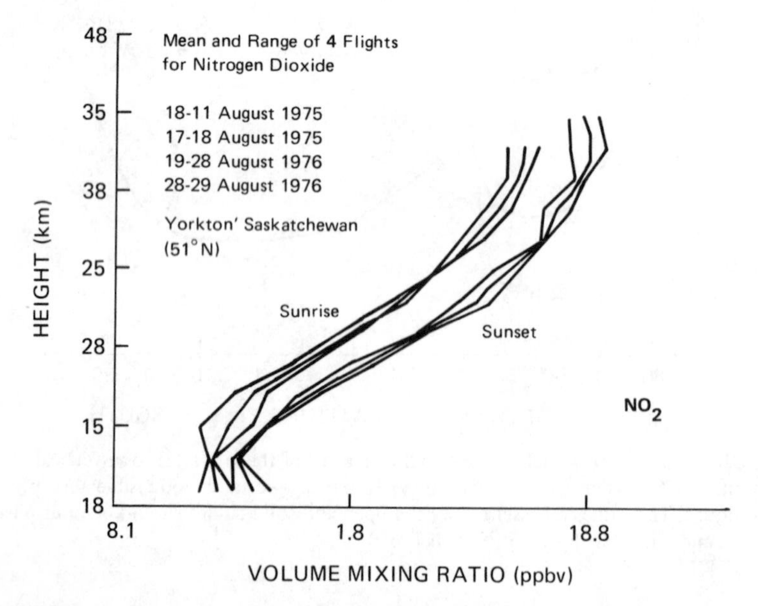

FIGURE D.38 The sunrise and sunset altitude profiles of NO_2 reported by Evans et al. (1978) from the Canadian stratosphere flight series. The upper and lower limits indicate the maximum observed deviations from the mean. The measurements were made using a balloon-borne visible absorption apparatus.

In closing this brief discussion of the NO_2 data base, we note a final point made by Noxon (1979, 1980) that NO_2 has been observed to change by a factor of 2 within the time span of a few days, as illustrated in Figure D.40.

Atomic Nitrogen (N)

There are no reported observations in the stratosphere of atomic nitrogen in either its ^2D or ^4S states. Although atomic resonance fluorescence could detect, in situ, N atom concentrations in the 10^5 cm^{-3} range, current predictions place its expected concentration well below that, and no experiments have been attempted.

The Nitrate Radical (NO_3)

The chemical link between the NO_x catalytic radicals NO and NO_2 and the higher oxides of nitrogen is believed to be NO_3, and yet there are very few data on this important intermediate. There are no daytime observations available (the only analytical technique used thus far for NO_3 is visible absorption--a technique identical to that used for NO_2). A single nighttime profile has been reported, obtained at 43°N latitude from a balloon-borne visible spectrophotometer using Venus as a light source. That profile is shown in Figure D.41.

If the data in Figure D.41 are integrated, one obtains a vertical column between 20 and 40 km of 3.5×10^{13} cm^{-2}. This figure is not inconsistent with the only other data available, that of Noxon, who estimates, based on ground-based visible absorption data, a column density of 10^{14} cm^{-2} in the spring and an upper limit in the summer of 4×10^{13} cm^{-2}.

While it may well turn out to be irrelevant due to the very different moisture level, total pressure, and heterogeneity of the troposphere, it should be noted that NO_3 currently represents an enigma to tropospheric NO_x studies in that dramatically less NO_3 is observed (when simultaneous NO_2 measurements are made) than one would predict given current models. The implication is that a large sink for NO_3 exists because the source is well established.

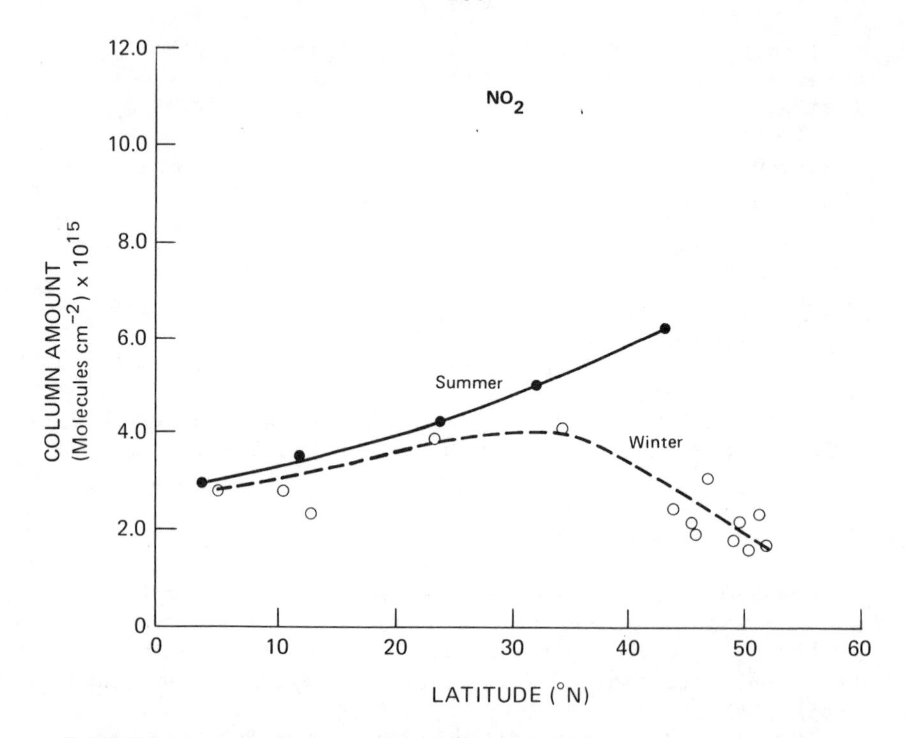

FIGURE D.39 Latitudinal and seasonal variations of the late afternoon vertical column of NO_2, as measured by Mankin and co-workers using aircraft-borne infrared absorption techniques (Coffey et al. 1981).

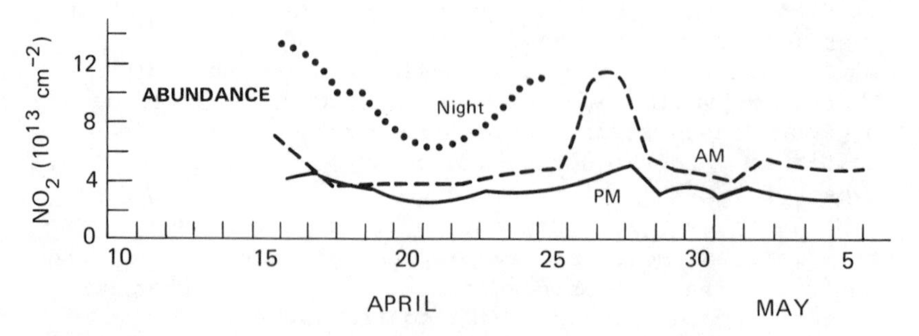

FIGURE D.40 The daily variation of the vertical column of the nighttime, late afternoon, and early morning NO_2 at 40°N in April and May 1976 as seen by Noxon et al. (1979), using ground-based visible absorption spectroscopy. The abundances should be multiplied by 1.25 (Noxon 1980).

Dinitrogen Pentoxide (N_2O_5)

Although dinitrogen pentoxide is recognized to be of major importance to our understanding of atmospheric NO_x chemistry, and has received increasing attention from the experimental community, there have been no new results reported beyond those presented in Hudson and Reed (1979). In that document, Evans et al. report a tentative detection of 2 ppbv at 30 km and Murcray reported an upper limit of 1.2 x 10^{15} cm^{-2} above 18 km in February.

Nitrous Acid (HONO)

The nitrous acid molecule, formed in the recombination reaction of OH and NO, is rapidly photolyzed in the stratosphere and thus is expected to be present at concentrations several orders of magnitude below the detection threshold of the high-resolution infrared absorption experiments. It does not possess a strong, well-defined electronic transition and is thus not amenable to resonance fluorescence techniques.

Thus, although it has been searched for in the IR absorption data, there are no reported observations, and the upper limits one would extract from the data are incapable of testing our understanding of hydrogen-nitrogen oxygen photochemistry.

Nitric Acid ($HONO_2$)

Nitric acid has, since the inclusion of reactive nitrogen compounds into stratospheric chemistry, been recognized as the dominant chemical "reservoir" for the oxides of nitrogen and has thus received considerable attention. Four analytical methods are currently available for the detection of $HONO_2$, two in situ and two remote.

In situ observations were first reported by Lazrus and Gandrud (1974) using a filter collection technique deployed on a balloon to determine the vertical profile of $HONO_2$ between the tropopause and 38 km. Those observations were taken in the spring season in three consecutive years: 1971, 1972, and 1973. Those results appear in Hudson and Reed (1979). Results using a second in situ technique, a rocket-borne ion-sampling method developed by Arnold and co-workers, have been recently reported (see Arnold et al. 1980).

Remote observations have been reported employing both infrared absorption (Fontanella et al. 1975 and H. Fischer, personal communication to D. Albritton for the WMO/NASA report, 1980) and infrared emission (Harries et al. 1976, Evans et al. 1978, and D.G. Murcray, University of Denver, personal communication to D. Albritton for the WMO/NASA report, 1980).

Although the data of Lazrus and Gandrud (1974), Harries et al. (1976), and Fontanella et al. (1975) have appeared in previous reports by both NASA and the NRC, we include those results with the more recent data to appraise the entire data base defining the mid-latitude, northern hemisphere vertical distribution of nitric acid. We present both the mixing ratio data in Figure D.42 (which is identical to the figure appearing in the WMO/NASA report) and the absolute concentration data in Figure D.43, the latter with an expanded abscissa (a format required for a detailed comparison with modeled distributions, as we will see in the sections that follow).

Given that four independent analytical techniques were employed by seven research groups, the consistency of the $HONO_2$ data is exceptional. It remains, of course, to employ the methods simultaneously for soundings of the same air mass and to standardize the deconvolution techniques in order to establish whether the scatter represents experimental uncertainty or atmospheric variability.

A particularly important point regarding the long path absorption sunrise-sunset data is that the chemical time constant of nitric acid is much longer than a diurnal period, and thus those data can be immediately interpreted in terms of the model calculations without reference to details of the diurnal dependence.

Figure D.43 constitutes the most important body of evidence available for testing our understanding of the nitric acid distribution in the stratosphere, and we will recall the figure later. However, there are other important experimental results that have been obtained for nitric acid.

First, the combined seasonal and latitude scans using aircraft-borne total column infrared measurements, summarized in Figure D.44, demonstrate a notable lack of seasonal dependence in the characteristic monotonic increase in $HONO_2$ from equator to pole, up to 40°N. This is, of course, in sharp contrast to the corresponding results for NO_2.

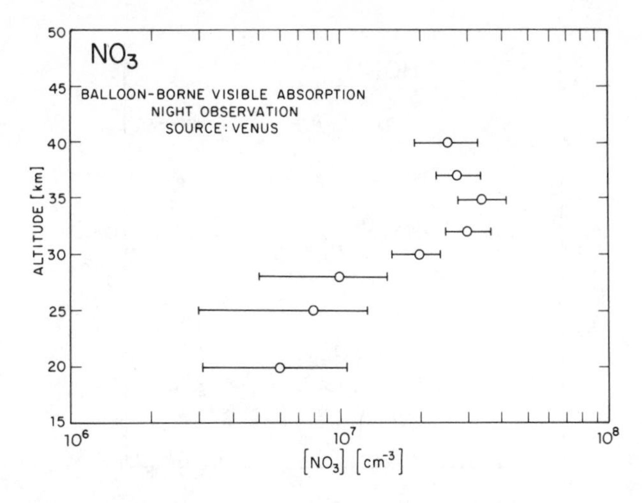

FIGURE D.41 Vertical distribution of NO_3 in the stratosphere at night obtained by balloon-borne absorption techniques in the visible using Venus as the source.

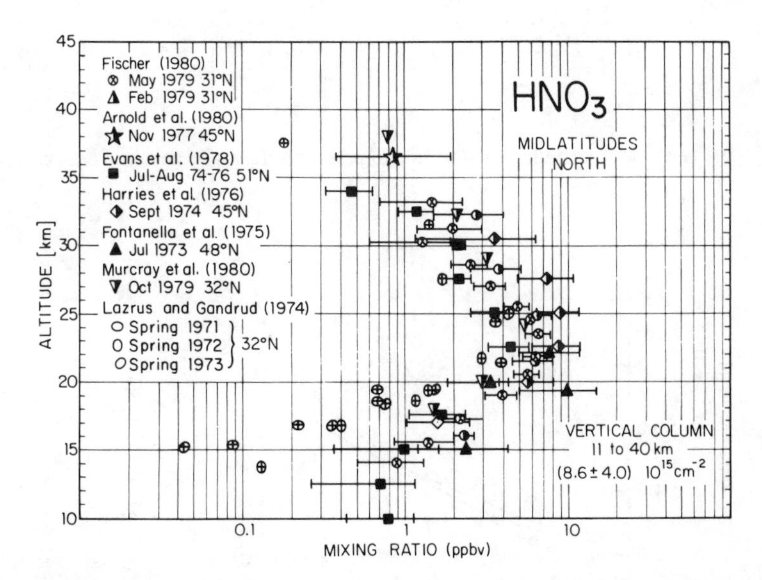

FIGURE D.42 In situ and remote measurements of the HNO_3 mixing ratio at northern mid-latitudes.

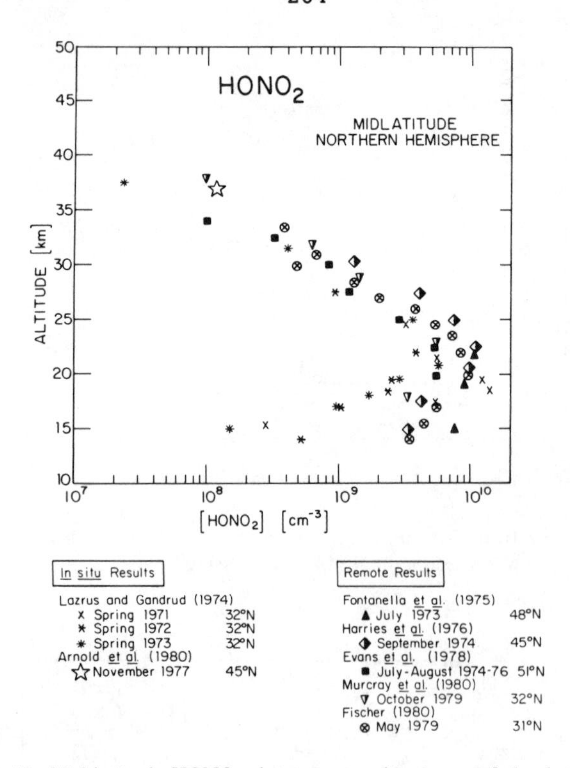

FIGURE D.43 Mid-latitude $HONO_2$ data expressed in terms of absolute concentration.

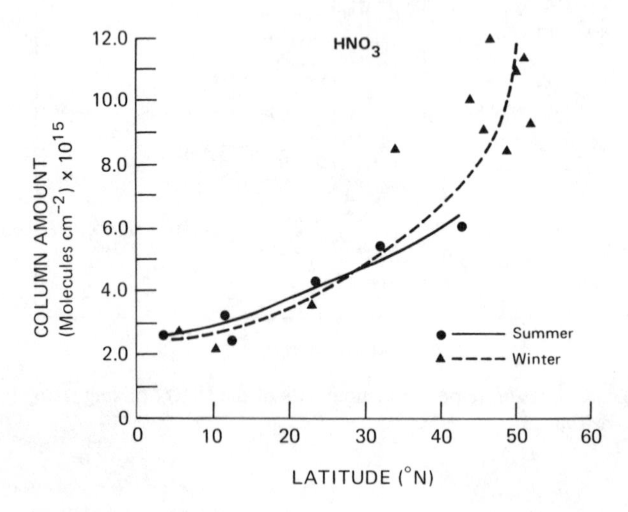

FIGURE D.44 Evidence for the lack of a seasonal variation in the vertical column density of HNO_3 at latitudes less than 40°N, as measured by Coffey et al. (1981) using infrared absorption.

The aircraft vertical column data substantiate the general morphology of the nitric acid total column concentration first reported by Murcray et al. (1974). There is a strong increase in the vertical column density with increasing latitude, in both the northern and the southern hemispheres. Above about 60°N latitude, there appears to be a pronounced seasonal variation. In the winter at high latitudes, the vertical column concentration of $HONO_2$ is distinctly larger than in early summer.

Group 5: Reactive Trace Constituents Containing Bromine

The Bromine Monoxide Radical (BrO)

There are no reported observations of BrO in the stratosphere. It should be amenable to detection by laser heterodyne techniques in the middle infrared, by mm-wave emission and by in situ chemical conversion resonance fluorescence. BrO is the pivotal radical in the bromine-ozone system.

Atomic Bromine (Br)

There are no reported observations of atomic bromine in the stratosphere.

The Bromine Dioxide Radical (BrO_2)

There are no reported observations of BrO_2 in the stratosphere.

Bromine Dioxide (OBrO)

There are no reported observations of OBrO in the stratosphere.

Hydrogen Bromide (HBr)

No direct observations of HBr have been reported. However, a recent publication by Berg et al. (1980)

report observations of total bromine using neutron activation techniques applied to an activated charcoal sampling matrix deployed from both aircraft and balloon platforms. Initial results from six aircraft flights and one balloon mission in the lower stratosphere are presented for latitudes between 16° and 67°N. Five total bromine values showed substantial variability ranging from 7 ± 4 ppt by volume to 40 ± 11 ppt. If the assumption that HBr dominates the total bromine budget in the middle and upper stratosphere is correct, these figures should reflect the HBr concentration in that region.

Hydrogen Oxybromide (HOBr)

There are no reported observations of HOBr in the stratosphere.

Bromine Nitrate ($BrONO_2$)

There are no reported observations of $BrONO_2$ in the stratosphere.

Group 6: Reactive Trace Constituents Containing Fluorine

Fluorine Monoxide (FO)

There are no reported observations of FO in the stratosphere.

Atomic Fluorine (F)

There are no reported observations of atomic fluorine in the stratosphere.

The Fluorine Dioxide Radical (FO_2)

There are no reported observations of FO_2 in the stratosphere.

Fluorine Dioxide (OFO)

There are no reported observations of OFO in the stratosphere.

Hydrogen Fluoride (HF)

Hydrogen fluoride has received considerable attention both because it possesses a strong IR absorption spectrum in the middle infrared and because it provides a very important check on the amount of fluorine released from the chlorofluorocarbons. Since the strength of the HF bond is sufficiently strong that no radical reacts with HF to any measurable degree, it is also of interest to compare ratios of HF and HCl in the same air mass.

Table D.7, taken from Hudson et al. (1982), summarizes the data, latitude, altitude range, method, and experimenter for each of the reported HF observations.

The two measurements that can be most directly compared are, as in the case of HCl, the high-resolution infrared absorption measurements of Farmer et al. (1980) and Buijs et al. (1980). The techniques are identical; the results are shown in Figure D.45. Although the slopes correlated reasonably well, the Buijs results are a factor of 2 greater than those of Farmer. It is unfortunate that such a large latitude discrepancy exists between the observations (although HF should not depend sensitively on latitude). It may well be that an adjustment for tropospheric height could remedy the disparity significantly.

Two additional profiles for HF have been obtained by Bangham et al. (1980) and Marche et al. (1980a), the former from balloon-borne observations at 32°N in emission and the latter from ground-based absorption measurements at 42°N. These two profiles cover different altitude regions in the stratosphere, but at the one common altitude of 30 km differ by about a factor of 5 (see Figure D.45). As was the case for HCl, however, the emission data values shown here are preliminary and, in particular, are likely to increase at the lower altitudes when a more rigorous analysis of the data is performed. Should this be the case, they may well be in agreement with the Buijs et al. (1980) data at lower altitudes.

The remaining remote sensing measurements are those of Zander et al. (1981), who report total column abundances above three different float altitudes from balloon flights

TABLE D.7 Summary of HF Measurements (Hudson et al. 1982)

Experimenter	Observation Date	Latitude	Altitude Range	Method	Reference
Zander	September 1974		above 27 km	IR absorption	Zander (1975)
Zander	May 1976	32°N	above 27 km	IR absorption	Zander (1980)
Buijs	May 1976	65°N	15-30 km	IR absorption	Buijs et al. (1980)
Mroz	February to November 1976	30°-33°N	15-37 km	In situ (filter)	Mroz (1977)
Farmer	March 1977	30°S	14-40 km	IR absorption	Farmer et al. (1980)
Zander	October 1978	32°N	above 30 km	IR absorption	Zander (1980)
Carli	April 1979	32°N	30-40 km	IR emission	Bangham et al. (1980)
Marche	May 1979	49°N	20-30 km	IR absorption (ground-based)	Marche et al. (1980)
Zander	September 1979	32°N	above 36 km	IR absorption	Zander (1980)

made over a period of four years. These measurements were made at successively higher altitudes and yield increasingly larger values for the total HF burden above the balloon, which, if interpreted as a profile, produce the result shown in Figure D.46. They may also be indicative of a long-term increase in the stratospheric HF, which Zander has observed in the course of the IR absorption studies. The possibility of such an increase renders even more difficult intercomparisons between measurements in a data base acquired over a period of six years and clearly demonstrates the need to establish a reliable baseline profile for HF against which future measurements can be assessed.

The in situ data of Mroz et al. (1977) shown in Figure D.45 are the average of four seasonal sets of measurements made in 1976. Although the shapes of the profiles are different, the total stratospheric burdens for HF that can be deduced from the data of Mroz and co-workers and of Farmer et al. (1980) appear to be in reasonably good agreement. However, the sampling technique used by Mroz and co-workers is stated to be sensitive to total fluoride, including COF_2 and COFCl: Depending on the model used, this implies that as much as one third of the collected material could have been in the form of these two gases. Thus, the HF in situ results are similar to those for HCl in that they are generally lower than the results obtained using remote sensing techniques.

HOW WELL DOES THE CURRENT DATA BASE ON STRATOSPHERIC REACTIVE TRACE SPECIES CONSTRAIN MODEL PREDICTIONS OF OZONE REDUCTION LEVELS?

Progressing beyond the demonstration that the basic tenets of a given ozone reduction theory have a reasonably high probability of being qualitatively correct requires a significant advance in both the quality of the data available and the manner in which those data are employed.

What we mean by "demonstration that the basic tenets . . . have a reasonably high probability of being qualitatively correct . . ." is that direct observations have verified that

- a given catalytic cycle enhancing the rate of ozone recombination occurs in the stratosphere, e.g., by observing NO, NO_2, O and O_3, or Cl, ClO, O and O_3;

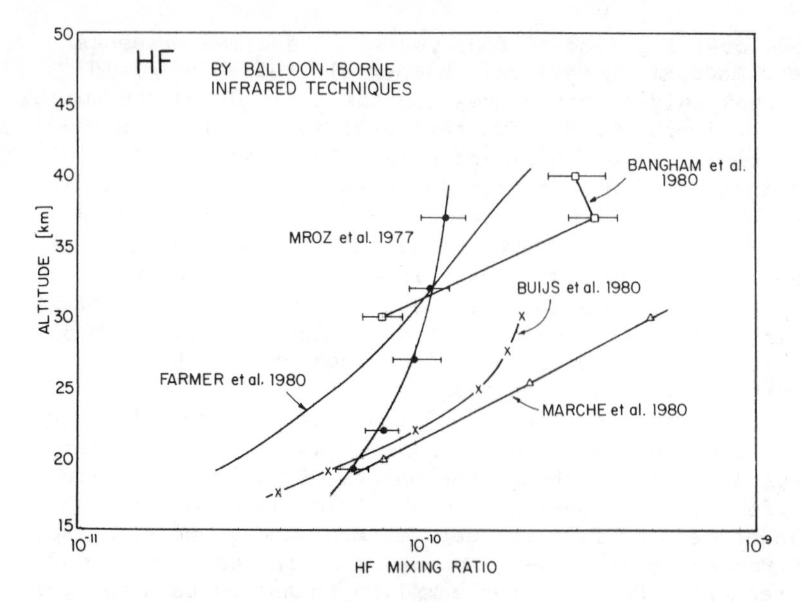

FIGURE D.45 Stratospheric HF profile measurements.

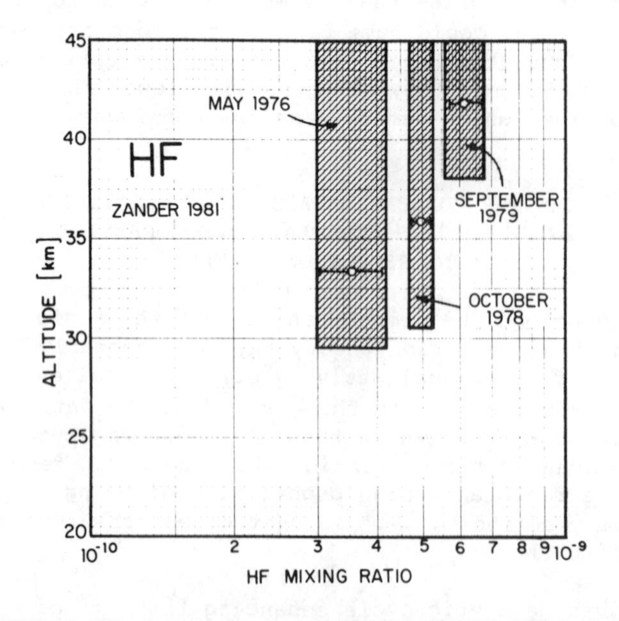

FIGURE D.46 HF mixing ratio reported by Zander (1981). The shaded areas were deduced from near and subhorizontal observations for each of the three flights indicated. Comparison with Figure D.45 indicates a marked change between 20 and 30 km.

• the radicals that make up a given catalytic cycle exist in approximately the model-calculated proportion to their reservoir terms;

• the source molecules that are believed to augment the concentration of a given family of reactants--e.g., N_2O or CH_3Cl, $CFCl_3$, CF_2Cl_2--penetrate the stratosphere approximately as calculated based on their assumed mechanism of destruction; and

• the observed total budget comprising source, reservoir, and free radical concentrations (e.g., CH_3Cl, CH_3CCl_3, $CFCl_3$, CF_2Cl_2; HCl, $ClONO_2$; and Cl, ClO) is in line with modeled predictions.

A review of the data base summarized earlier and treated in detail in Hudson et al. (1982) demonstrates that these criteria have been met for the nitrogen and chlorine systems throughout much of the stratosphere limited to the catalytic cycles:

$$\begin{array}{l} NO + O_3 \rightarrow NO_2 + O_2 \\ NO_2 + O \rightarrow NO + O_2 \\ \hline O + O_3 \rightarrow O_2 + O_2 \end{array} \qquad \begin{array}{l} Cl + O_3 \rightarrow ClO + O_2 \\ ClO + O \rightarrow Cl + O_2 \\ \hline O + O_3 \rightarrow O_2 + O_2 \end{array}$$

and for the hydrogen system in the middle stratosphere for the cycles

$$\begin{array}{l} OH + O_3 \rightarrow HO_2 + O_2 \\ HO_2 + O \rightarrow OH + O_2 \\ \hline O + O_3 \rightarrow O_2 + O_2 \end{array} \qquad \begin{array}{l} OH + O_3 \rightarrow HO_2 + O_2 \\ HO_2 + O_3 \rightarrow OH + O_2 + O_2 \\ \hline O_3 + O_3 \rightarrow 3O_2 \end{array}$$

The central issue is whether we have the experimental evidence to test

1. the quantitative change in ozone for a given change in the source molecule concentration given the set of reactions currently adopted in the best models of the stratosphere; and
2. whether that reaction set is complete with respect to those mechanisms that can directly affect odd oxygen.

We will discover that the data currently available in large measure fail on both counts, but there are important exceptions. Demonstrating precisely why the data fail is important not only to define the limits of our current understanding of the stratosphere, but also to serve as an important lesson for future measurement strategies, a topic discussed in the next section.

The approach adopted here is to trace the impact of the various assumptions about the rate constants through the model-predicted free radical concentrations to the predicted ozone depletion profiles, as summarized in Figure D.47.

In order to correlate the rate constant data, the free radical concentration, and the resulting ozone reduction profiles, we define the following six model cases, which encompass the major uncertainties in laboratory rate data.

Case 1: A set of rate constants identical to that recommended in Hudson and Reed (1979). See Table D.8. This case references all modeled distributions to those employed in the last NRC report.

Case 2: Rate constants similar to those in Case 1 with changes as tabulated in Table D.8. All of the changes are of minor significance for stratospheric modeling (although not for tropospheric modeling).

Case 3: Rate constants identical to those of Case 2, except a temperature-dependent rate constant for the reaction

$$OH + HONO_2 \rightarrow H_2O + NO_3$$

of $k = 1.5 \times 10^{-14} \exp(650/T)\ cm^3\ s^{-1}$ is used in place of the temperature independent rate,

$$k = 8.5 \times 10^{-14} \exp[(0 + 100)/T]\ cm^3\ s^{-1}$$

Case 4: Rate constants identical to those of Case 3, except the rate constant for the reaction

$$OH + HO_2HO_2 \rightarrow H_2O + NO_2 + O_2$$

is assumed to be 5 times that of Case 2.

Case 5: Rate constants identical to those of Case 4, except the rate constant for the reaction

$$OH + HO_2 \rightarrow H_2O + O_2$$

is assumed to be twice that of Case 2.

Case 6: Rate constants the same as those of Case 5, except with a "slow" formation rate for $ClONO_2$. It is assumed in this case that other isomeric forms are rapidly photodissociated following sunrise. See Hudson et al. (1982) for details.

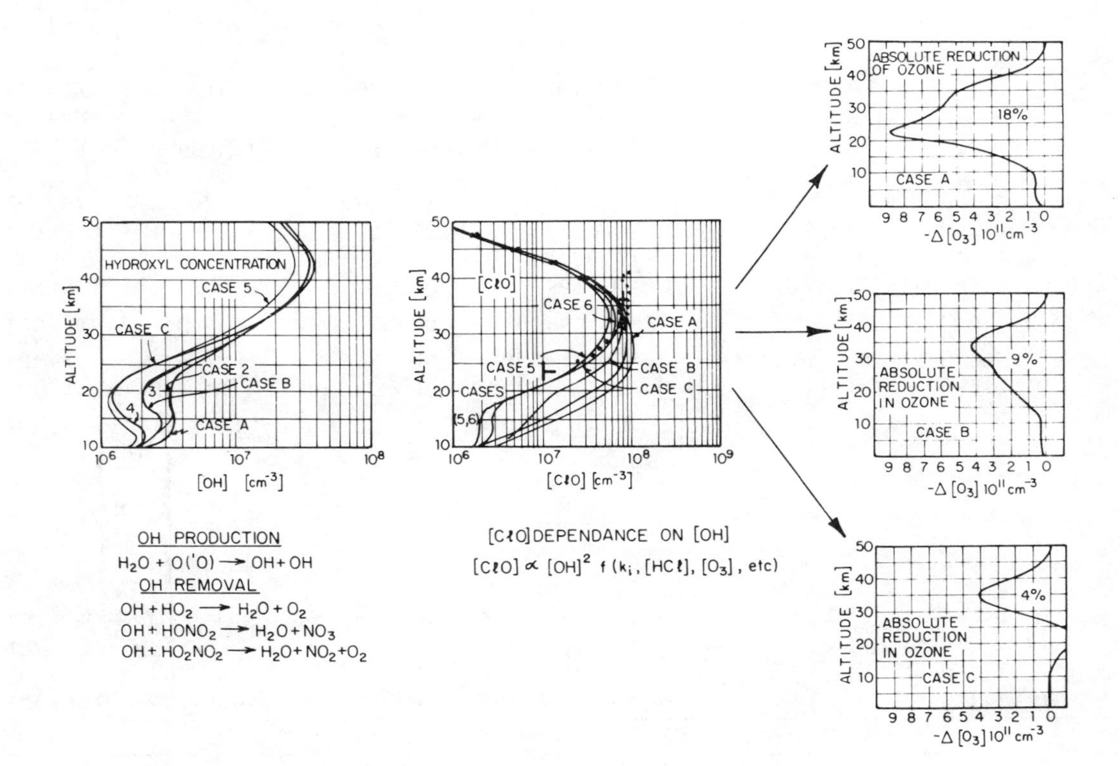

FIGURE D.47 Schematic representation linking the dependence of [ClO] on [OH] and the resulting altitude dependence of ozone depletion as a function of altitude.

TABLE D.8 Reaction Rates of Cases 1 and 2 (cm^{-3} s^{-1})

Reaction	Case 1: NASA 1049	Case 2
$OH + H_2O_2 \rightarrow H_2O + HO_2$	$1.0 \times 10^{-11} \exp[(-750 \pm 350)/T]$	$2.7 \times 10^{-12} \exp[(-145 \pm 100)/T]$
$OH + O \rightarrow H + O_2$	$4.0 \times 10^{-11} \exp[(0 \pm 300)/T]$	$2.3 \times 10^{-11} \exp[(-110 \pm 200)/T]$
$HO_2 + O \rightarrow OH + O_2$	$3.5 \times 10^{-11} \exp[(0 \pm 350)/T]$	$4.0 \times 10^{-11} \exp[(0 \pm 350)/T]$
$ClO + NO \rightarrow NO_2 + Cl$	$7.8 \times 10^{-11} \exp[(+250 \pm 100)/T]$	$6.5 \times 10^{-12} \exp[(+280 \pm 100)/T]$
$O(^1D) + \begin{Bmatrix} N_2O \\ H_2O \\ CH_4 \\ N_2 \\ O_2 \\ CO_2 \end{Bmatrix} \rightarrow$ Product	See NASA 1049	See JPL Publication 81-3
$HO_2 + HO_2 \rightarrow H_2O_2 + O_2$	$2.5 \times 10^{-12} \exp\left[\left(-0 {}^{+0}_{-1245}\right)/T\right]$	2.5×10^{-12}
$HO_2 + NO \rightarrow OH + NO_2$	$4.3 \times 10^{-12} \exp[(200 \pm 200)/T]$	$3.5 \times 10^{-12} \exp[(+250 \pm 100)/T]$
$OH + OH \rightarrow H_2O + O_2$	$1.0 \times 10^{-11} \exp[(500 \pm 400)/T]$	$4.5 \times 10^{-12} \exp[(-275 \pm 275)/T]$
$OH + HOCl \rightarrow H_2O + ClO$	$3.0 \times 10^{-12} \exp[(-800 \pm 500)/T]$	$3.0 \times 10^{-12} \exp\left[\left(-150 {}^{+350}_{-150}\right)/T\right]$
$Cl + CH_4 \rightarrow HCl + CH_3$	$9.9 \times 10^{-12} \exp[(-1359 \pm 150)/T]$	$9.6 \times 10^{-12} \exp[(-1350 \pm 150)/T]$
$Cl + HO_2 \rightarrow HCl + O_2$	$4.5 \times 10^{-11} \exp[(0 \pm 250)/T]$	$4.8 \times 10^{-11} \exp[(0 \pm 250)/T]$
$ClO + HO_2 \rightarrow HOCl + O_2$	5.2×10^{-12}	$4.6 \times 10^{-13} \exp[(+710 \pm 250)/T]$
$NO_3 + h\nu \rightarrow NO + O_2$ $\rightarrow NO_2 + O$	See NASA 1049, page 23	See discussion in this appendix
$OH + HONO_2 \rightarrow$ Products	$8.5 \times 10^{-14} \exp[(0 \pm 100)/T]$	Same as Case 1
$OH + HO_2NO_2 \rightarrow$ Products	5.0×10^{-13}	8×10^{-13}
$OH + HO_2 \rightarrow H_2O + O_2$	$4 \times 10^{-11} \exp[(0 \pm 250)/T]$	Same as Case 1

Briefly, these six cases can be summarized by noting that Case 1 corresponds approximately to the reaction and rate constant set used at the time of the previous NRC report. Case 2 catalogs all the refinements in reported rate constants that do not have a significant effect on the calculated distribution of any key reactive species in the stratosphere. Cases 3, 4, and 5 define the impact of, respectively, the faster low-temperature rate constant for $OH + HONO_2$, the faster overall rate for $OH + HO_2NO_2$, and the faster overall rate for $OH + HO_2$. Case 6 isolates the impact of assuming the formation of isomers other than $ClONO_2$ in the termolecular recombination of NO_2 and ClO.

Given that the most immediate concern of this report is an assessment of the ozone reduction resulting from the release of fluorocarbon compounds, we treat the chlorine-ozone question first.

Chlorine-Induced Destruction of Ozone

The [ClO] Profile

Figure D.48 compares the model calculated [ClO] profiles for each of the six cases defined above, with the corresponding altitude dependence of ozone reduction for steady state conditions given 1976 release rates of fluorocarbons and 1979 release rates for chloroform. These results were provided by D.J. Wuebbles and J.S. Chang of the Lawrence Livermore National Laboratory. Also cited in the figure are the integrated column reduction percentages for each of the six cases.

The most obvious conclusion to be drawn from Figure D.48 is that both the altitude distribution and the integrated reduction in ozone are exceedingly sensitive to the rate constants selected for the HO_x reactions. This results from the quadratic dependence of [ClO], the rate limiting radical in the dominant chlorine catalytic cycle:

$$Cl + O_3 \rightarrow ClO + O_2$$
$$ClO + O \rightarrow Cl + O_3$$
$$\overline{O_3 + O \rightarrow 2O_2}$$

on [OH], as noted in Appendix C. The second conclusion is that, for the diminished OH concentration in the lower

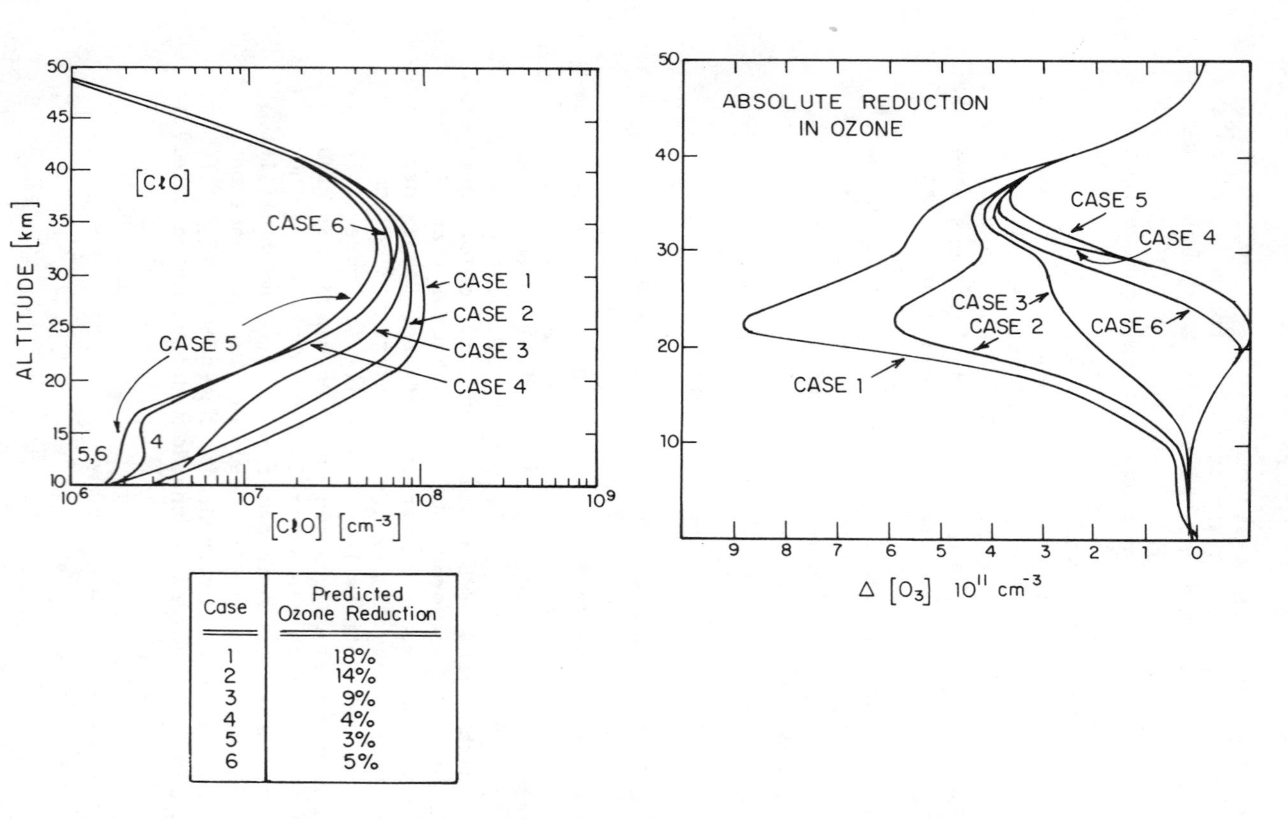

Case	Predicted Ozone Reduction
1	18%
2	14%
3	9%
4	4%
5	3%
6	5%

FIGURE D.48 Predictions of stratospheric ClO concentration and ozone reduction for the six cases described in the text.

stratosphere corresponding to Cases 4, 5, and 6, the chlorine-induced destruction of ozone is confined to the altitude region above 30 km with less than 10 percent of the integrated effect occurring at lower altitudes. The bimodal distribution in $\Delta[O_3]$ as a function of altitude for Cases 1, 2, and 3, which results from the contribution of the catalytic cycles (see Wuebbles and Chang 1981)

$$
\begin{array}{l}
Cl + O_3 \rightarrow ClO + O_2 \\
ClO + HO_2 \rightarrow HOCl + O_2 \\
OH + O_3 \rightarrow HO_2 + O_2 \\
HOCl + h\nu \rightarrow OH + Cl \\
\hline
O_3 + O_3 \rightarrow 3O_2
\end{array}
\qquad
\begin{array}{l}
Cl + O_3 \rightarrow ClO + O_2 \\
ClO + NO_2 \rightarrow ClONO_2 \\
NO + O_3 \rightarrow NO_2 + O_2 \\
ClONO_2 + h\nu \rightarrow Cl + NO_3 \\
NO_3 + h\nu \rightarrow NO + O_2 \\
\hline
O_3 + O_3 \rightarrow 3O_2
\end{array}
$$

below 30 km, disappears entirely for Cases 4, 5, and 6, which are characterized by 10 times lower [ClO] at 20 km. The isolation of chlorine-induced ozone destruction to the middle and upper stratosphere greatly simplifies the interpretation of ozone destruction by chlorine both because the chemical time constant for ozone (defined here as the ozone concentration divided by the rate of odd oxygen production) is much shorter than transport times in the middle and upper stratosphere, and thus local chemical production and destruction rates determine the ozone concentration. The cases characterized by low [ClO] also represent a significant decoupling of the chlorine system from hydrogen and nitrogen in the lower stratosphere.

Can we, based on the available ClO data, select which of the six cases most accurately reflects conditions in the real atmosphere? Figure D.49 displays Cases 1 through 6 superposed with the in situ data that comprise the envelope of observations critiqued earlier. The spread in the in situ observations is approximately ± 50 percent about the mean, but there is a clear indication that: (a) the measurements strongly favor the calculated distributions characterized by a rapid decrease in [ClO] below 30 km; and (b) the rapid decrease in [ClO] above 35 km predicted by all six model cases is not substantiated by the data.

Next, in Figure D.50, we superpose the mean of the nine in situ observations shown in Figure D.49 with the balloon-borne mm-wave emission data of Waters et al.

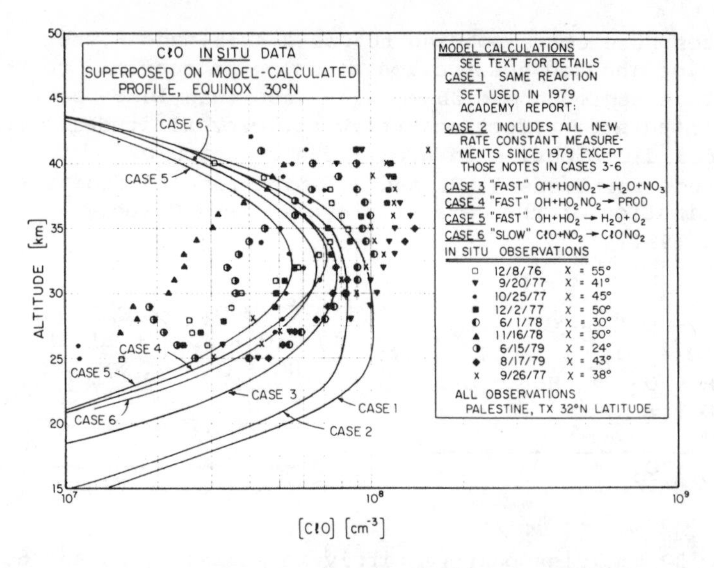

FIGURE D.49 ClO in situ data superposed on model-calculated profile, equinox 30°N.

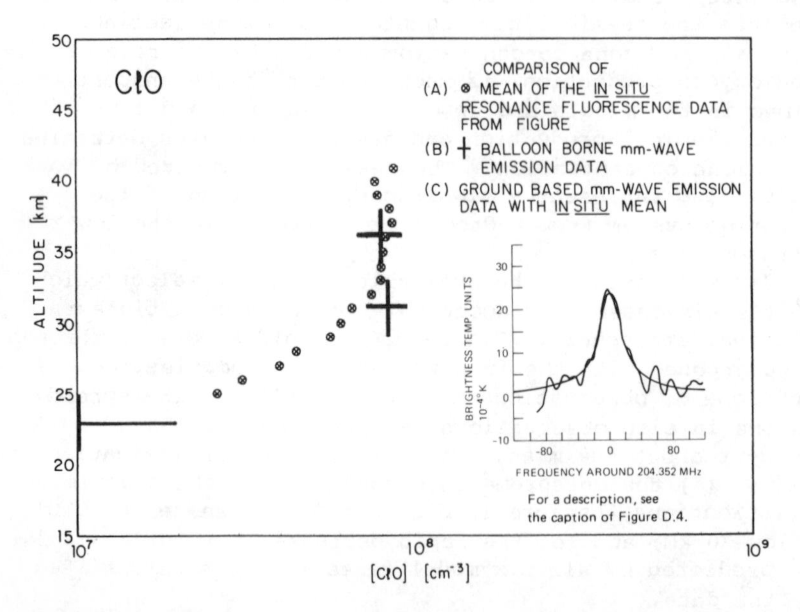

FIGURE D.50 Summary of the comparison between the ground-based mm-wave emission data of Parrish et al. (1980), the balloon-borne mm-wave emission data of Waters et al. (1981), and the in situ data of Anderson and co-workers.

(1981) and note with an inset the comparison between the mean of the in situ data and the ground-based mm-wave emission data of Parrish et al. (1981).

We conclude from Figure D.50 that to first order the three independent ClO detection methods provide very consistent results. A comparison between those data, represented by the balloon-borne observations that directly observe the distribution of [ClO] with altitude, and the six modeled cases is presented in Figure D.51.

Taken as a whole, the data clearly support the calculations that predict the minimum ClO concentration below 30 km, i.e., Cases 4 through 6. This is a conclusion of considerable importance, not only because, if accepted, it alters the predicted ozone reduction levels very significantly (see Figure D.48), but also because it seriously constrains any proposed mechanism involving chlorine radicals in the lower stratosphere. Such a proposed mechanism might involve either a catalytic cycle that is involved directly in the recombination of odd oxygen or a reaction linking the chlorine system to another family of reactants (e.g., bromine, nitrogen, hydrogen, etc.); in either case, the factor of 10 lower ClO concentration in the lower stratosphere seriously reduces the probability that such a mechanism can be of quantitative significance.

We are thus limited to Cases 4 to 6, which isolate the chlorine-induced destruction of ozone to altitudes above 30 km (where the ozone density is controlled predominantly by local chemical production and destruction). In this altitude regime, it is clear that observed [ClO] exceeds the calculated distribution by nearly a factor of 3 at 40 km. The first-order importance of this can be represented by an "overlap integral" between the altitude dependence of chlorine-induced ozone destruction, $\Delta[O_3]$ versus altitude, and the ratio of calculated to observed [ClO]. This is summarized in Figure D.52.

The implication of Figure D.52 is that a mechanism not currently included in the models exists that converts HCl (the dominant form of chlorine at 40 km) to the free radical form ClO. If that mechanism does not enter directly in a rate limiting process for ozone production or destruction, then Figure D.52 implies that the chlorine-induced destruction of ozone will deepen and shift to higher altitudes. If the missing mechanism does involve odd oxygen production or destruction directly, then one cannot conclude even the sign of the effect resulting from the inclusion of the mechanism.

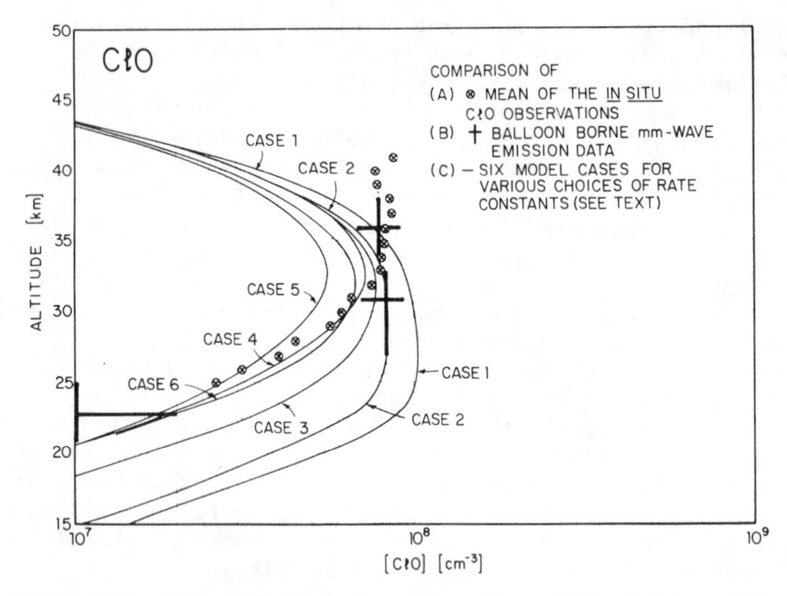

FIGURE D.51 Superposed balloon-borne observations of ClO and the six model-calculated cases defined in the text.

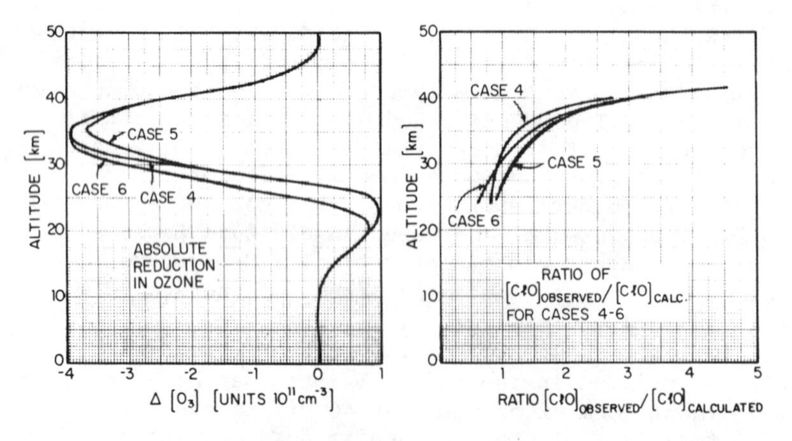

FIGURE D.52 Correlation between the model-calculated altitude dependence of ozone depletion (resulting from 1976/1979 release rates, as noted in the text) and the ratio of calculated to observed [ClO] in the stratosphere.

The [OH] Profile

Given the direct relationship between the ClO concentration profile and the predicted reduction in ozone, we consider next whether other observations substantiate or refute the selection of Cases 4 to 6 as those most appropriately representative of stratospheric photochemistry for the cases considered. The most obvious test is, of course, the correlation between calculated and observed [OH] because (a) the ClO concentration depends quadratically on [OH] below 30 km, as discussed in Appendix C; and (b) it is the $HO_x = OH + HO_2$ concentration that is altered by the various rate constant assumptions for

$OH + HONO_2 \rightarrow N_2O + NO_3$	Case 3
$OH + HO_2NO_2 \rightarrow$ products	Case 4
$OH + HO_2 \rightarrow H_2O + O_2$	Case 5

Figure D.53 summarizes the constraints placed on the six cases by the available OH data, represented here by the in situ balloon-borne data. We select those data because they are consistent with the ground-based total column measurements of OH and they provide the only information available on the shape of the [OH] distribution in the stratosphere. The conclusion, however, is disappointing. The observations are insufficient in number, (absolute) accuracy, and altitude coverage to distinguish between any of the available cases. In the most important region below 30 km, no data exist.

If we consider the other HO_x radical, HO_2, we find the same situation; the data are too scattered and of insufficient altitude coverage to test this critical question of lower stratospheric OH_x (Figure D.54).

A gap of considerable importance thus exists in the case linking low ClO concentrations in the region below 30 km to low OH concentrations, which are in turn explained by exhanced HO_x destruction via reactions of OH with nitric and pernitric acid.

We can, of course, search elsewhere for clues regarding the destruction of OH in the lower stratosphere, most notably in the nitrogen system. Before doing this, however, we turn to a brief review of the nitrogen-catalyzed destruction of odd oxygen and the perturbation of O_3 resulting from the doubling of N_2O.

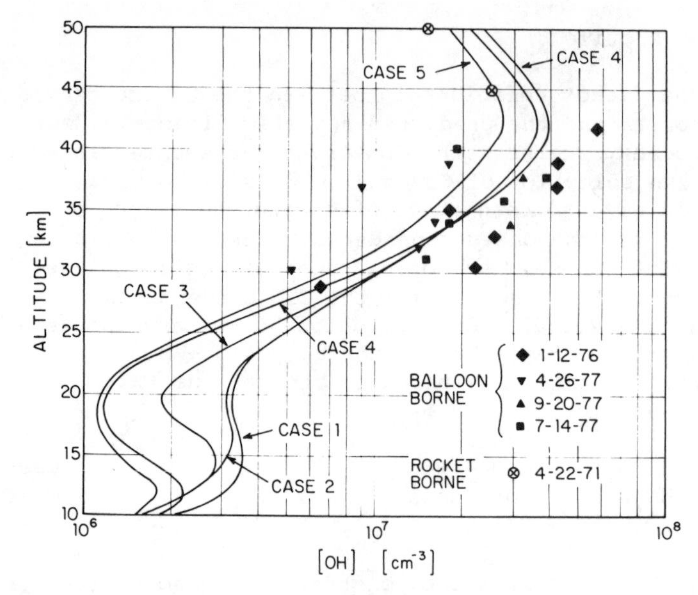

FIGURE D.53 Comparison between observed and calculated OH for the six model cases. Case 6 matches the profile for Case 5.

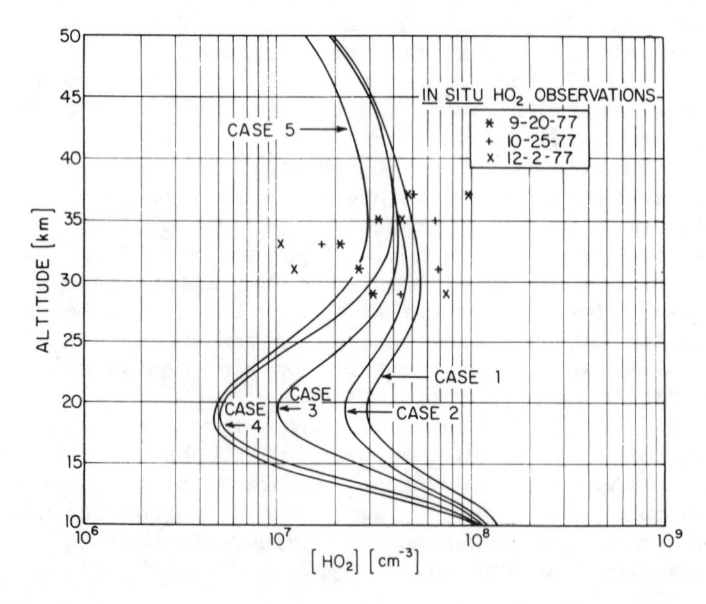

FIGURE D.54 Summary of the correlation between observed and calculated HO_2.

Nitrogen-Induced Destruction of Ozone

The $[NO_2]$ Profile

We first define, in Figure D.55, the response of ozone, as a function of altitude, to a doubling of N_2O for each of the six cases, previously defined. Also included in that figure is the corresponding altitude distribution of the rate limiting radical, NO_2. Cases 1 and 2 are characterized by (nearly) equal but opposite lobes in ΔO_3 that virtually cancel when integrated. This behavior underlines the important connection between the nitrogen and hydrogen catalytic systems. Positive values of ΔO_3 for Cases 1 and 2 below 25 km result from a decrease in the rate of O_x catalysis (not a production of odd oxygen!) by HO_x for increased levels of NO_x because the reaction

$$HO_2 + NO \rightarrow OH + NO_2$$

shifts the available HO_x from the O_x catalytic rate limiting form HO_2 to OH. This behavior is observed only under conditions in which there is sufficient HO_x to dominate the budget of O_x and the behavior is simply one of a deepening of the ozone destruction profile with increasing NO_2. Note that, as OH decreases, the concentration of NO_2 increases since NO_x is removed predominantly by the recombination of OH with NO_2 to form nitric acid.

In Figure D.56, we present the overlay of the 32°N latitude NO_2 data (corrected to midday conditions) with the six modeled $[NO_2]$ profiles. Several points are immediately apparent. First, the range in calculated NO_2 is significantly smaller than that for ClO and thus, while the spread in the NO_2 data is less than that for ClO, it is decidedly more difficult to extract a clear conclusion. The implication, however, is that Cases 1 to 3 correlate better with the data than do Cases 4, 5, and 6. This is obviously a point of considerable importance, not only as a clue to the question of lower stratospheric HO_x and the consistency of our picture of ClO concentration below 35 km, but also from the point of view of the total odd oxygen balance. As noted in Appendix C (Figure C.6a), approximately 70 percent of the total production rate of odd oxygen between 20 and 35 km is balanced by NO_x catalysis (rate limited by NO_2) for a reaction rate constant set corresponding to Case 5, so differences of even ± 50 percent are of major import.

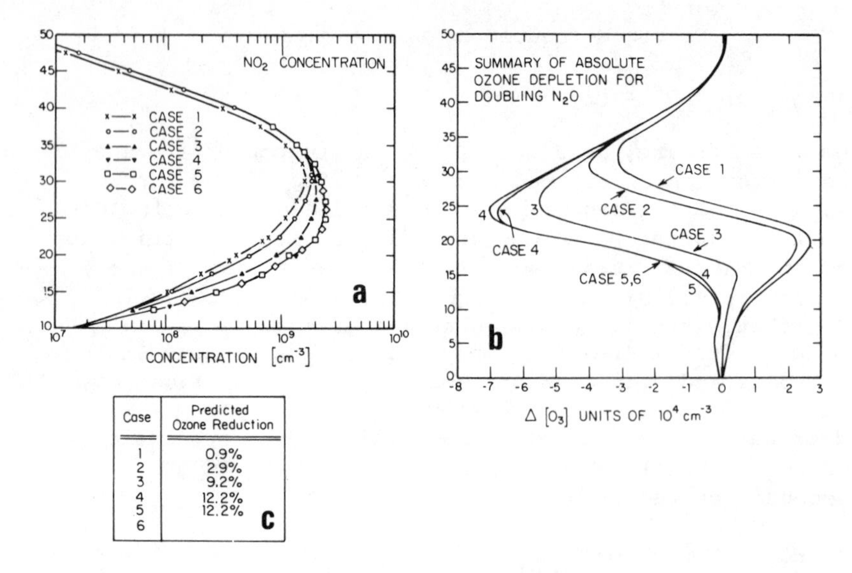

Case	Predicted Ozone Reduction
1	0.9%
2	2.9%
3	9.2%
4	12.2%
5	12.2%
6	

FIGURE D.55 (a) Calculated NO_2, (b) ozone reduction as a function of altitude for a doubling of NO_2, and (c) the integrated column reduction of ozone for each of the six cases defined in the text.

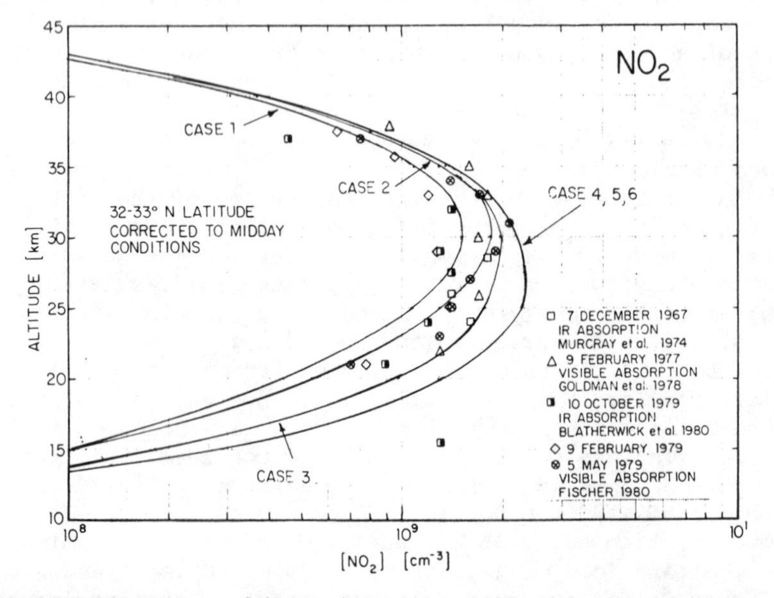

FIGURE D.56 Summary of the correlation between NO_2 observed at midday and the six modeled cases defined in the text. The model calculations here do not include the effect of the spherical earth on multiple scattering at large zenith angles. This effect is important below 30 km.

Examination of the correlation between observed and calculated middle-latitude NO is summarized in Figure D.57, using the most consistent data set available, that of Ridley and co-workers. Although Figure D.57 reflects the significantly larger data base available for NO than for NO_2 (note, also, that the in situ observations were done at midday and thus do not need to be "corrected" for diurnal differences), the superposition of observations and theory is inconclusive. One cannot discriminate, based on the best available data, between NO concentrations corresponding to column-integrated ozone reduction figures from 1 to 12 percent.

The [$HONO_2$] Profile

We consider next the correlation between calculated and observed $HONO_2$, singling out the mid-latitude data previously presented in Figure D.57. It is clear from Figure D.58 that the nitric acid concentration below 25 km is insensitive to the particular choice of rate constants partitioning reactive nitrogen among NO, NO_2, NO_3, N_2O_5, and $HONO_2$ because nitric acid dominates the reactive nitrogen budget. In the upper stratosphere, nitric acid is only weakly dependent on [OH], and thus there is little hope of using such data to constrain current models with respect to ozone reduction prediction. It is of considerable interest, however, to note the clear divergence between all six cases and the envelope of observations at altitudes above 25 km. This has been a persistent and unresolved feature and while the nitric acid data above 25 km are not as direct a check on the odd oxygen budget as are observations of NO and NO_2, such differences are very clearly of concern.

The Profile of the Ratio of [$O(^3P)$] to [O_3]

We conclude this section by comparing the calculated and observed ratio of [$O(^3P)$] to [O_3]. We display in Figure D.59 the calculated ratio and the mean of the $O(^3P)$ in situ observations and the most recent in situ ozone observations obtained in June (1978 and 1981) at Palestine, Texas, from Figure D.20. The in situ ozone data, discussed earlier, was obtained using three different methods that agree within the uncertainty of the techniques, which is less than 10 percent.

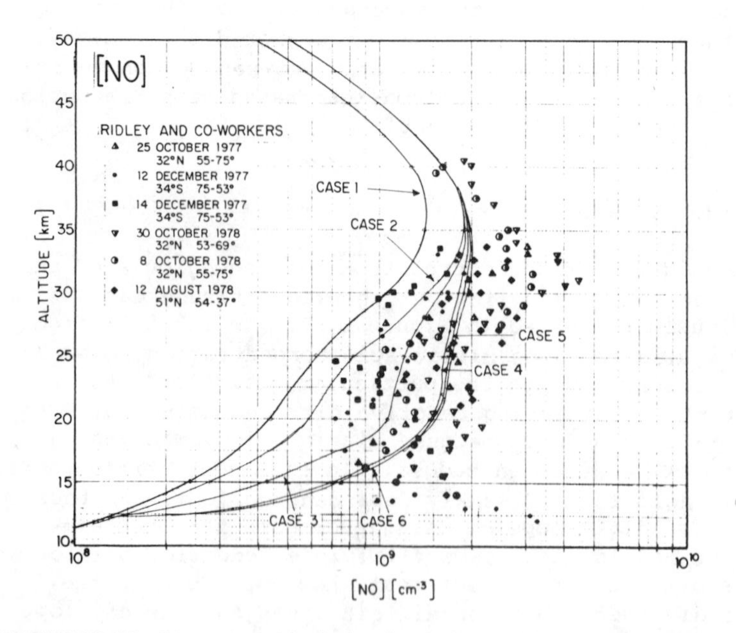

FIGURE D.57 Comparison of calculated and observed NO at mid-latitude.

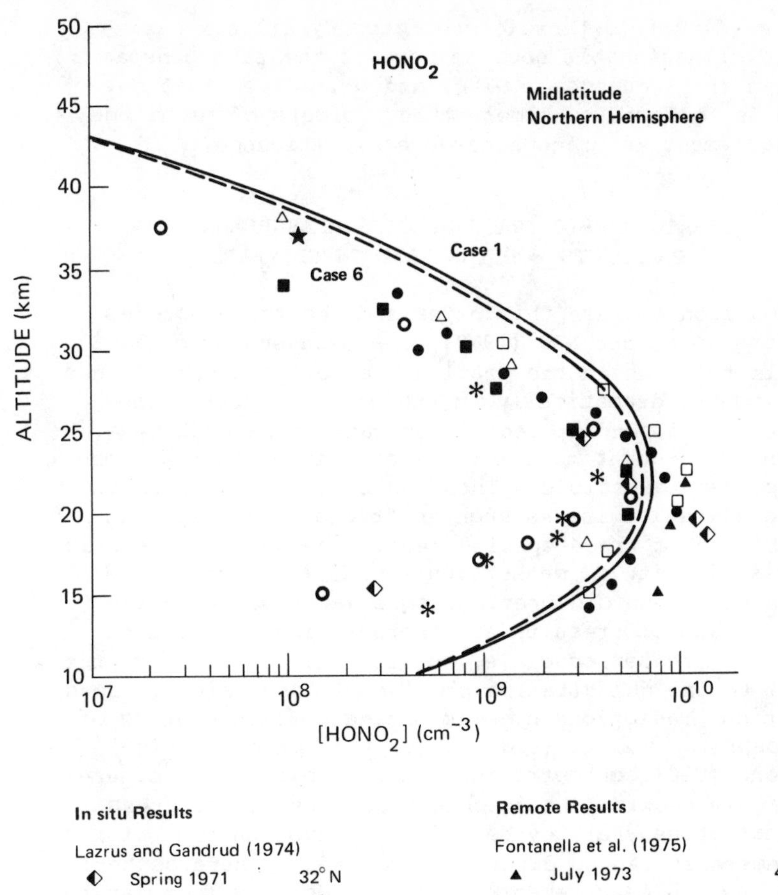

In situ Results

Lazrus and Gandrud (1974)

◐	Spring 1971	32° N
*	Spring 1972	32° N
o	Spring 1973	32° N

Arnold et al. (1980)

★	November 1977	45° N

Remote Results

Fontanella et al. (1975)

▲	July 1973	48° N

Harries et al. (1976)

□	September 1974	45° N

Evans et al. (1978)

■	July-August 1974-76	51° N

Murcray et al. (1980)

△	October 1979	32° N

Fischer (1980)

●	May 1979	31° N

FIGURE D.58 Comparison of calculated and observed $HONO_2$ at mid-latitude.

The $[O(^3P)]/[O_3]$ ratio predicted by all six cases is indistinguishable because none of the rate constants involve the exchange of $O(^3P)$ and O_3. Given that the ratio is followed over more than 2 orders of magnitude, the agreement is of considerable significance.

STRATOSPHERIC TRACE SPECIES MEASUREMENTS: PROSPECTS FOR THE NEXT THREE YEARS

A comparison between this paper and the trace species sections in either NRC (1979) or Hudson and Reed (1979) reveals that, while the total number of observations has not expanded dramatically in the past two years, the number of independent techniques has, in several key instances, brought much more clearly into focus a number of important questions. Those questions range from the more qualitative issues such as "typical" atmospheric variability of trace species (e.g., the newest generation of (six) in situ NO measurements exhibits much greater consistency than did previous observations, while the H_2O, OH, and ClO results are characterized by sets of data that are reproducible, but for which clear exceptions exist) to the quantitative problem of constraining ozone reduction predictions by eliminating certain classes of stratospheric models (e.g., those that predict "high" [OH] and [ClO] concentrations in the lower stratosphere).

In this section, we discuss prospects for progress in the next three years by defining the evolving set of problems that can be directly addressed by stratospheric trace species measurements. This is done by formulating a series of questions abstracted from earlier discussions. The emphasis is placed almost entirely on photochemical mechanisms involving the higher reactive trace species that either couple the chemical families (i.e., nitrogen, hydrogen, chlorine, etc.) together by radical-radical recombination steps or enter directly into the rate-determining processes for odd oxygen production/destruction. This bias eliminates in large measure the exceedingly important topic of satellite observations, a subject that has been recently reviewed in detail in Hudson et al. (1982).

Following a statement of the questions that must be addressed by stratospheric measurements in the near future, an appraisal of the prospects for making significant progress in the next three years is presented.

QUESTION 1: Is the cause of the large discrepancy between the observed and calculated [ClO] profile above 35 km an experimental problem or is there an important mechanism missing in the models that converts chlorine to the free radical ClO?

This discrepancy is critical because it coincides in altitude with the peak in the chlorine-induced odd oxygen destruction profile (see Figure D.52). It also has a great deal in common with a similar divergence between calculated and observed [ClO] below 30 km that existed at the time of the last NRC report, a situation summarized in Figure D.60.

In the past two years, both the balloon-borne microwave emission data and the ground-based mm-wave emission data have confirmed the in situ results below 30 km. Of perhaps greater importance was that a plausible mechanism for the cause of the discrepancy evolved out of laboratory measurements of the reaction rate constant data for OH + $HONO_2$ and OH + HO_2NO_2, as noted in the definition of Cases 3 and 4.

Without a resolution to Question 1, there will remain two schools of thought on quantitative predictions of fluorocarbon-induced ozone reduction, because the latter is a sensitive function of the vertical distribution of the rate limiting radical, ClO, in the chlorine-catalyzed destruction of ozone.

PROSPECTS: Several advances in ClO detection methodology will, with high probability, settle the remaining questions concerning the vertical distribution of ClO at mid-latitudes. Progress is forecast in three areas. First, the cross-calibration of the balloon-borne techniques (which within the next two years will include in situ chemical conversion-resonance fluorescence, mm-wave emission, and laser heterodyne radiometry) will define the experimental uncertainties of the three methods. It is already clear from the intercomparison of the mm-wave and in situ methods that agreement within the cited uncertainties of ±30 percent exists. Extensive laboratory simulation work, the use of redundant instruments on each flight, and the addition of several in-flight calibration checks coupled with more careful control of descent velocities will reduce the experimental uncertainties of the in situ methods to about ±10 or 15 percent. Similar figures are the planned objective of the remote techniques.

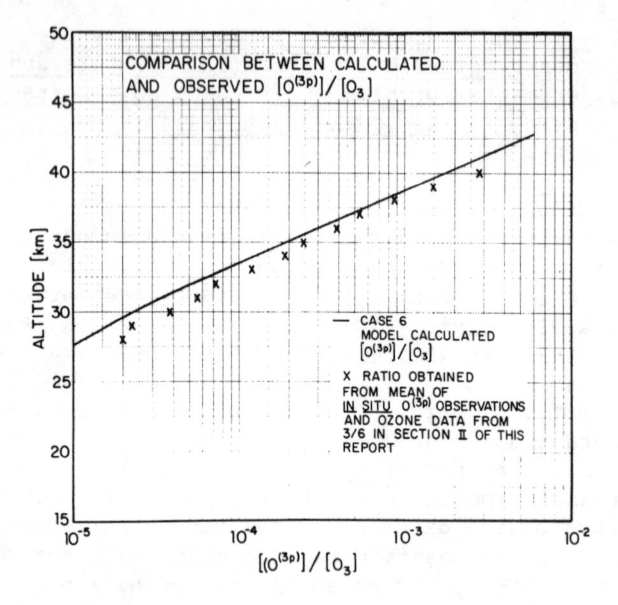

FIGURE D.59 Comparison between observed and calculated $[O(^3P)]/[O_3]$.

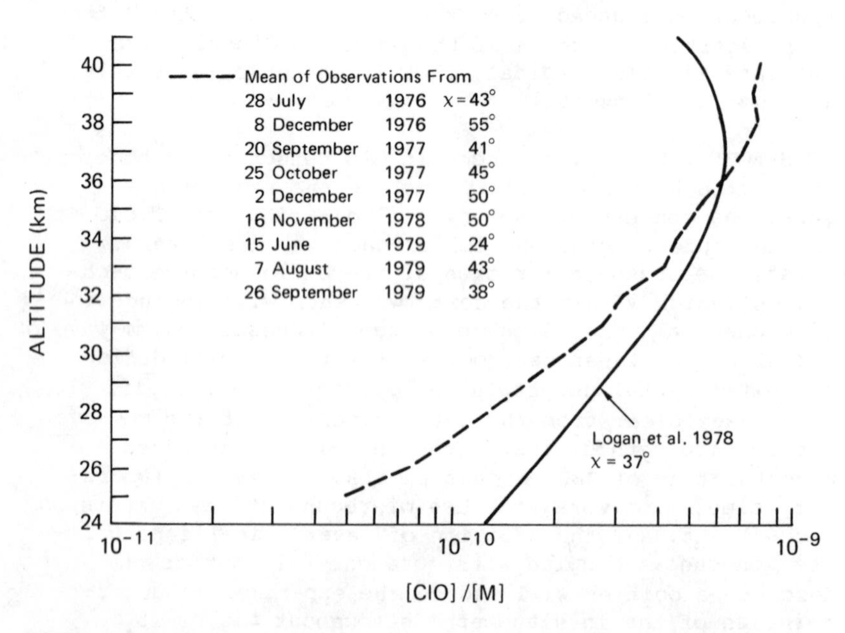

FIGURE D.60 Comparison between the mean of all in situ ClO observations, excluding July 14, 1977, and the calculated ClO distribution from Logan et al. (1978).

The ClO distribution above 35 km must be explored with the balloon-borne mm-wave emission measurements to determine, in particular, the total ClO column density above 40 km. This is most effectively done simultaneously with in situ observations obtained with multiple vertical scans using the reel down/reel up deployment technique currently under development.

It is important to note that all the measurement techniques provide data that are easiest to interpret in the low-pressure region of the upper stratosphere, so that if the middle and lower stratospheric profiles are correct, the probability is large that the high-altitude end of the profile is correct.

Second, the development of ground-based mm-wave emission techniques provides the means for obtaining much better temporal coverage to search for the occurrence of enhancements reported by the in situ methods--the only impediment currently preventing the initiation of that coverage is the serious attenuation of 204-GHz radiation by water vapor in the troposphere, which maximizes during the summer months, encompassing July when both high values were observed.

Careful delineation of the diurnal behavior of ClO should also be accomplished in the next five years using both the mm-wave emission technique and the multiple vertical scan in situ technique. These results should cast light on the question of isomer formation from the reaction of ClO with NO_2. This is not suggested as a substitute for direct laboratory data, but rather as a complementary approach to establish the temporal behavior of ClO as a function of altitude throughout the night and following sunrise.

QUESTION 2: Is the stratosphere most accurately characterized by "high" [OH] below 30 km, as represented by Cases 1 and 2 in Figure D.53, or by "low" [OH], as defined by Cases 4 and 5 in the same figure?

From the standpoint of understanding perturbations to stratospheric ozone, this question is of unequaled importance because an unequivocal answer will establish (a) whether the rate of ozone destruction in the lower stratosphere is controlled by catalytic cycles involving HO_x or NO_x radicals, and (b) whether the chlorine radicals Cl and ClO have any measureable impact on the odd oxygen budget below 30 km. Without direct observations of OH (with an excellent signal-to-noise ratio) in the region between 15 and 30 km, obtained simultaneously with measurements of H_2O, an intolerable gap will remain in the case linking chemical perturbation to ozone reduction in the stratosphere.

PROSPECTS: Two methods have been developed to extend previous OH measurements in the upper stratosphere to lower altitudes. A lidar method, employing a pulse laser with time-resolved detection, has been initially tested in the stratosphere and should yield the first balloon-borne remote measurements in the next five years. In addition, an in situ method employing a high repetition rate (20,000 Hz) tuneable laser has been developed to determine the OH concentration in situ throughout the stratosphere with approximately 1000 times the signal-to-noise ratio of the experiments reported earlier. The in situ method also promises to provide observations of HO_2 by using chemical conversion (the addition of NO) to convert HO_2 to OH, followed by laser-induced fluorescence detection of OH. This "simultaneous" detection of OH and HO_2 with the same absolute calibration will establish the sum of the two major HO_x species and the ratio with an altitude resolution of less than or equal to 0.5 km and a signal-to-noise ratio greater than or equal to 10 throughout the stratosphere.

Development of cryogenically cooled detection chambers for the measurement of H_2O by fragment fluorescence under daylight conditions should, within two years, provide the first data on the ratio of $[HO_x]$ to $[H_2O]$.

QUESTION 3: What is the mean distribution of NO_2 as a function of altitude between 15 and 45 km determined to an (absolute) accuracy of ±10 percent throughout the day at equatorial, lower mid-latitude, and upper mid-latitude locations?

Given our current picture of odd oxygen destruction rates, as summarized in Figure C.6a of Appendix C, catalytic destruction of odd oxygen by NO_x constitutes at least 70 percent of the ozone budget between the tropopause and 35 km. That catalytic cycle is rate limited by NO_2 at all altitudes, yet we do not have high-accuracy data on this critical radical as a function of altitude and latitude.

PROSPECTS: High-accuracy/precision NO_2 observations with excellent signal-to-noise ratios have not been reported, but a recent experiment employing photolytic conversion of NO_2 to NO followed by the chemiluminescent detection of NO holds promise of making a considerable contribution to this exceedingly serious shortcoming in our observational data base. A particularly attractive feature of the technique is that it provides a measurement of NO with the same absolute calibration, so highly precise ratios of NO to NO_2 should result. In addition, on-board NO and NO_2 calibrated samples should yield exceedingly accurate absolute results.

There are other methods currently under study in the laboratory for both NO and NO_2 including laser-induced fluorescence, double photon ionization, and double photon fluorescence. Those in situ methods may well yield the first "cause and effect" studies of the odd oxygen budget by correlating local fluctuations in NO_2 and O_3 in the middle stratosphere, where the loss rate of O_3 is controlled almost entirely by the NO_x catalytic cycle.

The infrared methods that are not plagued by the restriction of sunset-sunrise geometries may also, if cross-calibrated in the same air mass, yield important results that will address Question 3.

<u>QUESTION 4: Do the infrared techniques applied to the "reservoir" terms HCl and $HONO_2$ yield results within the stated experimental accuracies when applied simultaneously to the same air mass?</u>

An exceedingly important check on the models used for ozone reduction calculation comes from a comparison between calculated and observed concentrations of the chlorine and nitrogen compounds of intermediate lifetime (about a few months) in the middle and lower stratosphere. This is because these compounds, primarily HCl and $HONO_2$, dominate the total budget of reactive chlorine and nitrogen in the lower and middle stratosphere and are transport controlled and thus sensitive to model assumptions regarding vertical and horizontal transport.

<u>PROSPECTS</u>: An extensive series of cross-calibration flights, wherein the major infrared remote techniques for HCl and $HONO_2$ detection are used to interrogate the same air mass, is scheduled for the next two years. Experiments from both Europe and North America will be included, and a standard series of deconvolution programs will be applied to the data with careful comparison of the resulting profiles.

This flight series should make major advances toward narrowing the experimental uncertainties in the observation of both HCl and $HONO_2$. It has generally been found that a few carefully orchestrated observations are more effective for testing models than a large number of observations with questionable absolute calibration. It is equally true that the observation of a given trace reactant by as many independent methods (with comparably defensible absolute calibration) is essential for acceptance by the scientific community. This joint flight of the analytical techniques for HCl and $HONO_2$ is in response to that fact, and the results will be applicable to a broad range of molecules that can, at present, only be observed by remote IR methods.

QUESTION 5: What is the diurnal behavior of ClO, NO_2, NO, OH, and HO_2 as a function of altitude between the tropopause and 45 km?

Although high-quality profiles of the major radicals at midday are of first-order importance, there is a great deal to be learned from the temporal behavior, under carefully controlled conditions, of the highly reactive trace species following sunrise and sunset. It is crucial in these studies to achieve altitude resolution of 1 to 2 km and to watch several related species simultaneously.

PROSPECTS: The need to obtain simultaneous data on the five constituents with good altitude resolution is one of the most difficult analytical challenges currently facing the field. It will require the simultaneous deployment of three sophisticated experiments with repetitive vertical scans, concentrating on the sunset and sunrise periods. While this capability will probably be within reach in the next two to three years, considerable progress will almost certainly be made using mm-wave emission techniques to examine the diurnal behavior of ClO, IR emission techniques and chemiluminescence for NO and NO_2, and balloon-borne lidar or in situ laser-induced fluorescence measurements for OH and HO_2, separately deployed in each case.

QUESTION 6: What is the spectral distribution of solar radiation between 180 and 240 nm as a function of altitude down to the tropopause?

The absence of published high-resolution data of the solar flux as a function of altitude, solar zenith angle, and wavelength is a shortcoming of major importance. Without those direct measurements, the loss rate of the critical "source" terms (e.g., $CFCl_3$, CF_2Cl_2, and CH_3Cl) cannot be checked.

PROSPECTS: Within the next year, publication of the first high-resolution data on the penetration of solar flux in the 180- to 240-nm spectral interval should begin to eliminate a serious shortcoming on the question. If this does not clear up discrepancies in the loss rates of, for example, $CFCl_3$, then it may be necessary to consider the difficult observations of dissociation rates directly measured in situ. A discussion of the discrepancies between observed and calculated source molecules (CH_4, N_2O, CH_3Cl, $CFCl_3$, ethane) appears in Appendix C.

QUESTION 7: What is the vertical distribution of ClO, NO, NO_2, OH, and HO_2 between 15 and 45 km in the equatorial latitudes?

Far too much emphasis has been placed on the analysis of mid-latitude data as a result of the concentration of experimental results on this region. However, the dominant region of global ozone production exists at latitudes below 30°N, and it is of first-order importance to discover whether [ClO], for example, exhibits the behavior characterized by a rapid decrease below 30 km as it does at 32°N. There are comparably important examples in the HO_x and NO_2 systems.

PROSPECTS: Within two years, the new generation of techniques previously discussed should have provided the first high-quality soundings of these key radicals, hopefully with simultaneous observation of H_2O and O_3 with the OH and HO_2 experiments. It will require, perhaps, another two years to establish with considerable confidence the mean distribution of those radicals, but the large observed fluctuations in H_2O above the tropopause may yield valuable insight into the chemical linking between the NO_x, HO_x, and ClO_x families by studying the covariance between these radicals. Simultaneous in situ observations of ozone may yield exceedingly important insight into the odd oxygen budget from the same series of observations.

QUESTION 8: What is the altitude distribution of the important intermediates, HOCl, $ClONO_2$, HO_2NO_2, NO_3, N_2O_5, and $HONO_2$ in the stratosphere?

The reasons that these products of radical-radical recombination reactions are important are discussed throughout this report and need not be repeated. They present a particularly difficult analytical problem, however, because they are in general large polyatomic molecules that do not possess strong electronic transitions, yet their predicted concentrations fall below the detection threshold of long-path IR absorption techniques.

PROSPECTS: The first three molecules in this group constitute an exceedingly difficult triplet from the point of view of analytical techniques that can be applied to the stratosphere. Initial detection of $ClONO_2$ has been reported, but the detection is marginally possible with the best IR methods available, and no method has reported observation of HOCl and HO_2NO_2. Significant difficulties are predicted for progress on these molecules, but the options have not been exhausted. Double photon ionization methods and fragment fluorescence may be applicable, although the ubiquitous nature of the hydrogen, nitrogen, and oxygen fragments in pernitric acid will make such measurements difficult to interpret.

Initial measurements of NO_3 at night are encouraging. Attempts to detect N_2O_5 by thermal dissociation followed by detection of the NO_x products formed have been made in the laboratory, but have not shown sufficient promise to warrant stratospheric application.

It would, in addition, be exceedingly important if an unambiguous technique for detecting $HONO_2$ in situ could be developed. This would contribute significantly to the question of the NO_2, $HONO_2$, OH chemistry of the lower stratosphere.

QUESTION 9: What is the concentration of NO_2, NO, ClO, OH and O_3 simultaneously determined in an air mass characterized by the very low NO_2 concentration observed by Noxon northward of the high-latitude ledge features, described in Figure D.34?

The apparent intrusion of polar air to northern mid-latitudes in the spring represents the opportunity to test in an interesting way the nitrogen, hydrogen, and chlorine chemistry of the stratosphere.

PROSPECTS: The analytical techniques will be available within two years to explore in situ and simultaneously the concentration of NO, NO_2, ClO, OH, HO_2, and O_3 in the vicinity of the NO_2 "ledge" reported by Noxon, based on ground-based observations of nitrogen dioxide. A detailed understanding of the free radical concentration in such an event would be an exceedingly interesting perturbation experiment.

QUESTION 10: Is water vapor the constituent responsible for inducing the variability in free radical concentrations evident in virtually all the results reported in this paper?

Given the extreme sensitivity of $[H_2O]$ to the tropopause temperature and the large observed fluctuations of water above the tropical tropopause, it seems plausible that fluctuations in H_2O, which in turn cause fluctuations in OH and HO_2, constitute a starting point for observed local changes in NO, NO_2, and ClO. The mechanistic links are discussed both here and in Appendix C.

PROSPECTS: As the signal-to-noise ratio, absolute calibration, altitude resolution, and capability to make a large number of simultaneous observations improve in the next three to four years, a wealth of information about how fluctuations in local water vapor concentrations affect the HO_x, NO_x, and ClO_x chemistry of the stratosphere will evolve. Thus correlation experiments may best be carried out in the equatorial region, where fluctuations in H_2O may be the most dramatic. If local variability reported from aircraft observations well above the tropopause hold at higher altitudes, an entirely new class of correlation experiments will evolve. Such measurements hold great promise for establishing cause-and-effect links within the complex net of reactions linking the various families through radical-radical reactions.

QUESTION 11: Does the odd oxygen production/destruction budget balance, based on observed concentrations of the rate limiting free radicals?

Although transport times in the odd oxygen continuity equation obviate the possibility of applying a purely chemical test to the balance of local odd oxygen production and destruction in the lower stratosphere, it is essential that we continue to press the issue of improved analytical techniques for NO_2, HO_2, ClO, $O(^3P)$, and O_3 to quantify, as a function of altitude and latitude, the balance between production and destruction of odd oxygen. Although this approach cannot directly test cause-and-effect relationships with the odd oxygen budget and the approach is currently seriously diluted by large experimental uncertainties, it must be carefully pursued.

PROSPECTS: The next two years will bring considerably more accurate detection techniques for the major rate limiting radicals, NO_2, ClO, HO_2, OH, $O(^3P)$, and O_3 with cross-calibration against remote techniques and limited latitude coverage. Although such techniques can never prove completeness in our definition of ozone production and loss processes, the detailed accounting will provide important evidence suggesting the altitude dependence of proposed mechanisms.

REFERENCES

Ackerman, M. and C. Muller (1973) Stratospheric methane and nitrogen dioxide from infrared spectra. Pure and Applied Geophysics 106-108:1325-1335.

Ackerman, M., J.C. Fontanella, D. Frimout, A. Girard, N. Louisnard, and C. Muller (1975) Simultaneous measurements of NO and NO_2 in the stratosphere. Planetary and Space Science 23:651-660.

Aiken, A.C. and E.J.R. Maier (1978) Balloon-borne photoionization mass spectrometer for measurement of stratospheric gases. Review of Scientific Instruments 49:1034-1040.

Anderson, J.G. (1971) Rocket measurement of OH in the mesosphere. Journal of Geophysical Research 76:7820.

Anderson, J.G. (1975) Measurement of atomic oxygen and hydroxyl in the stratosphere. Pages 458-464, Proceedings, Fourth Conference on CIAP. Symposium No. 175. Washington, D.C.: U.S. Department of Transportation.

Anderson, J.G. (1980) Free radicals in the earth's stratosphere: A review of recent results. Pages 233-251, Proceedings of the NATO Advanced Study Institute on Atmospheric Ozone: Its Variation and Human Influences, edited by A.C. Aiken. October 1-13, 1979. U.S. Department of Transportation, Report No.

FAA-EE-80-20. Washington, D.C.: Federal Aviation Administration.

Anderson, J.G., J.J. Margitan, and D.H. Steoman (1977) Atomic chlorine and the chlorine monoxide free radical in the stratosphere: Three in situ observations. Science 198:501.

Anderson, J.G., R.E. Shetter, H.J. Grassel, and J.J. Margitan (1980) Stratospheric free chlorine measured by balloon-borne in situ resonance fluorescence. Journal of Geophysical Research 85:2869.

Arnold, F., R. Fabian, G. Henschen, and W. Joos (1980) Stratospheric trace gas analysis from ions: H_2O and HNO_3. Planetary and Space Science 28:581-585.

Bangham, M.J., A. Bonetti, R.H. Bradsell, B. Carli, J.G. Harries, F. Mencaraglia, D.G. Moss, J. Pollitt, E. Rossi, and N.R. Swann (1980) New measurements of stratospheric composition using submillimeter and infrared emission spectroscopy. (Unpublished manuscript available from A. Bonetti, University of Florence, Florence, Italy.)

Berg, W.W., P.J. Crutzen, F.E. Grabek, and S.N. Gitlin (1980) First measurements of total chlorine and bromine in the lower stratosphere. Geophysical Research Letters 7:937-940.

Blatherwick, R.D., A. Goldman, D.G. Murcray, F.J. Murcray, G.R. Cook, and J.W. Van Allen (1980) Simultaneous mixing ratio profiles of stratospheric NO and NO_2 as derived from balloon-borne infrared solar spectra. Geophysical Research Letters 7:471-473.

Buijs, H.L., G.L. Vail, G. Tremblay, and D.J.W. Kendall (1980) Simultaneous measurements of the volume mixing ratio of HF and HCl in the stratosphere. Geophysical Research Letters 7:205.

Burnett, C.R. (1976) Terrestrial OH abundance measurement by spectroscopic observation of resonance absorption by sunlight. Geophysical Research Letters 3:319.

Burnett, C.R. (1977) Spectroscopic measurements of atmospheric OH abundance. Bulletin of the American Physical Society 22:539.

Burnett, C.R. and E.B. Burnett (1981) Spectroscopic measurements of the vertical column abundance of hydroxyl [OH] in the earth's atmosphere. Journal of Geophysical Research 86:5185.

Campbell, M.J., J.C. Sheppard, and B.J. An (1979) Measurement of hydroxyl concentration in boundary layer air by monitoring CO oxidation. Geophysical Research Letters 6:175.

Chaloner, C.P., J.R. Drummond, J.T. Houghton, R.F. Jarnot, and H.K. Roscoe (1978) Infrared measurements of stratospheric composition I. The balloon instrument and water vapor measurements. Proceedings of the Royal Society of London, Series A 364:145-159.

Chance, K.V., J.C. Brasunas, and W.A. Traub (1980) Far infrared measurement of stratospheric HCl. Geophysical Research Letters 7:704.

Coffey, M.T., W.G. Mankin, and A. Goldman (1981) Simultaneous spectroscopic determination of the latitudinal, seasonal and diurnal variability of stratospheric N_2O, NO and HNO_3. Journal of Geophysical Research 86(C8):7331-7341.

Crutzen, P.J., I.S.A. Isaksen, and J.R. McAfee (1978) The impact of the chlorocarbon industry on the ozone layer. Journal of Geophysical Research 83:345.

Davis, P.D., W.S. Heaps, and T. McGee (1976) Direct measurements of natural tropospheric levels of OH via an aircraft borne tunable dye laser. Geophysical Research Letters 3:331.

Davis, P.D., W.S. Heaps, D. Philen, M. Rodger, T. McGee, A. Nelson, and A.J. Moriarty (1979) An airborne laser induced fluorescence system for measuring OH and other trace gases in the parts-per-quadrillion to parts-per-trillion range. Review of Scientific Instruments 50:1505.

Drummond, J.W., J.M. Rosen, and D.J. Hoffman (1977) Balloon-borne chemiluminescent measurements of NO to 45 km. Nature 265:319-320.

Ehhalt, D.H., U. Schmidt, and L.E. Heidt (1977) Vertical profiles of molecular hydrogen in the troposphere and the stratosphere. Journal of Geophysical Research 82:5907-5911.

Evans, W.F.J. and E.J. Llewellyn (1970) Molecular oxygen emission in the airglow. Annales de Geophysique 26:167.

Evans, W.F.J., E.J. Llewellyn, and A.V. James (1969) Balloon observations of the temporal variation of the infrared atmospheric oxygen bands in the airglow. Planetary and Space Science 17:933.

Evans, W.F.J., H. Fast, J.B. Kerr, C.T. McElroy, R.S. O'Brien, D.I. Wardle, J.C. McConnell, and B.A. Ridley (1978) Stratospheric constituent measurements from project stratoprobe. Pages 55-60, Proceedings of the WMO Symposium on the Geophysical Aspects and Consequences of Change in the Composition of the Stratosphere, WMO Publication 511. Geneva: World Meteorological Organization.

Eyre, J.R. and H.K. Roscoe (1977) Radiometric measurements of HCl. Nature 226:243.

Farmer, C.B., O.F. Raper, B.D. Robbins, R.A. Toth, and C. Muller (1980) Simultaneous spectroscopic measurements of stratospheric species: O_3, CH_4, CO, CO_2, N_2O, HCl, and HF at northern and southern mid-latitudes. Journal of Geophysical Research 85 (C3):1621.

Fischer, H., F. Fergg, and D. Rabus (1982) Radiometric measurements of stratospheric H_2O, HNO_3 and NO_2 profiles. Proceedings of the International Radiation Symposium, Fort Collins, Colorado, August 1980. (To be published in the Journal of the National Cancer Institute in 1982.)

Fontanella, J.C., A. Girard, L. Gramont, and N. Louisnard (1975) Vertical distribution of NO, NO_2 and HNO_3 as derived from stratospheric absorption infrared spectra. Applied Optics 14:825-839.

German, K.R. (1975) Radiative and predissociative lifetimes of the $V^1 = D_1D_1$ and 2 levels of the $A^2\Sigma^+$ state of OH and OD. Journal of Chemistry and Physics 63:5252-5255.

German, K.R. (1976) Collision and quenching cross sections in the $A^2\Sigma^+$ state of OH and OD. Journal of Chemistry and Physics 64:4065-4068.

Girard, A., J. Besson, R. Giraudet, and L. Gramont (1978/1979) Correlated seasonal and climate variations of trace constituents in the stratosphere. Pure and Applied Geophysics 117:381-394.

Goldman, A., F.G. Fernald, W.J. Williams, and D.G. Murcray (1978) Vertical distribution of NO_2 in the stratosphere as determined from balloon measurements of solar spectra in the 4500 Å region. Geophysical Research Letters 5:257-260.

Harries, J.E., D.G. Moss, N.R.W. Swann, G.F. Neill, and P. Gildwarg (1976) Simultaneous measurements of H_2O, NO_2 and HNO_3 in the daytime stratosphere from 15 to 35 km. Nature 259:300-301.

Heaps, W.S., T.J. McGee, R.D. Hudson, and L.O. Caudill (1981) Balloon-borne Lidar Measurements of Stratospheric Hydroxyl and Ozone, NASA Publication X-963-81-27. Greenbelt, Md.: National Aeronautics and Space Administration.

Horvath, J.J. and C.J. Mason (1978) Nitric oxide mixing ratios near the stratopause measured by a rocket-borne chemiluminescent detector. Geophysical Research Letters 5:1023-1026.

Hudson, R.D. and E.I. Reed, eds. (1979) The Stratosphere: Present and Future. NASA Reference Publication 1049. Greenbelt, Md.: National Aeronautics and Space Administration; N80-14641-14648. Springfield, Va.: National Technical Information Service.

Hudson, R.D. et al., eds. (1982) The Stratosphere 1981: Theory and Measurement. WMO Global Research and Monitoring Project Report No. 11. Geneva: World Meteorological Organization. (Available from National Aeronautics and Space Administration, Code 963, Greenbelt, Md. 20771.)

Kerr, J.B. and C.T. McElroy (1976) Measurement of stratospheric nitrogen dioxide from the AES stratospheric balloon program. Atmosphere 14:166-171.

Kley, D., J.W. Drummond, and A.L. Schmeltekopf (1980) On the structure and microstructure of stratospheric water vapor. Pages 315-327, Atmospheric Water Vapour, edited by A. Deepak, T.D. Wilkerson, and L.H. Ruhnke. New York, N.Y.: Academic Press, Inc.

Lazrus, A.L. and B.W. Gandrud (1974) Distribution of stratospheric nitric acid vapor. Journal of Atmospheric Science 31:1102.

Lazrus, A.L., B.W. Gandrud, J. Greenberg, J. Bonelli, E. Mroz, and W.A. Sedlacek (1977) Midlatitude seasonal measurements of stratospheric and chlorine vapor. Geophysical Research Letters 4:587.

Loewenstein, M., W.J. Borucki, H.F. Savage, J.G. Borucki, and R.C. Whitten (1978a) Geographical variations of NO and O_3 in the lower stratosphere. Journal of Geophysical Research 83:1875-1882.

Loewenstein, M., W.J. Starr, and D.G. Murcray (1978b) Stratospheric NO and HNO_3 observations in the northern hemisphere for three seasons. Geophysical Research Letters 5:531-534.

Logan, J.A., M.J. Prather, S.C. Wofsy, and M.B. McElroy (1978) Atmospheric chemistry: Response to human influence. Philosophical Transactions of the Royal Society 290:187.

Logan, J.A., M.J. Prather, S.C. Wofsy, and M.B. McElroy (1981) Tropospheric chemistry: A global perspective. Journal of Geophysical Research 80:7210.

Marche, P., A. Barbe, C. Secroun, J. Corr, and P. Jouve (1980a) Ground-based spectroscopic measurements of HCl. Geophysical Research Letters 7:869.

Marche, P., A. Barbe, C. Secroun, J. Corr, and P. Jouve (1980b) Mesures des acides fluorhydrique et chlorhydrique dans l'atmosphere par spectroscopic infraroque a partir du sol. Comptes Rendus Hebdomadaires des Seances de l'Academie des Sciences, Series B 290(14):369-371.

Mauersberger, K., R. Finstad, S. Anderson, and P. Robbins (1981) A comparison of ozone measurements. Geophysical Research Letters 8:361.

Menzies, R.T. (1978) Remote measurements of ClO in the stratosphere. Geophysical Research Letters 6:151.

Menzies, R.T., C.W. Rutledge, R.A. Zanteson, and P.D. Spears (1981) Balloon-borne laser heterodyne radiometer for measurements of atmospheric trace species. Applied Optics 20:536.

Mihelcic, D., D.H. Ehhalt, G.F. Kulessa, J. Klomfass, M. Trainer, U. Schmidt, and H. Röhrs (1978) Measurements of free radicals in the atmosphere by matrix isolation and electron paramagnetic resonance. Pure and Applied Geophysics 116:530.

Molina, M.J. and F.S. Rowland (1974) Stratospheric sink for chlorofluoromethanes: Chlorine-atom catalyzed destruction of ozone. Nature 249:810-812.

Mroz, E.J., A.J. Lazrus, and J. Bornelli (1977) Direct measurements of stratospheric fluoride. Geophysical Research Letters 4:149.

Murcray, D.G., A. Goldman, W.J. Williams, F.H. Murcray, J.N. Brooks, J. Van Allen, R.N. Stocker, J.J. Kosters, D.B. Barker, and D.E. Snider (1974) Recent results of stratospheric trace-gas measurements from balloon-borne spectrometers. Pages 184-192, Proceedings of the Third CIAP Conference, Report No. DOT-TSC-OST-74-15. Washington, D.C.: U.S. Department of Transportation.

Murcray, D.G., A. Goldman, F.H. Murcray, F. J. Murcray, and W.J. Williams (1979) Stratospheric distribution of $ClONO_2$. Geophysical Research Letters 6:857.

National Research Council (1979) Stratospheric Ozone Depletion by Halocarbons: Chemistry and Transport. Panel on Chemistry and Transport, Committee on Impacts of Stratospheric Change, Assembly of Mathematical and Physical Sciences. Washington, D.C.: National Academy of Sciences.

Naudet, J.P., D. Huguenin, P. Rigaurd, and D. Cariolle (1981) Stratospheric observations of NO_3 and its experimental and theoretical distribution between 20 and 40 km. Planetary and Space Science 29:707.

Noxon, J.F. (1968) Day airglow. Space Science Reviews 8:92.

Noxon, J.F. (1978) Stratospheric NO_2 in the Antarctic winter. Geophysical Research Letters 5:1021-1022.

Noxon, J.F. (1979) Stratospheric NO_2. II. Global behavior. Journal of Geophysical Research 84:5067-5076.

Noxon, J.F. (1980) Correction. Journal of Geophysical Research 85:4560-4561.

Noxon, J.F., E.C. Whipple, Jr., and R.S. Hyde (1979) Stratospheric NO_2. I. Observational method and behavior at mid-latitude. Journal of Geophysical Research 84:5047-5065.

Parrish, A., R.L. de Zafra, P.M. Solomon, J.W. Barrett, and E.R. Carlson (1981) Chlorine oxide in the stratospheric ozone layer: Ground-based detection and measurement. Science 211(4487):1158-1161.

Patel, C.K.N., E.G. Burkhardt, and C.A. Lambert (1974) Spectroscopic measurements of stratospheric nitric oxide and water vapor. Science 184:1173-1176.

Perner, D., D.H. Ehhalt, H.W. Pätz, U. Platl, E.P. Roth, and A. Volz (1976) OH radicals in the lower troposphere. Geophysical Research Letters 3:460.

Ridley, B.A. and D.R. Hastie (1981) Stratospheric odd-nitrogen: NO measurements at 51°N in summer. Journal of Geophysical Research 86(C4):3162-3166.

Ridley, B.A. and L.C. Howlett (1974) An instrument for nitric oxide measurements in the stratosphere. Review of Scientific Instruments 45:742-746.

Ridley, B.A. and H.I. Schiff (1981) Stratospheric odd-nitrogen: Nitric oxide measurements at 31°N in autumn. Journal of Geophysical Research 86:3167-3172.

Roscoe, H.K., J.R. Frummond, and R.F. Jarnot (1981) Infrared measurements of stratospheric composition III. The daytime changes of NO and NO_2. Proceedings of the Royal Society of London A 375:507.

Roy, C.R., I.E. Galbally, and B.A. Ridley (1980) Stratospheric odd nitrogen. II. Measurements of nitric oxide in the southern hemisphere. Quarterly Journal of the Royal Meteorological Society 106:887-894.

Schmidt, U. (1978) The latitudinal and vertical distribution of molecular hydrogen in the troposphere. Journal of Geophysical Research 83:941-946.

Schmidt, U., G. Kulessa, and E.P. Roth (1980) The atmospheric H_2 cycle. Pages 307-322, Proceedings of the NATO Advanced Study Institute on Atmospheric Ozone: Its Variation and Human Influences, edited by A.C. Aiken. October 1-13, 1979. U.S. Department of Transportation, Report No. FAA-EE-80-20. Washington, D.C.: Federal Aviation Administration.

Sze, N.D. and M.K.W. Ko (1981) The effects of the rate of OH + HNO_3 and $HONO_2$ photolysis on the stratospheric chemistry. Atmospheric Environment 15:1301.

Wallace, L. and D.M. Hunten (1968) Dayglow of the oxygen A band. Journal of Geophysical Research 73:4813.

Wang, C.C., L.I. Davis, Jr., P.M. Zelzer, and R. Munoz (1981) Improved airborne measurements of OH in the atmosphere using the technique of laser induced fluorescence. Journal of Geophysical Research 86:1181.

Waters, J.W., J.J. Gustincic, R.K. Kakar, H.K. Roscoe, P.N. Swanson, T.G. Phillips, T. deGraauw, A.R. Kerr, and R.J. Mattauch (1979) Aircraft search for millimeter-wavelength emission by stratospheric ClO. Journal of Geophysical Research 84:7034-7040.

Waters, J.W., J.C. Hardy, R.F. Jarnot, and H.M. Pickett (1981) Chlorine monoxide radical, ozone and hydrogen peroxide: Stratospheric measurements by microwave limb sounding. Science 214:61.

Weinstock, E.M., M.J. Phillips, and J.G. Anderson (1981) In situ observations of ClO in the stratosphere: A review of recent results. Journal of Geophysical Research 86:7273.

Wuebbles, D.J. and J.S. Chang (1981) A study of the effectiveness of the Cl_x catalytic ozone loss mechanism. Journal of Geophysical Research 86:9869.

Zander, R. (1975) Presence de HF dans la stratosphere superieure. Comptes Rendus Hebdomadaires des Seances de l'Academie des Sciences, Series B 281(12):213-214.

Zander, R. (1981) Recent observations of HF and HCl in the upper stratosphere. Geophysical Research Letters 8(4):413-416.

Zander, R., H. Lecbert, and L.D. Kaplan (1981) Concentration of carbon monoxide in the upper stratosphere. Geophysical Research Letters 8(4):365-368.

Appendix E

TREND ANALYSIS OF TOTAL OZONE

Hans A. Panofsky
Department of Meteorology
Pennsylvania State University

INTRODUCTION

Total ozone has been measured at Arosa for almost 50 years by the Dobson spectrophotometer. The number of Dobson stations in the global observing network increased slowly at first, but more rapidly since the late 1950s. Measurements made with instruments developed in the USSR were included in the network system after 1958, but their reliability approached that of the Dobson measurements only since about 1972. Most of the observing stations are on continents and in the northern hemisphere.

Analysis of these records by Angell and Korshover (1981) reveals considerable differences in long-term variations from region to region; but, on the average, trends appear to be mostly positive in the 1960s and near zero in the 1970s. As we will see, more recent analyses based on sophisticated statistical models suggest positive trends in the 1970s, which are, however, not significantly different from zero.

Whereas it is relatively easy to estimate trends from Dobson data, it is much more difficult to ascribe such trends to specific causes. There may have been trends in variables such as dust that influence the measured ozone but not necessarily the actual ozone; there may have been real changes of ozone due to changes in circulation of the atmosphere; there may have been changes of temperature or of various trace elements that influence the ozone budget; and there may have been influences of solar variation. Further, the Dobson network may not be representative of global averages.

As we shall see, some causes of trends in Dobson-measured ozone can be evaluated by sophisticated statistical analysis; for other causes we can only make some not very well educated guesses.

Ultraviolet satellite measurements of total ozone trend largely avoid the problem of spatial representativeness (not completely, because the dark polar cap remains unobserved). However, satellite records so far are relatively short and have other deficiencies. Estimates of total ozone have been made by backscattered ultraviolet (BUV) on the Nimbus 4 satellite beginning in April 1970. However, this instrument initially suggested global averages 3 to 4 percent less than the total Dobson averages. This difference increased further over 81 months due to instrumental drift. In addition, failure of a solar panel in June 1972 reduced the number of observations after that date, introducing a problem of spatial representativeness of the trends over the whole 81 months. The Solar Back-Scatter (SBUV) instrument on Nimbus 7, operational since 1978, did not have these problems (see Hudson et al. 1982). But the period of its operation so far is too short to make reliable trend estimates.

In principle, however, satellites should eventually lead to better trend measurements than Dobson instruments for two reasons: (1) there is better spatial representation, and (2) only one instrument is used, whereas the different Dobson instruments have separate idiosyncrasies and errors. Gradual instrumental drift could be estimated by comparison with a well-calibrated Dobson network.

Pittock (see Hudson et al. 1982, p. 3-46) estimates that eventually satellites will be able to detect "true global trends" with standard deviations of the order of 1 or 2 percent from observations by the same satellite over 10 years. Such trends, of course, could have man-made as well as natural causes. So far, satellites, just as Dobson instruments, have not suggested any significant trend in total ozone.

Most recent physical models also suggest very little ozone change due to human interference. Therefore there is, at present, no real disagreement between statisticians and physical modelers.

The main disagreement remaining concerns the extent to which statistical analysis can be used as an early warning system for the future.

For the Dobson network, there are two main problems: first, how representative is the Dobson network of the global ozone distribution? This problem has been attacked by use of satellite data. The results are somewhat controversial, as we shall see.

The second difficulty involves the importance of "natural" variations over long periods. This question has been attacked by studying the relatively long Arosa record. Again, the results are controversial. For satellite early warning, only the second problem is important.

STATISTICAL ANALYSIS OF THE DOBSON NETWORK

Three separate groups of statisticians (Bloomfield et al. 1981, Reinsel et al. 1981, St. John et al. 1981) have analyzed monthly averages at 36 Dobson stations with records over 10 years long. Their studies differ in detail but have many common features: the results also are quite similar.

All of the groups originally based their analyses on the "hockey stick" or "boomerang" approach. They argued that since there was no obvious cause for the upward trend in the 1960s, the observed upward trend must be due to "red noise"--natural long-period variations. Therefore the "true" long-range trend up to 1970 was assumed to be zero. Given this zero trend, estimates for the Dobson mean trend in the 1970s were made by various statistical procedures. Such trends were generally positive (Figure E.1) but did not differ significantly from zero.

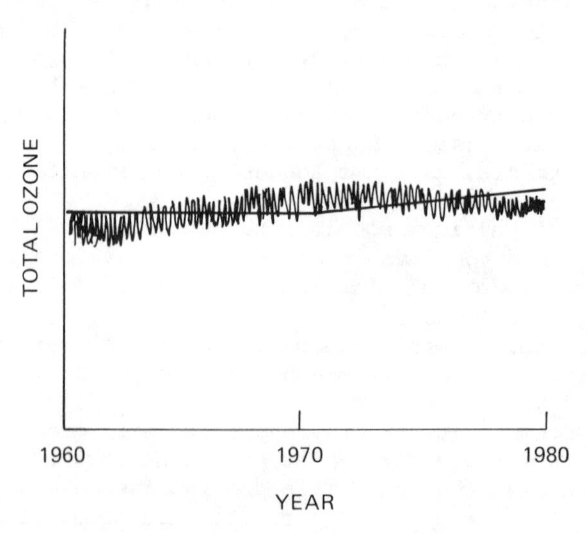

FIGURE E.1 Typical ozone variations with time, and "hockey stick" fit (solid line).

Bloomfield et al. (1981) also produced separate trend analyses for the 1960s and 1970s (no hockey stick assumed), and obtained a near-zero trend for the 1970s.

Further possibilities tested by Bloomfield et al. (1981) were based on assumed relationships between ozone and nuclear tests, and between ozone and solar activity. Although the trends in the 1970s derived from this analysis did change somewhat, they still are not significantly different from zero.

With 36 individual trends, it has been possible to assess the uncertainty in the mean trend of the 36 stations, based on the 10- to 20-year records. Statistically, this variability from station to station was split into (1) random effects at each station; (2) variation among stations in each region (the regions are large, e.g., North America is a single region); and (3) variation among regions. However, it should be noted that these three sets of differences cannot be associated with physical causes on a one-to-one basis.

For example, "random" errors are introduced in the monthly averages by missing data, by the effect of large variations with periods of the order of a week, by local air pollution episodes or clouds, or simply by observational error.

Variations in trends within regions are primarily due to different weather at the stations in the same region. Different weather implies real differences in ozone, and also differences in observational accuracy; e.g., clouds, which interfere with the accuracy of the ozone observations. Also, rates of deterioration of instrumental parts may be different, and calibrations may have been performed at different points in the solar cycle.

Weather differences are even more important in different regions; further, each region uses different secondary standards for instrument calibration.

All groups agree that the standard deviation of average trend derived from these stations, due to a combination of all these factors, is about 0.6 percent per decade. However, the question of trend "bias" due to the assumption of the "hockey stick" model has been raised frequently and requires further analysis.

SPATIAL BIAS

In principle, it is possible that trend estimates from the Dobson network would differ systematically from global trends; e.g., if ozone moved from land areas to ocean areas for an extended period, there might be an indicated ozone change but no real global trend.

Three techniques have been used to estimate the bias of global trend estimated from the Dobson network. In the first technique, Hasebe (1980) has estimated total ozone values at grid points over the world from the Dobson network by a process called "optimum interpolation." However, since there are huge areas of no ozone stations, particularly in the southern hemisphere, the interpolated (and sometimes extrapolated) values in such areas are extremely uncertain. Hasebe computes global trends from these grid-point values. These do not differ significantly from those estimated by Angell and Korshover (1981). In any case, it is unlikely that Hasebe's techniques for estimating global trends are necessarily superior to the simpler methods used by the earlier authors.

In a second technique, Moxim and Mahlman (1980) compared global trends with Dobson location trends as computed from a "simple" three-dimensional numerical global model of the atmosphere, containing ozone. They find differences between <u>one-year</u> trends computed from monthly averages between global and Dobson location ozone of the order of 1 percent.

Finally, several groups of authors have attempted estimates of spatial bias by use of the BUV satellite observations, described in the introduction. Of course, it is difficult to evaluate the accuracy of these comparisons, since the BUV instrument deteriorated after several years, and reliable trends over periods longer than three years or so could not be computed.

London and Ling (1981) compared global total ozone with Dobson location ozone and found only small mean differences, but large standard deviations of these averages. However, they did not compare trends.

Reinsel et al. (1982) compared statistics of ozone trends all over the world with data for April 1970 to May 1975 with those of ozone trends in small areas surrounding a select group of 36 Dobson stations. They found no statistically significant differences in this one small sample covering only a few years. The authors concluded from this result that there are no differences between global trends and trends derived from Dobson station data. Meteorologists are doubtful about this conclusion and suggest standard deviations of global trends due to Dobson location bias could well be of the order of 1 percent per decade. This order of magnitude is also suggested by satellite records analyzed by A.J. Miller (see Hudson et al. 1982), and is somewhat smaller than what would have been expected from Moxim and Mahlman's model study. In fact, the results of Reinsel et al. do not differ significantly from an assumption of a bias of decadal average Dobson trend of 1 percent. Hence, meteorologists generally suggest a spatial bias in the Dobson network, of the order of 1 percent per decade, but realize that this is an extemely uncertain number.

LOW-FREQUENCY VARIATIONS

Bishop and Hill (1981) used a partially inhomogeneous Arosa record to estimate the trend uncertainty due to low-frequency variations (not directly analyzable with the record of the Dobson network) from the variation among decadal trends. Their result was an uncertainty of the order of 0.8 percent per decade.

The Arosa record used by Bishop and Hill is remarkable for the absence of fluctuations with periods of the order of a century, periods that are often quite apparent in temperature records. In fact, the homogenized Arosa ozone record for the period 1932-1980 (Dütsch, personal communication to J. London, University of Colorado, 1981) shows noticeable long-period variations (Figure E.2). The figure also shows the sensitivity of 10-year trends to the starting date of each decade. For example, the

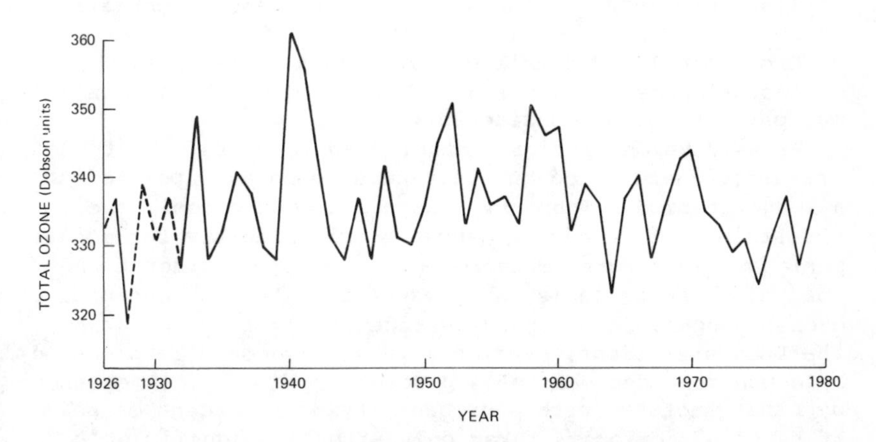

FIGURE E.2 Homogenized values for total ozone at Arosa (Hudson et al. 1982).

trend for 1936-1945 is almost zero. In contrast, that for 1940-1949 is -33 Dobson units. Hence, the estimate of 0.8 percent for long-period variability is quite uncertain. Further, this estimate is based on the record from only one station.

A further indication of the magnitude of uncertainties due to long-period fluctuations is the sensitivity of the trends in the 1970s to the statistical model used. Thus, the "hockey stick" approach and the hypotheses of separate unknown trends in the 1960s and 1970s produce trends in the 1970s differing by 1.5 percent. E.L. Scott (University of California, Berkeley, personal communication, 1981) also has criticized the dependence of the global trend estimates on the particular statistical model chosen. For these reasons, we suggest that the uncertainty of global decadal ozone trends due to long-period natural variations with unknown causes is at least of the order of 1 percent; but we consider this estimate also as very uncertain.

SUMMARY AND RECOMMENDATIONS

Because of the various controversies, it is not possible to arrive at a definite understanding of the uncertainties of global decadal ozone trends derived from Dobson stations or satellites. The statisticians using the "hockey stick" models would estimate the standard deviations of the decadal trends derived from Dobson stations to be of the order of 1 percent. Most meteorologists and at least one statistician would prefer standard deviations of the order of 2 percent or larger.

Once long-lived satellites with good reliability characteristics are available, the standard deviations may perhaps be cut considerably.

As we consider periods longer than 10 years, the trends are better determined but they have to be extrapolated to a longer period. These two factors almost cancel; the uncertainty of trend may decrease only slightly for longer periods. Therefore, meteorologists would consider trend analysis an unreliable early warning system for man-made ozone changes; an observed average trend of 0 percent at the Dobson stations, over, say 20 years, could be produced by a man-made decrease of 4 percent or larger, compensated by other factors, with a probability of 5 percent or so. If remedial action is taken only after a "significant" decrease of ozone, existing fluorocarbons could continue to decrease the ozone further to intolerable levels due to their long lifetimes. Some statisticians would consider a zero trend as an indication that the trend due to fluorocarbons must have been much smaller than 4 percent. Our best hope for using total ozone trends as early-warning systems rests on improved and well-calibrated records of total ozone measured from long-lived satellites.

REFERENCES

Angell, J.K. and J. Korshover (1981) Update of ozone variations through 1979. Pages 393-396, Proceedings of the Quadrennial International Ozone Symposium, August 4-9, 1980. Boulder, Colo.: National Center for Atmospheric Research.

Bishop, L. and W.J. Hill (1981) Analyzing Stratospheric Ozone for the Natural and Man-made Trend Variability. Pages 304-305, Summaries of Conference Presentations. (Available from the Society for Industrial and Applied Mathematics, Philadelphia, Pa.)

Bloomfield, P., M.L. Thompson, G.S. Watson, and S. Zeger (1981) Frequency Domain Estimation of Trends in Atmospheric Ozone. Technical Report No. 182, Department of Statistics, Princeton University. (Submitted for publication to Journal of Geophysical Research)

Hasebe, Fumio (1980) A global analysis of the fluctuations of total ozone II nonstationary annual oscillation, quasi-biennial oscillation, and long-term variations in total ozone. Journal of the Meteorological Society of Japan 58:104-117.

Hudson, R.D., et al., eds. (1982) The Stratosphere 1981: Theory and Measurement. Geneva: World Meteorological Organization. (Available from National Aeronautics and Space Administration, Code 963, Greenbelt, Md. 20771)

London, J. and X. Ling (1981) The geographic bias in determining average variations of total ozone from ground-based observations. Pages 337-339, Proceedings of the Quadrennial International Ozone Symposium, August 4-9, 1980. Boulder, Colo.: National Center for Atmospheric Research.

Moxim, W.J. and J.D. Mahlman (1980) Evaluation of various total ozone sampling networks using the GFDL 3-D tracer model. Journal of Geophysical Research 85 (C8):4527-4539.

Reinsel, G., G.C. Tiao, M.N. Wang, R. Lewis, and D. Nytchka (1981) Statistical analysis of stratospheric ozone data for detection of trend. Atmospheric Environment 15:1569-1578.

Reinsel, G., G.C. Tiao, and R. Lewis (1982) A statistical analysis of total ozone data from the Nimbus BUV satellite experiment. Journal of Atmospheric Sciences 39.

St. John, D.S., S.P. Bailey, W.H. Fellner, J.M. Manor, and R.D. Snee (1981) Time series analysis for trends in total ozone measurements. Journal of Geophysical Research 86:7299-7311.

Appendix F

DETECTION OF TRENDS IN THE VERTICAL DISTRIBUTION OF OZONE

A. Barrie Pittock
CSIRO Division of Atmospheric Physics
Mordialloc, Australia

UMKEHR METHOD

Observations of the vertical distribution of ozone by the Umkehr method can be made with any Dobson spectrophotometer in the total ozone network. However, only about 18 stations in the network currently make regular Umkehr observations (see Table F.1). Of these, only three are in the southern hemisphere. Six stations have records extending back more than 20 years.

In addition to calibration drift, the Umkehr method is subject to a number of sources of error and bias, notably the effects of tropospheric and stratospheric dust (Dave et al. 1981, De Luisi et al. 1975), and a meteorological bias due to the inability to make observations under cloudy conditions (Pittock 1970).

Long-term trends in ozone concentrations at various altitudes as observed by the Umkehr technique are thus subject to major uncertainties in addition to the possibilities of random and systematic errors. The major uncertainties are due to changes in atmospheric concentrations of dust especially from volcanic eruptions, possible trends in cloudiness, and a serious problem of geographical representativeness. Global mean concentrations of ozone at particular altitudes are, however, rather meaningless since the vertical distribution of ozone varies markedly with latitude and season. Umkehr-derived vertical distributions are most accurate in the middle stratosphere (around 30- to 45-km altitude) if adequate allowance can be made for stratospheric dust. Such allowance using optical depth measurements has been proposed (Dave et al. 1981) and may enable meaningful estimates of trends in the middle stratosphere at northern middle latitudes to be made.

TABLE F.1 Umkehr Stations in Operation as of 1980

Station	Initial Year (Since 1958)	Months of Missing Data
Europe		
Arosa	1961	6
Belsk	1963	46
Cairo	1978	13
Lisbon	1967	39
North America[a]		
Boulder	1978	1
Edmonton	1974	18
Japan		
Kagoshima	1958	88
Naha	1976	31
Sapporo	1958	96
Tateno	1958	23
India		
Mount Abu	1964	30
New Delhi	1965	57
Poona	1975	27
Srinagar	1976	14
Varanasi	1964	72
Australia		
Aspendale	1958	0
Brisbane	1959	0
Macquarie Island	1964	0

[a]Goose Bay and Churchill have reported old data, but no data for 1979 or 1980.

Statistical analysis of Umkehr data (Bloomfield et al. 1982, Penner et al. 1981) leads to estimates of the "revealed" uncertainties--that is, those detectable from the scatter or range of measurements--to which must be added estimates of the unrevealed uncertainties--that is, those due to lack of geographic coverage or to poorly estimated global trends in stratospheric dust. Combined, these analyses suggest standard deviations of global trend estimates of around 5 percent per decade in the 30- to 45-km altitude range, and somewhat less if the trend estimate is for the north temperate latitudes only (Hudson et al. 1982). An ozone depletion of about 10 percent per decade should thus be detectable at the 95 percent confidence level in the middle stratosphere of the north temperate zone. Trends of this magnitude are not revealed by the existing data, which are summarized for northern middle latitudes in Figure F.1. Umkehr data for the Australian stations over the last several years are only now becoming available and have not yet been included in global trend estimates.

BALLOON-BORNE OZONESONDES

Balloon-borne ozonesondes are currently flown on a regular basis at a small number of stations in Western Europe, North America, Japan, and India, and at one station in the southern hemisphere (see Table F.2).

These observations have high vertical resolution but are adjusted by a single factor to give absolute agreement with the total ozone amount measured by a nearby Dobson spectrophotometer. This single-factor adjustment is a major source of uncertainty both in individual profiles and in trend determinations because the sondes perform with lower efficiencies at low ambient pressures (high altitude). This deficiency is compensated for by using a standard pressure-dependent correction factor as well as the single-factor adjustment. However, individual sondes may differ in performance from the standard pressure-dependent correction factor, and subtle changes in the manufacture or preparation of instruments could introduce a secular trend in this performance. Actual trends in ozone concentrations above balloon burst altitude can also cause fictitious trends in the profiles at lower altitudes to appear via the correction factor to the Dobson total amount (see Pittock 1977b).

Another problem with ozonesonde measurements is that polluted tropospheric air may contaminate the intake system causing artificially low readings especially at low altitudes. Using the adjustment factor may then result in overestimates of ozone at high altitudes.

For all the above reasons, the expected random error in individual ozone soundings is least between the tropopause and about 25 km altitude (standard deviation about 4 percent). The expected random error is higher (about 8 percent) in the troposphere and at 30 km and above (Hudson et al. 1982).

Another major uncertainty in estimates of global trends is due to the poor spatial coverage of the ozonesonde network. This uncertainty is considerably reduced if the analysis is confined to north temperate latitudes where the spatial coverage is relatively good. However, satel-

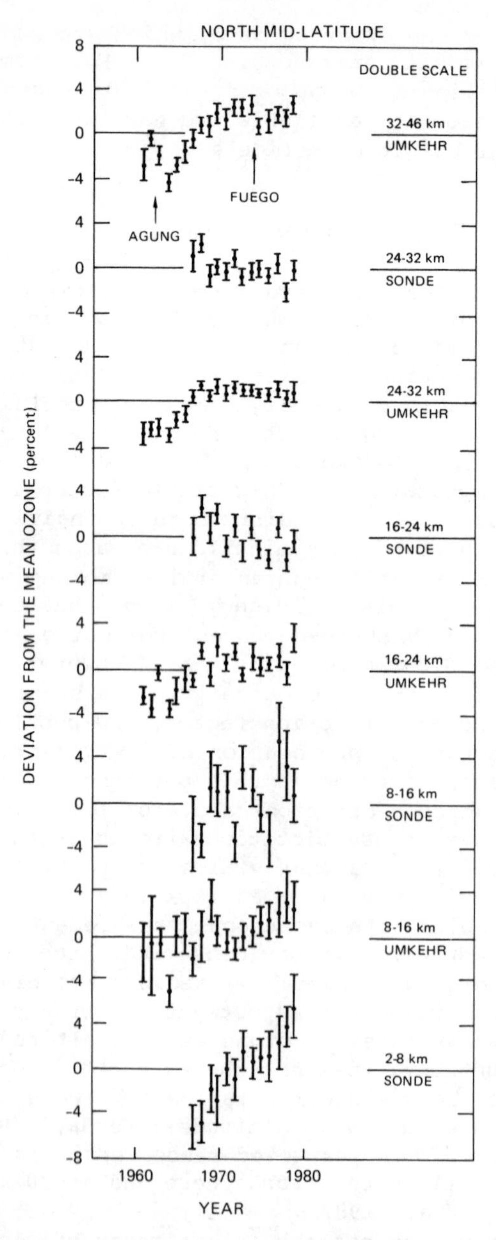

FIGURE F.1 Observed ozone variations for different layers in the troposphere and stratosphere at middle northern latitudes. The vertical bars represent approximate 95 percent confidence intervals (Angell and Korshover 1981).

TABLE F.2 Balloon-borne Ozonesonde Stations in Operation as of 1980

Station	Initial Year (Since 1958)	Months of Missing Data
North America		
Churchill	1973	0
Edmonton	1973	0
Cold Lake	1977	26
Goose Bay	1969	0
Palestine	1977	18
Toronto	1976	26
Wallops Island	1970	26
Europe		
Biscarrosse	1976	0
Hohenpeissenberg	1966	0
Legionowo	1980	0
Lindenberg (Tempelhof)	1967	11
Payerne	1968	0
Uccle	1965	20
North Polar		
Resolute	1966	5
Tropics		
Natal (Brazil)	1979	2
Australia		
Aspendale	1965	6
Japan		
Kagoshima	1968	19
Sapporo	1968	35
Tateno	1968	23

lite measurements and theoretical considerations reveal asymmetries between the northern and southern hemispheres; data from the north temperate latitudes cannot be extrapolated elsewhere.

At 30 km, which is about as high as ozonesondes regularly reach, these considerations and statistical analyses of the data (Pittock 1977a, Hudson and Reed 1979, Hudson et al. 1982) lead to estimates of standard deviations of estimated global ozone trends per decade of about 5.5 percent. Thus an ozone change of about 11 percent per decade could be detected at the 95 percent confidence level. A change in ozone concentration at 30 km in the north temperate zone could be detected from the ozonesonde data with confidence if it were to exceed about 8 percent per decade. No such global trend or trend in the north temperate zone has yet been detected.

Trends in the ozone content of the upper troposphere are also of interest since the same theoretical models that predict ozone reduction in the middle and upper stratosphere predict ozone increases in the upper troposphere due to NO_x emissions from the surface and/or aircraft exhausts.

Estimates of revealed and unrevealed errors in trend determinations in the 2- to 8-km layer from ozonesonde data suggest that a change of about $\pm$18 percent at this level worldwide could be detected at the 95 percent confidence level (Hudson et al. 1982). Using data from the north temperate zone only, it should be possible to detect a trend of $\pm$14 percent per decade at the 95 percent confidence level. As most surface and aircraft emissions of NO_x occur in the north temperate zone, this is a more sensible place to look for early evidence of a tropospheric trend, especially as tropospheric effects of pollutants will have a shorter lifetime than those in the stratosphere due to removal by meteorological processes.

A linear regression analysis of the ozonesonde data at 2-8 km in the north temperate zone (see Figure F.1) reveals a trend during the 1970s of about +7 percent per decade (Liu et al. 1980, Angell and Korshover 1981). The revealed uncertainties as indicated by the error bars in Figure F.1 suggest that this trend might be statistically significant. However, consideration of estimates of possible unrevealed errors, as discussed above, increases the uncertainty to a standard deviation of about 7 percent per decade. Thus, with 95 percent confidence, the trend lies between -7 percent and +21 percent per decade. The most probable value of the trend thus differs from zero by about one standard deviation and has a probability of only two chances in three of being real. The data therefore are quite suggestive of an increase, but the level of confidence in the result is not high. Careful checking and stratification of the ozonesonde data and application of refined statistical techniques may result in a reduction of the uncertainty in this trend estimate.

SATELLITE METHODS

The most extensive sets of data on ozone concentrations in the middle and upper stratosphere so far obtained by satellites have been obtained with the backscattered ultraviolet (BUV) and the infrared limb emission techniques (Hudson et al. 1982). The usefulness of satellite data for trend analysis depends on obtaining long and essentially homogeneous time series of data. This requires continuity with the same type of, or closely comparable, instruments and inversion algorithms (the methods used to transform measured radiation intensities to the ozone distributions that give rise to them) and regular calibration by in-flight and "ground-truth" methods. Allowance must also be made for possible natural fluctuations in ozone concentration, due particularly in the upper stratosphere to possible solar cycle variations.

As indicated by Panofsky (see Appendix E) in discussing total ozone measurements, the BUV data from the NIMBUS-4 satellite, which commenced operation in April 1970, suffered from instrumental drift and also from a loss of spatial coverage after June 1972. Additional problems of spatial representativeness arose from interference with observations of vertical distributions caused by high-energy charged particles in the vicinity of the South Atlantic magnetic anomaly. The drift problem has forced almost total reliance on Umkehr, rocket, and balloon-borne ozonesondes for validation and assessment of instrument performance.

The solar backscattered ultraviolet (SBUV) instrument on NIMBUS-7, which commenced operation in November 1978, was designed to overcome these problems. Shorter data sets are available from the Limb Radiance Inversion Radiometer (LRIR) on NIMBUS-6 from June 1975 through January 1976, and the Limb Infrared Monitor of the Stratosphere (LIMS) on NIMBUS-7 from October 1978 through May 1979. Other data were obtained by the Stratospheric Aerosol and Gas Experiment (SAGE) on the AEM2 satellite from February 1979 to the present, an early BUV type experiment on OGO-4 in 1967-1968, and a later BUV instrument on AE-5, commencing in November 1975 and still operational. Data from these last two BUV instruments are not yet available.

According to Heath, in as-yet unpublished work (quoted in Hudson et al. (1982) and in *Science*, September 4, 1981, pp. 1088-1089 and submitted for publication in

Science), comparison of vertical profiles of ozone concentration from the NIMBUS-4 BUV instrument in 1970 and 1971 with those from the SBUV instrument on NIMBUS-7 in the corresponding months of 1978 and 1979 suggests that there has been some ozone depletion in the layer between 2 and 6 mbar (approximately 35- to 45-km altitude). At the altitude of maximum depletion, around 38- to 40-km altitude, the decrease averaged about 5 percent per decade (see Figure F.2).

Best estimates of the revealed and unrevealed errors in global mean ozone concentrations at 40 km from the NIMBUS-7 SBUV instrument alone suggest that a trend, due to whatever cause, of ±1.4 percent per decade could be detected at the 95 percent confidence level with 10 years of observations (Hudson et al. 1982). The actual uncertainty in estimates of ozone reduction from the combination of NIMBUS-4 and NIMBUS-7 data is difficult to quantify but is certainly likely to be much greater than 1.4 percent per decade owing to the problems with NIMBUS-4 outlined above and the necessity to allow for a solar cycle effect above 35 km.

The changing sensitivity of the NIMBUS-4 BUV instrument has been taken into account by Heath using comparisons with near overpass Umkehr observations, assuming that the Umkehr network did not itself drift in calibration. The solar cycle effect was taken into account by assuming that natural concentrations of stratospheric ozone vary in phase with solar activity and that the amplitude of this effect increases monotonically with increasing altitude. Thus the smooth curves (dashed lines) in Figure F.2, which match the observations (solid lines) at the 10.0- and 0.7-mbar levels, were taken to represent the solar cycle effect, and the difference from the observations at intermediate levels was taken to represent the ozone decrease tentatively attributed by Heath to destruction by chlorofluorocarbons.

Any departure of the real solar cycle effect from Heath's interpolated monotonic curves would lead to an error in the hypothesized ozone depletion profile. Heath's assumed solar cycle effect is in broad agreement with Dutsch (1979), but not with the theoretical calculations of Penner and Chang (1978), nor with those of Brasseur and Simon (1981), which are based on recent solar UV flux data. Neither theoretical study supports a monotonic variation of the solar cycle effect in the range between 30 and 50 km. According to Brasseur and Simon (1981), the magnitude of the solar cycle effect at

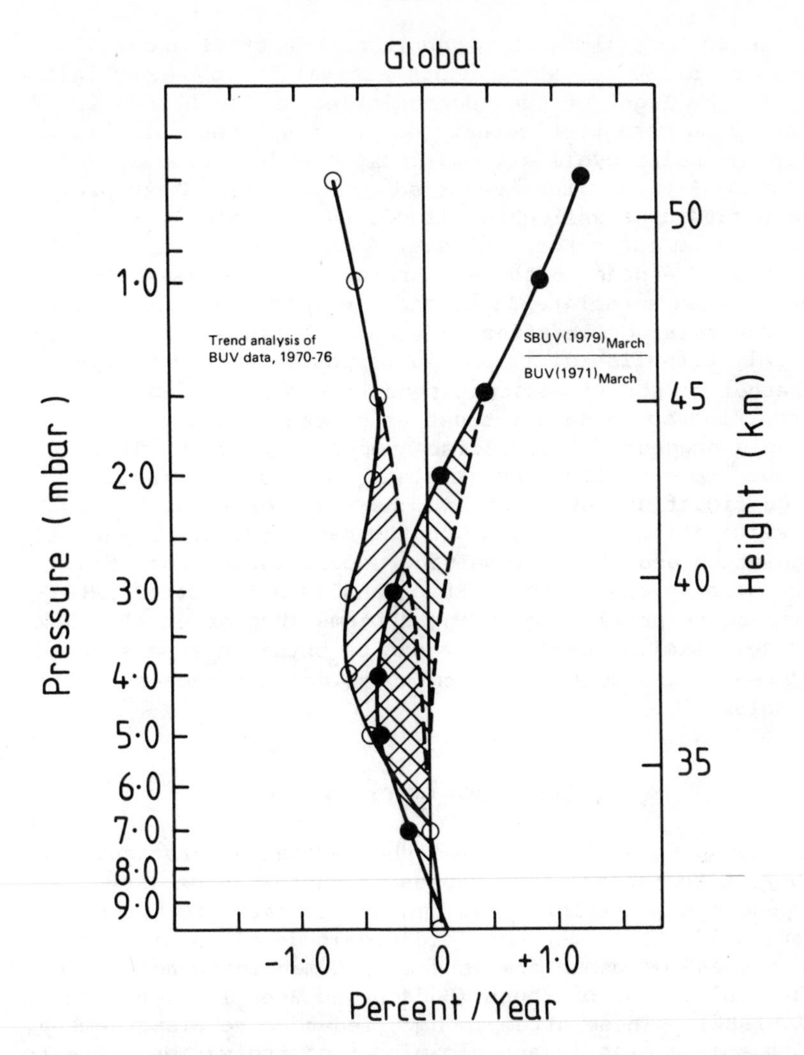

FIGURE F.2 Inferred long-term ozone variations in the stratosphere from satellite observations, according to Heath (1981). Solid lines represent observed ozone variations. Dashed lines are assumed effect of solar cycle variations only. Hatched area represents decrease in ozone tentatively attributed by Heath to destruction by chlorofluoromethanes.

38 to 40 km is about 8 percent, and is still about 7 percent at 30 km, where Heath assumes it to be negligible.

In the light of the uncertainties of the NIMBUS-4 BUV data, and more particularly of the controversial allowance for the solar cycle effect, the ozone depletion around 35- to 45-km altitude reported by Heath cannot at present be regarded as well established. Nevertheless, it is clear from the error analysis of the NIMBUS-7 system that with only another 5 to 10 years of homogeneous well-calibrated satellite data, and provided that the nature of the solar cycle effect at these altitudes can be more firmly established, it should be possible to determine whether or not significant reduction of ozone is occurring at these altitudes as current photochemical models suggest. The precise number of years of data needed to establish the existence of a statistically significant depletion within the range of theoretical possibilities will depend on the magnitude of the actual depletion and of the remaining uncertainties regarding the solar cycle effect. Error analyses for other satellite ozone profile measuring systems (Hudson et al. 1982) suggest similar sensitivity can be obtained from several systems using quite different physical approaches to the problem.

QUESTIONS OF CAUSALITY

In the absence of a detailed theoretical understanding of many of the alternative causes of ozone trends (see Appendix A), statisticians and others have attempted to assign limits to the possible magnitude of natural ozone trends and of ozone trends due to human influences other than chlorofluorocarbons (Hudson and Reed 1979, Hudson et al. 1982). These attempts have led to some disagreement, with some statisticians claiming that the variance due to long-term natural ozone variability can be estimated from total ozone data going back only a couple of decades, or in the case of one or two stations, 40 or 50 years. Other scientists who are familiar with long-term variability in other climatic variables claim that much longer records are necessary to obtain reasonable estimates of long-term natural variability. In part this disagreement rests on differing ideas about the possible nature of natural long-term climatic variability: the statisticians believe it is essentially a manifestation of a partially cumulative short-term random variability, whereas many climatologists maintain that climatic regimes may change with time, perhaps discontinuously, such that their statistical properties measured over intervals of a few decades may not be entirely representative of a longer time-span (e.g., see Flohn 1975). Such considerations may well apply to the stratosphere as well as to tropospheric climate.

The simplistic notion that natural variability of ozone must have some upper limit, so that the detection of a real trend in ozone greater than this limit must imply human influence, is an appealing one. Drawing the line on the basis of revealed statistical variations in the data, or of an intuitive "feel" for natural variability is, however, hazardous. It would be preferable if a physical approach could be adopted in which quantitative estimates were made of the various alternative causal mechanisms such as those outlined in Appendix A. Only when such an admittedly difficult and demanding course is followed will it be possible to ascribe with confidence particular causes to any observed real trend in ozone. Simultaneous measurements of other relevant variables such as temperatures, circulation parameters, the solar spectrum, and various other trace constituents and pollutants, will obviously aid the diagnostic process, and variation of effects with height and latitude provide additional means of discrimination between alternative causal mechanisms, notably solar effects, effects of the global CO_2 increase, and those due to chlorofluorocarbons or NO_x emissions. The present statistically based arguments and criteria cannot be regarded as scientifically satisfactory, but must be seen as necessary interim procedures to help in the process of making decisions in the face of uncertainty (Pittock 1980).

CONCLUSIONS

Current observations of the vertical distribution of ozone are severely limited as tools for the detection of ozone depletion due to (a) very poor spatial coverage by the balloon-borne ozonesonde and Umkehr observing networks, and (b) the short duration of continuous and homogeneous global data coverage by satellite.

The rate of ozone depletion due to human influences is expected to vary with latitude and altitude, with maximum rates of depletion calculated to occur around 35 to 45 km in altitude, and some ozone increase possible in the upper troposphere at northern middle latitudes. Since spatial coverage by balloon-borne ozonesonde and Umkehr methods is best in the northern mid-latitude zone, a focus on data analysis in this zone seems appropriate.

Two possible human effects on the vertical distribution of ozone have been reported to date. One is a possible increase in upper tropospheric ozone concentration in the north temperate zone, of about 7 percent during the 1970s (Liu et al. 1980, Angell and Korshover 1981). Given the various sources of uncertainty (Hudson et al. 1982), the probability, based on observations, that this effect is real (i.e., different from zero) is about 2 in 3. If it is real, this effect is attributed to an increase in NO_x concentrations in the upper troposphere due to emissions from aircraft and surface combustion, and should not be present to an appreciable extent in the southern hemisphere.

The second reported human effect is a claimed ozone depletion of the order of 5 percent per decade in the 38- to 40-km layer (Hudson et al. 1982) deduced from NIMBUS-4 BUV and NIMBUS-7 SBUV data over the time interval 1970 to 1979. Considering the problems experienced with the NIMBUS-4 instrument, and the uncertain but critical allowance for a possible solar cycle effect at these altitudes, this reported ozone depletion cannot at present be regarded as well established.

This last uncertainty highlights the question of causality in assessing the probability of observational data reflecting ozone depletion of human origin. The question of the influence of the 11-year solar cycle on ozone concentrations above about 25-km altitude is particularly important. A definitive description of this solar cycle influence in the middle and upper troposphere, which Brasseur and Simon (1981) estimate as having an amplitude of about 5 to 10 percent in these layers, is critical to early detection of ozone depletion at altitudes where photochemical theory indicates that the effects of chlorofluorocarbons should be greatest. Unless the solar cycle effect on ozone can be definitively described theoretically, it may be necessary to wait for accurate observations at critical altitudes over at least one whole solar cycle (11 years) in order to confidently infer that ozone depletion is due to pollution, even though the error limits in satellite vertical distribution measurements are small enough that a real trend in ozone concentration at 40-km altitude may be detected earlier.

A suitable set of satellite data is not currently available prior to 1978, and the Umkehr data are not only limited in spatial coverage but subject to uncertainty due to the need to allow for the effects of varying aerosol concentrations.

Given a resolution of the solar cycle effect, satellite-based observations of ozone concentrations in the 35- to 45-km region seem to provide the best hope for early detection of ozone depletion effects.

RECOMMENDATIONS

In order to obtain conclusive evidence for or against the reality of significant depletion of ozone by pollutants of human origin, the following actions are recommended.

1. Satellite methods. Highest priority should be given to the maintenance of one or more continuously operating and well-calibrated homogeneous satellite systems for the determination of the vertical distribution of ozone. This should include independent ground-truth obtained from the Dobson spectrophotometer network and balloon- or rocket-borne ozonesondes. Data obtained by more than one independent satellite system operating simultaneously, using different physical principles (e.g., backscattered UV and limb-scanning systems), would add greatly to confidence in any conclusions reached.

2. Focus on zones. Attention should be focused on those altitudes and latitudes where theoretical effects of pollution are greatest and nonsatellite data coverage is best. This implies a focus on altitudes in the range of 35 to 45 km, and the upper troposphere in the north temperate zone. Observations in the south temperate zone would provide a useful check, especially as upper tropospheric effects are expected to be negligible in the southern hemisphere.

3. Solar cycle effect. A definitive description of the effect on stratospheric ozone of the 11-year solar cycle, especially at 35 to 45 km, is urgently needed. Efforts should be directed to

(a) theoretical analysis of the solar cycle effects,
(b) monitoring of solar ultraviolet radiation, solar protons, and any other solar outputs likely to affect ozone concentrations, and

(c) analysis of zonally representative ozone profile data at 35 to 45 km over at least one whole 11-year solar cycle, including further attempts to refine existing Umkehr data from the north temperate zone with proper allowance for variable aerosol effects.

4. Umkehr method. The spatial coverage by the Umkehr method should be increased, especially in the north and south temperate zones. Since many Dobson spectrophotometers are already in place that could but do not at present make Umkehr measurements, this should not be unduly difficult to achieve. Effort must also be made to make proper allowance for the effects of varying concentrations of tropospheric and stratospheric particulate matter using actual particle concentrations as suggested by Dave et al. (1981).

5. Ozonesondes. The balloon-borne ozonesonde network in the north temperate zone should be maintained and if possible improved, and that in the south temperate zone (currently one station only) increased. More refined statistical techniques and critical data analysis should be applied to the existing north temperate zone ozonesonde data.

6. Tropospheric ozone. More theoretical work is needed on the distribution of ozone in the troposphere, including especially the effect of NO_x from aircraft and surface emissions and the chronological evolution of these effects using emission data.

7. Monitoring other variables. In view not only of the solar cycle effect, but also of the effect of changing temperature (due to increasing carbon dioxide concentrations) and variations in atmospheric circulation, other relevant meteorological variables and chemical constituents must be monitored in order both to test photochemical theory and reaction rates more critically and to enable a useful reduction in background variance due to causes other than pollution (e.g., see Bloomfield et al. 1981, Pittock 1973).

REFERENCES

Angell, J.K. and J. Korshover (1981) Update of ozone variations through 1979. Pages 393-396, Proceedings of the Quadrennial International Ozone Symposium, August 4-9, 1980. Boulder, Colo.: National Center for Atmospheric Research.

Bloomfield, P., M.L. Thompson, G.S. Watson, and S. Zeger (1981) The association of ozone with meteorological variables. Pages 306-313, Proceedings of the Quadrennial International Ozone Symposium, August 4-9, 1980. Boulder, Colo.: National Center for Atmospheric Research.

Bloomfield, P., M.L. Thompson, G.S. Watson, and S. Zeger (1982) Frequency Domain Estimation of Trends in Stratospheric Ozone. Technical Report No. 182, Department of Statistics, Princeton University. (Submitted for publication to Journal of Geophysical Research.)

Brasseur, G. and P.C. Simon (1981) Stratospheric chemical and thermal response to long-term variability in solar UV irradiance. Journal of Geophysical Research 86(C8):7343-7362.

Dave, J.V., C.L. Mateer, and J.J. De Luisi (1981) An examination of the effect of haze on the short Umkehr method for deducing the vertical distribution of ozone. Pages 222-229, Proceedings of the Quadrennial International Ozone Symposium, August 4-9, 1980. Boulder, Colo.: National Center for Atmospheric Research.

De Luisi, J.J., B.M. Herman, R.S. Browning, and R.K. Sato (1975) Theoretically determined multiple-scattering effects of dust on Umkehr observations. Quarterly Journal of the Royal Meteorological Society 101:325-331.

Dutsch, H.U. (1979) The search for solar cycle-ozone relationships. Journal of Atmospheric and Terrestrial Physics 41:771-785.

Flohn, H. (1975) History and intransitivity of climate. Pages 106-118, The Physical Basis of Climate and Climate Modelling. GARP Publication No. 16. Geneva: World Meteorological Organization.

Heath, D. (1981) Secular changes in stratospheric ozone from satellite observations (1970-1979). Unpublished manuscript. (Submitted to Science.)

Hudson, R.D. and E.I. Reed (1979) The Stratosphere: Present and Future. NASA 1049. Washington, D.C.: National Aeronautics and Space Administration.

Hudson, R.D., et al., eds. (1982) The Stratosphere 1981: Theory and Measurements. WMO Global Research and Monitoring Project Report No. 11. Geneva: World Meteorological Organization. (Available from National Aeronautics and Space Administration, Code 963, Greenbelt, Md. 20771.)

Liu, S.C., D. Kley, M. McFarland, J.D. Mahlman, and H. Levy II (1980) On the origin of tropospheric ozone. Journal of Geophysical Research 85:7546-7552.

Penner, J.E. and J.E. Chang (1978) Possible variations in atmospheric ozone related to the eleven-year solar cycle. Geophysical Research Letters 5:817-820.

Penner, J.E., L.P. Golen, and R.W. Mensing (1981) A time series analysis of Umkehr data from Arosa. UCRL Reprint 85420. Livermore, Cal.: Lawrence Livermore Laboratory.

Pittock, A.B. (1970) On the representativeness of mean ozone distributions. Quarterly Journal of the Royal Meteorological Society 96:32-39.

Pittock, A.B. (1973) Global meridional interactions in stratosphere and troposphere. Quarterly Journal of the Royal Meteorological Society 99:424-437.

Pittock, A.B. (1977a) Climatology of the vertical distribution of ozone over Aspendale (38°S, 145°E). Quarterly Journal of the Royal Meteorological Society 103:575-584.

Pittock, A.B. (1977b) Ozone sounding correction procedures and their implications. Quarterly Journal of the Royal Meteorological Society 103:809-810.

Pittock, A.B. (1980) Monitoring, causality and uncertainty in a stratospheric context. Pure and Applied Geophysics 118:643-661.

Appendix G

THE ASSOCIATION OF DNA DAMAGE WITH CANCER-INITIATING EVENTS

Some of the reasons for associating cancer initiation with damage to DNA are as follows:

1. Many carcinogenic agents are electrophilic in nature (attracted to negatively charged particles) and react with cellular macromolecules. There is an association between the mutagenicity of compounds (their reactivity with DNA) and their carcinogenicity, although one must remember the necessity for activating inactive mutagenic compounds. In animal carcinogenic test systems, the animals contain the necessary activating enzymes. Within families of active metabolites, such as the diolepoxides of benzo(a)pyrene, there is a very close association between mutagenicity and carcinogenicity, and a similar association is found for nitrosamines in liver-cell-activated mutagenicity and liver carcinogenicity.

2. The disease xeroderma pigmentosum is associated with a very high skin cancer prevalence, and the cancers are on the sun-exposed areas of the body. The cells of individuals with this disease are defective in one or more mechanisms that repair ultraviolet damage to DNA. Defects in repair are associated with a 10^3- to 10^4-fold higher skin cancer prevalence in xeroderma pigmentosum individuals than in the average population.

3. If the thymidine in the DNA of cells in culture is substituted by the analog bromodeoxyuridine, the cells become very sensitive to UV-B because bromodeoxyuridine has a much higher absorption coefficient in the UV-B than does thymidine. Substitution with bromodeoxyuridine is a way to sensitize cells to UV-B and the sensitization may be detected either in terms of cell killing, neoplastic

transformation, or the ability of UV-B to make single strand breaks in the DNA. In this system, there is a close association between DNA damage and neoplastic transformation. (Under normal conditions, cells in culture do not grow indefinitely, but are inhibited when they grow to sufficient numbers that they begin to contact one another. This contact inhibition may be destroyed by radiation so that the cells continue to proliferate and make so-called "transformed" foci. In many instances, such foci give rise to tumors when the cells from a focus are transplanted into appropriate mouse strains.)

4. Certain species of fish grow in clones. Cells from one member of the clone may be removed, irradiated in vitro, and injected back into other members of the clone. When this is done for thyroid cells of the species _Poecilia_ _formosa_, thyroid tumors develop in the recipients. If, however, visible light exposure follows the ultraviolet exposure, that is to say, the cells are subjected to photoreactivation, no tumors develop. Since photoreactivation is diagnostic for pyrimidine dimers (Chapter 3) in DNA, these experiments not only imply damage to DNA in tumor production but also implicate a specific photoproduct--pyrimidine dimers.

5. The action spectrum for neoplastic transformation is similar to that for affecting DNA in mammalian cells (Chapter 3).

Appendix H

PARTICIPANTS IN THE WORKSHOP ON BIOLOGICAL EFFECTS OF INCREASED SOLAR ULTRAVIOLET RADIATION

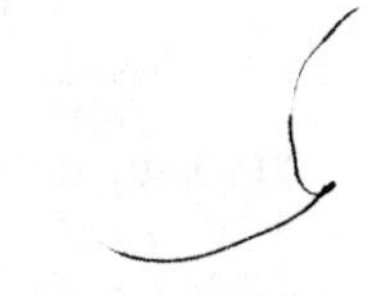

The purpose of the workshop was for the Committee on Biological Effects of Increased Solar Ultraviolet Radiation to hear the most recent data and their interpretation from scientists active or current in the workshop topic. The committee used this information in developing conclusions and research recommendations; however, the report and the judgments it contains are solely the responsibility of the committee. Participation in the workshop should not be taken as evidence that an individual endorses the conclusions or recommendations of this report.

RICHARD B. SETLOW* (Chairman), Biology Department, Brookhaven National Laboratory

R. HILTON BIGGS, Institute of Food Science and Agriculture, University of Florida
JOHN CALKINS, Department of Radiation Medicine, University of Kentucky
EDWARD deFABO, National Cancer Institute Frederick Research Center
JAY DONIGER, Laboratory of Biology, National Cancer Institute
DAVID ELDER, University of Pennsylvania Medical School
MORTIMER M. ELKIND, Division of Biological and Medical Research, Argonne National Laboratory
P. DONALD FORBES, The Skin and Cancer Hospital, Temple University
JAMES P. FRIEND,* Department of Chemistry, Drexel University

*Indicates members of the Committee on Biological Effects of Increased Solar Ultraviolet Radiation.

SAXON GRAHAM, Department of Social and Preventive Medicine, State University of New York at Buffalo

MAUREEN M. HENDERSON,* Health Sciences Center, University of Washington

JOHN R. HUNTER, Southwest Fisheries Center National Marine Fisheries Service

JOHN JAGGER,* Department of Biology, University of Texas at Dallas

RICHARD M. KLEIN,* Department of Botany, University of Vermont

ALBERT M. KLIGMAN, Department of Dermatology, University of Pennsylvania

KENNETH E. KOPECKY, National Animal Disease Center

MARGARET L. KRIPKE, National Cancer Institute Frederick Research Center

JOHN A.H. LEE, Department of Epidemiology, University of Washington (Dr. Lee could not attend the workshop but instead provided the committee with a paper prepared especially for the workshop entitled "Melanoma Epidemiology Since the Academy Reports of 1979.")

SIDNEY LERMAN, Department of Ophthalmology, Emory University School of Medicine

JOHN A. PARRISH,* Department of Dermatology, Massachusetts General Hospital/Harvard University

ELIZABETH L. SCOTT, Department of Statistics, University of California at Berkeley

JOSEPH SCOTTO, Biometry Branch, National Cancer Institute

HOWARD H. SELIGER,* Department of Biology, Johns Hopkins University

WILLIAM B. SISSON,* Department of Animal Range Science, U.S. Department of Agriculture/New Mexico State University

LAWRENCE B. SLOBODKIN, Department of Ecology and Evaluation, State University of New York at Stony Brook

ARTHUR SOBER, Department of Dermatology, Massachusetts General Hospital

ROBERT S. STERN, Department of Dermatology, Beth Israel Hospital

DANIEL STRICKLAND, Department of Epidemiology, University of Washington

ROBERT B. WEBB, Division of Biological and Medical Research, Argonne National Laboratory (Dr. Webb could not attend the workshop but did consult with the committee and provide background material.)

ROBERT C. WORREST, Environmental Research Laboratory, U.S. Environmental Protection Agency, Corvallis

Appendix I

BIOGRAPHICAL SKETCHES OF COMMITTEE MEMBERS AND CONSULTANTS

COMMITTEE MEMBERS

ROBERT E. DICKINSON is deputy director of the Atmospheric Analysis and Prediction Division at the National Center for Atmospheric Research in Boulder, Colorado. He received a Ph.D. in 1966 in meteorology from the Massachusetts Institute of Technology and has done research in atmospheric dynamics, radiative transfer, and climate modeling. He is a member of the NRC Committee on Atmospheric Sciences and was a member of the NRC panels that authored the reports cited as NRC (1976b, 1979b).

JAMES P. FRIEND is R.S. Hanson Professor of Atmospheric Chemistry at Drexel University, Philadelphia, Pennsylvania. He received an S.B. from the Massachusetts Institute of Technology in 1951 and an M.A. in 1953 and a Ph.D. in 1956 in chemistry from Columbia University. Dr. Friend is an expert in global cycles and geochemistry of trace substances in the atmosphere and climate impact assessments. He has worked for the Perkin-Elmer Corporation and Isotopes, Inc., and was a professor of atmospheric chemistry at New York University. Dr. Friend was a member of the three previous NRC committees that prepared the reports cited as NRC (1975; 1976a,b; 1979a,b).

MAUREEN M. HENDERSON is associate vice president for Health Sciences at the University of Washington, Seattle. She received an MB.BS (Dunelm) in 1949 and a D.P.H. (Dunelm) in 1956. Dr. Henderson is a physician epidemiologist with a speciality in the epidemiology of chronic diseases. She has taught at St. Bartholemew's Hospital, London, the University of Maryland, and Johns Hopkins University and has served on numerous advisory and review committees for the government. She is a member of the National Cancer Advisory Board and of the Council of the Institute of Medicine. She served on the NRC committees that prepared the reports cited as NRC (1976a, 1979a).

DONALD M. HUNTEN is professor of planetary science at the Lunar and Planetary Laboratory of the University of Arizona. He received a Ph.D. in physics from McGill University in 1950. His research interests are the upper atmosphere of earth and other planets and spectroscopic instrumentation. He is a member of the National Academy of Sciences and was a member of the NRC committee that prepared the report cited as NRC (1975).

JOHN JAGGER is a professor in the School of General Studies at the University of Texas at Dallas. He received a B.S. in 1949 in physics and a Ph.D. in 1954 in biophysics from Yale University. He is an expert on the effects of UV, especially UV-A, on bacteria, including photoreactivation, photoprotection, and effects on cell growth and membrane function. Dr. Jagger has worked at the Oak Ridge National Laboratory and has taught at the Southwest Center for Advanced Studies, the University of Tennessee, Pennsylvania State University, and the University of Kyoto, Japan. He is a former editor of the journal *Photochemistry and Photobiology*. He has been a member and president of the NRC's U.S. National Committee for Photobiology.

RICHARD M. KLEIN is professor of botany at the University of Vermont, Burlington. He received a B.S. in 1947, an M.S. in 1948, and a Ph.D. in 1951 in botany from the University of Chicago. His expertise is in the field of plant physiology, especially effects of UV radiation on plant growth and development and potential economic impacts. He has also worked at the New York Botanical Garden.

CHARLES H. KRUGER, JR., is professor of mechanical engineering at Stanford University, where he conducts research in the dynamics of high-temperature gases and combustion processes. He received a Ph.D. in 1960 in mechanical engineering from the Massachusetts Institute of Technology. He has been a member of the Hearing Board of the San Francisco Bay Area Air Quality Management District since 1969, serving as chairman between 1971 and 1977, and is a member of the NRC Environmental Studies Board. In 1970 he was awarded a medal from the American Institute for Aeronautics and Astronautics.

MICHAEL B. McELROY is Abbott Lawrence Rotch Professor of Atmospheric Sciences at the Center for Earth and Planetary Physics at Harvard University. He received a Ph.D. in theoretical physics from Queen's University, Belfast, in 1962 and was a physicist on the staff of Kitt Peak National Observatory before joining the faculty at Harvard. His research interests are in the physics and chemistry of planetary atmospheres.

JOHN A. PARRISH is associate professor of dermatology at the Harvard Medical School and assistant dermatologist at Massachusetts General Hospital in Boston. He received a B.A. in 1961 in political science from Duke University and an M.D. in 1965 from the Yale University School of Medicine. He has done both clinical work and research in photomedicine, especially the hazardous and therapeutic effects of light, including UV, on human skin. He is director of the Wellman Laboratories and the Photomedicine Research Unit at Massachusetts General Hospital. He served on the NRC's U.S. National Committee for Photobiology, is a councilor for the American Society for Photobiology and a member of the Photobiology Task Force for the American Academy of Dermatology.

HOWARD H. SELIGER holds a joint appointment as professor of biology in the Biology Department and the School of Hygiene and Public Health at the Johns Hopkins University, Baltimore, Maryland. He received a B.S. in 1943 from the City College of New York, an M.S. in 1948 from Purdue University, and a Ph.D. in 1954 in physics from the University of Maryland. Dr. Seliger's fields of expertise include the study of mechanisms of bioluminescence and chemiluminescence, photoecology, marine phytoplankton ecology, and radiobiology. He was a physicist in the Radioactivity Section of the National Bureau of Standards before joining the Johns Hopkins University in 1960. He is the immediate past president of the American Society for Photobiology.

RICHARD B. SETLOW is chairman of and senior biophysicist in the Biology Department at the Brookhaven National Laboratory, Upton, New York. He recevied an A.B. in 1941 from Swarthmore College and a Ph.D. in 1947 in physics from Yale University. He is an expert in the field of molecular biophysics, in particular, the effects of UV radiation on biological systems, and has done extensive work on the DNA action spectrum and on DNA repair systems. Dr. Setlow has worked at the Oak Ridge National Laboratory, has taught at Yale University, the University of Tennessee, and the State University of New York at Stony Brook, and has directed the University of Tennessee Oak Ridge Graduate School of Biomedical Sciences. He is a member of the National Academy of Sciences and has served on numerous NRC committees, including the committee that prepared the report cited as NRC (1975) and the Panel to Review Statistics on Skin Cancer of the Committee on National Statistics.

WILLIAM B. SISSON holds a joint appointment as plant physiologist at the USDA Jornada Experimental Range and on the graduate faculty of the Department of Animal and Range Sciences at New Mexico State University, Las Cruces. He received a B.S. in 1970 from Humbolt State University, an M.S. in 1972 from Texas Tech University, and a Ph.D. in 1975 from Utah State University. Dr. Sisson is a plant physiologist and is an expert on the effects of UV on plants and ecosystems, especially photosynthesis and carbon metabolism. He has also done research at Utah State University.

CONSULTANTS

JAMES G. ANDERSON is Robert P. Burden Professor of Atmospheric Chemistry at the Center for Earth and Planetary Physics at Harvard University. He received a Ph.D. in physics and astrogeophysics from the University of Colorado in 1970. His current research interests are in the development of techniques for measuring trace species in the troposphere and stratosphere and the application of these techniques to in situ observations of species important for ozone chemistry in the stratosphere.

RALPH J. CICERONE is director of the Atmospheric Chemistry and Aeronomy Division of the National Center for Atmospheric Research in Boulder, Colorado. He received a Ph.D. in physics and electrical engineering from the University of Illinois in 1970 and currently conducts theoretical research on the photochemistry of stratospheric ozone and experimental research on atmospheric trace gases. He is an editor of the

Journal of Geophysical Research, a member of the NRC Committee on Atmospheric Sciences as well as co-chairman of that committee's Panel on Atmospheric Chemistry, a member of the Upper Atmosphere Committee of the American Meteorological Society, and a Fellow of the American Geophysical Union and of the AAAS.

JENNIFER A. LOGAN is research associate in atmospheric chemistry at the Center for Earth and Planetary Physics at Harvard University. She received a Ph.D. from the Massachusetts Institute of Technology in physical chemistry in 1975. Her current research focuses on theoretical modeling of the chemistry of the stratosphere and troposphere.

HANS A. PANOFSKY is Evan Pugh Research Professor of Atmospheric Sciences at Pennsylvania State University. He was awarded a Ph.D. in astronomy by the University of California in 1941. His current research interests are in dynamic meteorology and micrometeorology. He was a member of the NRC panels that prepared the reports cited as NRC (1975, 1976a, 1979a).

A. BARRIE PITTOCK is principal research scientist in the Division of Atmospheric Physics of the Commonwealth Scientific and Industrial Research Organization (CSIRO) near Melbourne, Australia. He holds a Ph.D. in physics from Melbourne University and was a Fulbright scholar at the National Center for Atmospheric Research in Boulder, Colorado in 1963-1964. Dr. Pittock is a member of the National Committee on Atmospheric Science of the Australian Academy of Science and of its subcommittee on the World Climate Research Program. His current research is in monitoring and causality of stratospheric change, solar influences on weather and climate, natural variations in climate, and human impact on climate.

STEVEN C. WOFSY is associate professor of atmospheric chemistry at the Center for Earth and Planetary Physics at Harvard University. He was awarded a Ph.D. in chemistry by Harvard in 1971. He is currently conducting research in tropospheric and stratospheric chemistry, including theoretical modeling of stratospheric ozone chemistry.